W0193212

Web Semantics and Ontology

David Taniar, Monash University, Australia

Johanna Wenny Rahayu, La Trobe University, Australia

IDEA GROUP PUBLISHING

Hershey • London • Melbourne • Singapore

Acquisitions Editor:	Michelle Potter
Development Editor:	Kristin Roth
Senior Managing Editor:	Amanda Appicello
Managing Editor:	Jennifer Neidig
Copy Editor:	Nicole Dean
Typesetter:	Jennifer Neidig
Cover Design:	Lisa Tosheff
Printed at:	Yurchak Printing Inc.

Published in the United States of America by
    Idea Group Publishing (an imprint of Idea Group Inc.)
    701 E. Chocolate Avenue
    Hershey PA 17033
    Tel: 717-533-8845
    Fax: 717-533-8661
    E-mail: cust@idea-group.com
    Web site: http://www.idea-group.com

and in the United Kingdom by
    Idea Group Publishing (an imprint of Idea Group Inc.)
    3 Henrietta Street
    Covent Garden
    London WC2E 8LU
    Tel: 44 20 7240 0856
    Fax: 44 20 7379 0609
    Web site: http://www.eurospanonline.com

Copyright © 2006 by Idea Group Inc. All rights reserved. No part of this book may be reproduced, stored or distributed in any form or by any means, electronic or mechanical, including photocopying, without written permission from the publisher.

Product or company names used in this book are for identification purposes only. Inclusion of the names of the products or companies does not indicate a claim of ownership by IGI of the trademark or registered trademark.

Library of Congress Cataloging-in-Publication Data

Web semantics and ontology / David Taniar and Johanna Wenny Rahayu, editors.
    p. cm.
  Summary: "This book provides an overview of current research and development activities in the area of web semantics and ontology, giving an in-depth description of different issues, including modeling, using ontologies in enterprise systems, querying and knowledge discovering of ontologies"--Provided by publisher.
  Includes bibliographical references and index.
  ISBN 1-59140-905-5 (hardcover) -- ISBN 1-59140-906-3 (softcover) -- ISBN 1-59140-907-1 (ebook)
  1. Semantic Web. 2. Ontology. I. Taniar, David. II. Rahayu, Johanna Wenny.

TK5105.88815.W42 2006
025.04--dc22
British Cataloguing in Publication Data
A Cataloguing in Publication record for this book is available from the British Library.

All work contributed to this book is new, previously-unpublished material. The views expressed in this book are those of the authors, but not necessarily of the publisher.

Web Semantics and Ontology

Table of Contents

Section II. Ontologies and Enterprise Systems

Section III. Ontologies-Based Querying and Knowledge Discovery

Section IV. Applications and Policies

Preface

The chapters of this book provide an excellent overview of current research and development activities in the area of Web Semantics and ontology. They supply an in-depth description of different issues in Web Semantics and ontology, including modelling of Web Semantics and ontologies, using ontologies in enterprise systems, querying and knowledge discovering of ontologies, and adopting policies and building applications. Each chapter contains a thorough study of the topic, systematic proposed work, and a comprehensive list of references.

Following our call for chapters in 2005, we received more than 30 chapter proposals. Each proposed chapter was carefully reviewed and eventually, 12 chapters were accepted for inclusion in this book. This book brought together academic, researchers, and practitioners from many different countries, including Argentina, Australia, Belgium, Canada, France, Greece, India, Italy, Spain, and USA. Their research and industrial experience, which are reflected in their work, will certainly allow readers to gain an in-depth knowledge of their areas of expertise.

Intended Audience

Web Semantics and Ontology gives readers comprehensive knowledge on the current issues of Web Semantics and ontologies. The book describes the basic need arised from Web Semantics, the underpinning background of Web Semantics, the infrastructures, languages, and tools for Web Semantics; and real-world application domains of Web Semantics. This book is intended for individuals who want to enhance their knowledge of issues relating to modelling, adopting, querying, discovering knowledge, and building ontologies and Web Semantics. Specifically, these individuals could include:

- **General public interested in the Internet technology:** General publics who are interested in the Web technology will find this book useful as it covers current issues and practice of Web Semantics and ontology. This book can be used as a reference book on Web Semantics and ontology.

- **Information technology researchers:** Researchers who are primarily interested in current issues of Web technologies will find this book useful, as it presents issues and state of the art of Web Semantics. The topics that might give them particular interest include ontology, enterprise systems, modelling, knowledge discovery, queries, policies, and other issues.

- **Information technology students and lecturers:** The chapters in this book are grouped into four parts to cover important issues in the area. This will allow students and teachers in Web Semantics fields to effectively use the appropriate materials as a reference or reading resources. These categories are: (1) ontology modelling; (2) enterprise systems; (3) retrieval and knowledge discovery; and (4) policies and applications. Since this book covers the issues of Web Semantics and ontology comprehensively, it can be used as a textbook at a graduate level.

- **Web software developers:** Software developers will find this book useful particularly in the area of practical Web development involving OWL, XML, RDF, metadata, and UML. The final part of this book on applications would be useful for developers in learning on how a large scale application is built.

Prerequisites

The book as a whole is meant for anyone professionally interested in Web Semantics and ontology and who in some way wants to gain an understanding on how the issues in modelling and using ontologies in enterprise systems, as well as information retrieval and knowledge discovery of Web Semantics. Each chapter may be studied separately, or in conjunction with other chapters. As each chapter may cover topics different from other chapters, the prerequisites for each may vary. However, we assume the readers have at least a basic knowledge of:

- Web information systems and its applications,
- Information modelling,
- Enterprise information systems, and
- Queries and knowledge discovery.

Overview of Web Semantics and Ontology

Over the last years there has been a steady shifting from the Internet as we know it — unstructured, or at best, semistructured, to a more structured Web, referred to as the *Web Semantic*. Web Semantics is used to denote the next evolution step of the Web, which establishes a layer of machine understandable data. The data is suitable for automated agents, sophisticated search engines, and interoperability services, which provide a previously not reachable grade of automation. The ultimate goal of Web Semantic is to allow machines the sharing and exploitation of knowledge in the Web way, i.e., without central authority, with few basic rules, in a scalable, adaptable, extensible manner. In other words, Web Semantics is the key of the next generation of Web information system, where information is given a well-defined meaning, better enabling people and programs to work in cooperation with each other.

The emergence of Web technology has made global information sharing possible. Sharing of knowledge is motivated by Semantic Web whereby there is a necessity to make content searching more efficient and meaningful by providing contextual and structural information about the presented contents. This becomes possible through the establishment of an appropriate standard to define the conceptual level of a metalanguage, and such a standard is known as an *Ontology*, which is described as sharable conceptualization of specific domain of interest in a machine-understandable format.

Web Semantics aids the efficiency of searching in the World Wide Web, by providing more information about the presented texts. Although the applications are many, the most common examples are intelligent search engine with data mining capabilities, integrated decision support system applications, integrated enterprise applications, and so forth. The structuring of information on the Web will bring the Internet to the next era. The infrastructure to build such structured Semantic Web has been established, including the suitable language for representation of the data (XML), translation to HTML for presentation/display purposes (XSL), specification of metadata (RDF, XML-Schema, DTD), and finally the establishment of an appropriate standard to define the conceptual level of the metadata languages or the ontology (OWL, DAML-OIL).

Ontologies are the backbone that keeps the Semantic Web together, because they enforce an agreement at least on how the structure of information should be defined. In cases where different organizations cannot agree on the same structure, they would still use the same way of defining their structure as formulated in the ontology. This will enable the development of a tool that can translate between the different structures, thus enabling communication. While the collection of data or information on the Internet can be seen as a static data repository with possible regular incremental update, the user applications and requirements that utilize the collection of data for their individual purposes will change over time. For this reason, ontologies need to be dynamic and adaptable to cater for the diversity of users' needs and requirements, and the complexity of different applications that need to be integrated.

The new era of Web Semantic has enabled users to extract semantically relevant data from the Web. Web ontology plays an important role in the Semantic Web as it defines shared uniform structures which define how Web information is grouped and classified regardless of the implementation language or the syntax used to represent the data. However, as Web ontology grows and evolves, there are many issues to be addressed, including how it may be adopted in large organizations, how it can be queries, how the security may be guaranteed, etc.

Organization of This Book

The book is divided into four major sections:

I. Web Semantics and Ontologies Modelling

II. Ontologies and Enterprise Systems

III. Ontologies-Based Querying and Knowledge Discovery

IV. Applications and Policies

Each section in turn is divided into several chapters:

Section I focuses on modelling of Web Semantics and ontology. This section includes chapters on ontology extraction using views, patters, and modelling language. Section I consists of three chapters.

Chapter I, contributed by *Wouters*, *Rajagopalapillai*, *Dillon*, and *Rahayu*, investigates the use of materialized ontology view as an alternative efficient version in utilizing a whole large ontology. They describe the formalism of the materialized view for Web Semantic with conceptual and logical extensions.

The view provides the required conceptual and logical semantics to develop ontological bases. They also present a schemata transformation methodology to materialize Web Semantic view using an ontology extraction methodology framework where materialized ontology views are instantiated.

Chapter II, presented by *Kamthan* and *Pai*, focuses on patterns, which are refined from past experience due to recurring problems. They describe a process of creating an ontology in the language OWL for Web Application Patterns, called OWAP. The features during OWAP design, implementations and testing are also described.

Chapter III, presented by *Caliusco*, *Maidana*, *Galli*, and *Chiotti*, introduces contextual ontology, where an ontology is presented with its context definition. This contextual ontology needs to be expressed in a language at run-time, particularly for the analysis and design phase of a Web domain. They present a metamodel for modelling explicit and formal contextual ontologies to model contextual ontologies.

Section II concentrates on enterprise systems, covering major issues of ontology management in large-scale enterprise systems, enterprise information systems, and semantic annotation systems for text-based document systems. This section also consists of three chapters: Chapters IV, V, and VI.

Chapter IV, presented by *Lee*, *Goodwin*, and *Akkiraju*, describes their work on developing an enterprise-scale ontology management system that provides APIs and query languages, and scalability and performance that enterprise applications demand. They describe the design and implementation of the management system that programmatically supports the ontology needs of enterprise applications in a similar way a database management system supports the data needs of applications.

Chapter V, presented by *Rifaieh* and *Benharkat*, concentrates on studying the application of context and ontology which can serve as a formal background for reaching a suitable enterprise information system. They focus on formalism for contextual ontologies based combining description logics and modal logics. This in turn helps to overcome an ontology-phobia. They also show some examples the usefulness of these contextual ontologies for resolving the semantic sharing problems in some enterprise information systems.

Chapter VI, contributed by *Reeve* and *Han*, focuses on semantic annotation to be a key component in Semantic Web. They propose semi-automatic semantic annotation systems for text-based Web documents. This semantic annotation provides services supporting annotation, including ontology and knowledge base access and storage, information extraction, programming interfaces, and end-user interfaces.

Section III focuses on querying and knowledge discovery of ontology. It consists of three chapters covering acquiring new knowledge in ontology, dynamic knowledge discovery, and metadata and Web Semantic mining.

Chapter VII, presented by *Dey* and *Abulaish*, presents a text-mining based ontology enhancement and query processing system. The system supports ontology enhancement by identifying, defining, and adding new precise and imprecise concepts descriptions mined from text documents. They adopt a fuzzy reasoning method for query processing.

Chapter VIII, presented by *Castano, Ferrara*, and *Montanelli*, focuses on dynamic knowledge discovery, which is a capability of each node in a P2P or Grid network of finding knowledge in the system about information resources matching. They describe the models and techniques for ontology metadata management and ontology-based dynamic knowledge discovery in open distributed systems. They also describe the HELIOS peer-based system.

Chapter IX, written by *Aufaure, Le Grand, Soto*, and *Bennacer*, presents a state-of-the-art review of techniques covering metadata and ontologies, Semantic Web information retrieval, and automatic semantic extraction. They also describe open research areas and major current research programs in this domain.

Finally, **Section IV** presents applications and policies. This section consists of three chapters, Chapters X, XI, and XII. These chapters present applications in geospatial and pervasive computing, as well as policies for the security management.

Chapter X, written by *Winter* and *Tomko*, presents a review of the ways of georeferencing in Web resources. They present a case study which investigates the possibilities of translating the semantics of georeferences in Web resources to landmarks in route directions. They also show that interpreting goereferences in Web resources enhances the perceivable properties of described features.

Chapter XI, presented by *Tsounis, Anagnostopoulos, Hadjiethymiades*, and *Karali*, focuses on pervasive computing which creates an environment that seamlessly integrate devices with computing and communication capabilities. Since it poses an interoperability issues, they argue that the use of Web Semantic technology, like ontologies, may resolve these issues.

Finally, **Chapter XII**, written by *Clemente, Pérez, Blaya*, and *Skarmeta*, focuses on policies. They argue that by appropriately managing policies, a system can be continuously adjusted to accommodate variations imposed by constraints and environmental conditions. They present an evaluation of the use of ontology languages to represent policies for distributed systems.

How to Read This Book

Each chapter in this book has a different flavor from any others due to the nature of an edited book, although chapters within each part have a broad topic in common. A suggested plan for a first reading would be to choose a particular part of interest, and read the chapters in that part. For more specific seeking of information, readers interested in ontological views and extractions, ontological representation using OWL, ontological modelling language may read the first three chapters. Readers interested in looking at ontological management and adaptation in enterprise systems, as well as annotation systems may study the chapters in the second part. Readers, who are interested in ontological queries, metadata, knowledge discovery, and Semantic Web mining, may go directly to the third part. Finally, those interested in applications in geospatial Semantic Web, pervasive computing, and security management, may go directly to the final part of this book.

Each chapter opens with an abstract that gives the summary of the chapter, an introduction and closes with a conclusion. Following the introduction, the background and related work are often presented in order to give readers adequate background and knowledge to enable them to understand the subject matter. Most chapters also include an extensive list of references. This structure allows a reader to understand the subject matter more thoroughly by not only studying the topic in-depth, but also by referring to other work related to each topic.

What Makes This Book Different?

Web Semantics is a growing area in the broader field of Web technology. A dedicated book on important issues in Web Semantics and ontology is still difficult to find. Most books narrowly focus on one particular aspect of Web Semantics, such as RDF, etc. This book is therefore different in that it covers an extensive range of topics including ontological modelling, enterprise systems, querying and knowledge discovery, and wide range of applications.

This book gives a good overview of important aspects in the development of Web Semantics. The four major aspects covering ontological modelling, enterprise systems, Semantic Web mining, and applications, described in four parts of this book respectively, form the comprehensive foundations of Web Semantics and ontology.

The uniqueness of this book is also due to the solid mixture of both theoretical aspects as well as practical aspects of Web Semantics and ontology development. The application chapter presents a case study on geospatial Web Semantics. Other potential applications in pervasive computing environment are also presented. Throughout the book, languages and tools for Web Semantics and ontology are described. These include OWL, XML, metadata, RDF, etc. Theoretical issues, including security management and annotation, are also covered. Issues of adopting ontology in enterprise systems are also comprehensively discussed.

A Closing Remark

We would like to conclude this preface by saying the this book has been compiled from extensive work done by the contributing authors who are researchers and industry practitioners in this area and who particularly have expertise in the topic area addressed in their respective chapters. We hope that readers benefit from the works presented in this book.

David Taniar, PhD
Johanna Wenny Rahayu, PhD
Melbourne, Australia
October 2005

Acknowledgments

The editors would like to acknowledge the help of all involved in the collation and review process of the book, without whose support the project could not have been satisfactorily completed.

We would like to thank all the staff at Idea Group Inc., whose contributions throughout the whole process from inception of the initial idea to final publication have been invaluable. In particular, our thanks go to Kristin Roth, who kept the project on schedule by continuously monitoring our progress on every stage of the project, and to Mehdi Khosrow-Pour and Jan Travers, whose enthusiasm initially motivated us to accept their invitations to take on this project.

We are also grateful to our employers — Monash University and La Trobe University, for supporting this project. We acknowledge the support of the School of Business Systems at Monash and the Department of Computer Science and Computer Engineering at La Trobe in giving us archival server space for the reviewing process.

A special thank goes to Mr. Eric Pardede of La Trobe University, who assisted us in the entire process of the book: from collecting and indexing the proposals, distributing chapters for reviews and re-reviews, constantly reminding reviewers and authors, liaising with the publisher, to many other housekeeping duties which are endless.

In closing, we wish to thank all of the authors for their insights and excellent contributions to this book in addition to all those who assisted us in the review process.

David Taniar, PhD
Johanna Wenny Rahayu, PhD

Section I

Web Semantics and Ontologies Modelling

Chapter I

Ontology Extraction Using Views for Semantic Web

Carlo Wouters, La Trobe University, Australia

Rajugan Rajagopalapillai, University of Technology, Sydney, Australia

Tharam S. Dillon, University of Technology, Sydney, Australia

Wenny Rahayu, La Trobe University, Australia

Abstract

The emergence of Semantic Web (SW) and the related technologies promise to make the Web a meaningful experience. Conversely, success of SW and its applications depends largely on utilization and interoperability of well-formulated ontology bases in an automated heterogeneous environment. This creates a need to investigate utilization of an (materialized) ontology view as an alternative version of an ontology. However, high level modeling, design and querying techniques still proves to be a challenging task for SW paradigm, as, unlike classical database systems, ontology view definitions and querying have to be done at high-level abstraction. In order to address such an issue, in this chapter, we describe an abstract view formalism for

Copyright © 2006, Idea Group Inc. Copying or distributing in print or electronic forms without written permission of Idea Group Inc. is prohibited.

SW (SW-view) with conceptual and logical extensions. SW-views provides the needed conceptual and logical semantics to engineer ontology bases using three levels of abstraction, namely (1) conceptual, (2) logical/schema and (3) instance levels. We first provide the view model and it formal properties including a set of conceptual operators, which enable us to do ontology extraction at the conceptual level. Later, we provide a schemata transformation methodology to materialize SW-views under the Ontology Extraction Methodology (OEM) framework.

Introduction

Meaning of data is emerging as the main area of interest in the awake of meaningful Web era, which is the Semantic Web (SW) paradigm (W3C-SW, 2005a). As envisage by Berners-Lee (1998), SW is emerging as the new medium for the decentralized, automated global information sources for the new 21st century information-driven economies (Aberer et al., 2004). This is highly visible in the exponential increase of new research directions in engineering ontologies in a wide spectrum of domains ranging from traditional enterprise data to time-critical medical information and infectious decease databases. For such vast ontology bases to be successful and to support autonomous computing, in a meaningful distributed environment, the preliminary design and engineering of such ontologies should follow strict software engineering disciplines. Furthermore, supporting technologies for ontology engineering such as data extraction, integration and organization have be matured to provide adequate modeling and design mechanism to build, implement and maintain successful techniques. For such purpose, Object-Oriented (OO) paradigm seems to be an ideal choice as it has been proven in many other complex applications and domains (Dillon & Tan, 1993; Graham, Wills, & O'Callaghan, 2001).

OO conceptual models have the power in describing and modeling real-world data semantics and their interrelationships in a form that is precise and comprehensible to users (Dillon & Tan, 1993; Graham et al., 2001). But the existing OO modeling languages (such as UML [OMG-UML™, 2003a]) provide insufficient modeling constructs for engineering SW models and applications. This is mainly due to lack of inherent support for semistructured schema-based data descriptions and constraints in OO modeling languages and the shortcomings of many semistructured data models in providing visual modeling and higher levels of abstraction semantics (such as conceptual models) that are easily understood by humans. Due to this, in the Semantic Web paradigm, most modeling and design constructs are modeled at a lower level of abstraction, namely schema or data description language levels.

Copyright © 2006, Idea Group Inc. Copying or distributing in print or electronic forms without written permission of Idea Group Inc. is prohibited.

Regrettably, high level modeling, design and querying techniques still proves to be a challenging task for the SW paradigm as many requirements for such tasks require management and organization of heterogeneous vocabularies or ontologies at higher levels of abstraction. Conversely, in SW, formulation of data semantics are not provided by one or more fixed schema/(s), but an ontology. Such challenges present a motivation to investigate the use of views in the SW paradigm.

Since the introduction of view formalism in the relational data model (Date, 2003; Elmasri & Navathe, 2004), motivation for views has changed over the last two decades. At present view formalisms are used in Rajugan, Chang, Dillon, and Ling (2005a): (a) user access and user access control (UAC) applications, (b) defining user perspectives/profiles, (c) designing data perspectives, (d) dimensional data modeling, (e) providing improved performance and logical abstraction (materialized views) in data warehouse/OLAP and Web-data cache environments, (f) Web portals and profiles, and (g) Semantic Web (SW) (W3C-SW, 2005a) paradigms for sub-ontology or ontology views (Volz, Oberle, & Studer, 2003b; Wouters, Dillon, Rahayu, Chang, & Meersman, 2004b). From this list, it is very apparent that the applications and usefulness of views are realized more than their originally intended purpose (the 2-Es; data Extraction and Elaboration [Figure 1]), with extensive research being carried out by both researchers and industry to improve their design, construction and performance. Yet, the view concept is still a data language and model dependent low-level construct (implementation). Here we first briefly look at the history of the view mechanisms available today and some of the proposals for new view mechanisms supporting new semistructured data paradigms and SW.

Earlier we have shown that there are some important benefits in the database area by using views. The first major benefit, being able to view information in a different way without touching the actual structure (the adaptability aspect), is arguably even more important for Internet applications, as most of the users viewing the information are not the information authors, and have only read access. In general, it can be said that the information over the Internet has many different types of users, and it is harder to predict who these users will be while making the data available. This prevents an author to take into account all the users, and how they would like to view the information (i.e., what parts they consider relevant). The first identified benefit clearly is very important to ontologies and the Semantic Web. The second major benefit, enabling certain types of applications using views (the extendibility aspect), is also relevant to the Semantic Web, and once ontology views are commonplace, the same evolution as in database area can be expected.

For the purpose of this chapter, we need to make a distinction between the concept of abstract view definitions (addressed in this chapter) for SW and the

Copyright © 2006, Idea Group Inc. Copying or distributing in print or electronic forms without written permission of Idea Group Inc. is prohibited.

view definitions in SW languages such as Resource Description Framework (RDF) (W3C-RDF, 2004) and the Ontology Web Language (OWL, previously known as DAML+OIL) (W3C-OWL, 2002). Though expressive, SW-related technologies and languages suffer from visual modeling techniques, fixed models/schemas and evolving standards. In contrast, higher-level OO modeling language standards (with added semantics to capture Ontology domain specific constraints) are well-defined, practiced, and transparent to any underlying model, language syntax and/or structure. They also can provide well-defined models that can be transferred to the underlying implementation models with ease. Therefore for the purpose of this chapter, an abstract view for SW is a view, where its definitions are captured at a higher level of abstraction (namely, conceptual), which in turn can be transformed, mapped, and/or materialized at any given level of abstraction (logical, instance, etc.) in a SW-specific language and/or model.

To address such an issue, in this chapter, we propose a view formalism for SW (SW-view). The proposed view formalism provides: (1) conceptual and logical semantics with extensions, (2) an OO-based design methodology to design SW architectural constructs, and (3) an extensive set of *conceptual operators* (Rajugan, Chang, Dillon, & Ling, 2005b) that can be applied in Ontology Extraction Methodology (OEM) (Wouters et al., 2004b).

In this chapter, we present an SW-view formalism that can adapt to changing data model and language requirements in the SW paradigm. It is independent of data language and models, where view definitions are captured using any higher-level modeling language such as UML or XML Semantic (XSemantic) nets (Feng, Chang, & Dillon, 2002). Such flexibility is achieved by providing three levels of abstraction for view definition, namely at the conceptual, logical, and documentary levels. In addition, to support data extraction and elaboration, we provide an extensive set of conceptual operators with corresponding restrictive operator set for ontology extraction. The design methodology for SW-view is based on visual OO conceptual modeling techniques discussed extensively in Dillon and Tan (1993). Thus SW-view is a view formalism with built-in design methodology oriented towards semistructured data models and SW.

An overview of the chapter organization is as follows: In section 2, a discussion on view formalisms for different data models are given, followed by a brief discussion on benefits of views and in ontologies in section 3. Section 4 provides a detailed discussion on the view formalism for SW (SW-views), formal definitions, modeling issues and the conceptual operators. Section 5 presents discussion on applying SW-view formalism in the area of Ontology Extraction. It is shown that for ontologies, the formalism can still be applied. The Ontology Extraction Methodology uses a restricted version of the view. In section 6, a practical example is given of how views for Semantic Web and ontology extraction can be utilized in real-world scenarios. Especially, the automated

Copyright © 2006, Idea Group Inc. Copying or distributing in print or electronic forms without written permission of Idea Group Inc. is prohibited.

extraction of ontology views is used. Section 7 concludes the chapter with some discussion on our future research direction.

Related Work

We can group the existing view models into four categories, namely: (a) classical (or relational) views (Date, 2003; Elmasri & Navathe, 2004; Rajugan, Chang et al., 2005a; Rajugan, Chang, Dillon, & Ling, 2005c; Wouters et al., 2004b), (b) object-oriented view models, (c) semistructured (namely XML) view models, and (d) view models for SW. An extensive set of literature can be found in both academic and industry forums in relation to various view-related issues such as (1) models, (2) design, (3) performance, (4) automation, and (5) turning/ refinement, mainly supporting the 2-Es; data Extraction and Elaboration. A comprehensive discussion on existing view models can be also found in Rajugan, Chang et al. (2005c). Here, we focus only on view models for semistructured data.

Since the emergence XML (W3C-XML, 2004), the need for semistructured data models that have to be independent of the fixed data models and data access violates fundamental properties of classical data models. Many researchers attempted to solve semistructured data issues by using graph-based (Zhuge & Garcia-Molina, 1998) and/or semistructured data models (Abiteboul, Goldman, McHugh, Vassalos, & Zhuge, 1997; Liefke & Davidson, 2000). Again, the actual view definitions are only available at the lower level of the implementation and not at the conceptual and/or logical level. One of the early discussions on XML views was by Abiteboul (1999) and later more formally by Cluet et al. (2001). They proposed a declarative notion of XML views. Abiteboul et al. pointed out that a view for XML, unlike classical views, should do more than just provide different presentation of underlying data (Abiteboul, 1999). These concepts, which are implemented in the Xyleme project (Lucie-Xyleme, 2001), provide one of the most comprehensive mechanisms to construct an XML view to date. But, in relation to conceptual modeling, these view concepts provide no support. Other view models for XML include (a) the MIX (Mediation of Information using XML) view system (Ludaescher, Papakonstantinou, Velikhov, & Vianu, 1999), (b) an intuitive view model for XML using Object-Relationship-Attribute model for Semi-Structured data (ORA-SS) (Chen, Ling, & Lee, 2002). This is one of the first view models that supports some form of abstraction above the data language level and (c) a layered view model for XML (Rajugan, Chang et al., 2005c), with three levels of abstraction, namely conceptual, logical, and document level.

In related work in the Semantic Web (W3C-SW, 2005b) paradigm, some work has been done in views for SW (Volz, Oberle, & Studer, 2003a; Volz et al.,

Copyright © 2006, Idea Group Inc. Copying or distributing in print or electronic forms without written permission of Idea Group Inc. is prohibited.

2003b), where the authors proposed a view formalism for RDF document with support for RDF (W3C-RDF, 2004) schema (using an RDF schema supported query language called RQL). This is one of the early works focused purely on RDF/SW paradigm and has sufficient support for logical modeling of RDF views. The extension of this work (and other related projects) can be found at KAON (2004). RDF is an object-attribute-value triple, where it implies that the object has an attribute with a value (Feng, Chang, & Dillon, 2003). It only makes intentional semantics and not data modeling semantics. Therefore, unlike views for XML, views for such RDF (both logical and concrete) have no tangible scope outside its domain. In related area of research, the authors of the work propose a logical view formalism for ontology (Wouters, Dillon, Rahayu, Chang, & Meersman, 2004a; Wouters et al., 2004b) with limited support for conceptual extensions, where materialized ontology views are derived from conceptual/ abstract view extensions.

Another area that is currently under development is the view formalism for SW metalanguages such as OWL. In some SW communities, OWL is considered to be a conceptual modeling language for modeling Ontologies, while some others consider it to be a crossover language with rich conceptual semantics and RDF-like schema structures (Wouters et al., 2004a). It is outside the scope of this chapter to provide argument for or against OWL being a conceptual modeling language. Here, we only highlight one of view formalism that is under development for OWL, namely views for OWL in the "User Oriented Hybrid Ontology Development Environments" (HyOntUse, 2003) project.

Views, Databases, and Ontology

The main benefits of views have evolved since it was first introduced in relational database systems. This is mainly because the concept of views has been widely used in various advanced applications and database systems. In this chapter, we can categorize the benefits of views which are relevant to our proposed method from two perspectives:

• **Adaptability aspect:** The concept of views provides a mechanism to generate and present data in different structures and formats without the necessity to redefine the underlying structure of the stored data. This mechanism enables us to create user- or domain-oriented virtual data subsets which are relevant to some specific requirements. The fact that the view is only invoked on top of the stored interconnected relations means

Copyright © 2006, Idea Group Inc. Copying or distributing in print or electronic forms without written permission of Idea Group Inc. is prohibited.

that data integrity rules and constraints are all handled at the underlying data level. This provides further flexibility to the created views.

- **Extendibility aspect:** The concept of views has enabled a number of new and advanced applications to be built efficiently. These are normally those applications that deal with the storage and manipulation of a large amount of data and yet there is substantial need to analyze and view the information in different fashions. These applications include data warehousing (Mohania, Karlapalem, & Kambayashi, 1999; Roussopoulos, Kotidis, Labrinidis, & Sismanis, 2001), mediators in bioinformatics databases (Do & Rahm, 2004) and XML document repository (Chan, Dillon, & Siu, 2002; Cluet et al., 2001). While the basic idea of views is adopted in these applications, each of them has applied additional rules and mechanisms to make it feasible in the new application domains.

Database is a very well-defined area, where there are clear standards of what can and cannot be realized in a (traditional) database. Although there are extensions (e.g., active, deductive, spatial, and temporal databases), there is still a clear understanding of the basic principles of this area. Although the differences are many, the motivations of why databases and ontologies are used are very similar. Both serve to structure the vast amounts of information available. Databases transferred the unstructured text documents to structured tables and enabled applications to use this data. A similar approach is intended for ontologies, but then applied to the unstructured information on the Web. Because of the characteristics of the World Wide Web, databases can not successfully be applied to it. The major inhibitors for the database approach are:

- The Web is dynamic/ad hoc. Information is constantly changing, as well as the intended structures. Information can be very dynamic for databases, but the structures have to be static, and once established, should hardly change.
- The Web is distributed. Distributed databases is not, by far, the established area that databases is, and is still being researched. As per definition, all the up-to-date information is spread all over the world, and this is a major hurdle.
- The accepted standard for information, and partly responsible for the major success of the Internet, is HTML. However, this language offers few capabilities to structure information. Converting such documents to databases would not improve the structure or the ability of the information to be used in applications.

Copyright © 2006, Idea Group Inc. Copying or distributing in print or electronic forms without written permission of Idea Group Inc. is prohibited.

On the other hand, there is no consensus on the exact definition of an ontology, and different approaches assign varying levels of functionality to an ontology. For instance, OWL (W3C-OWL, 2002) lets you create instances as part of the ontology, while the DOGMA approach (Spyns, Meersman, & Mustafa, 2002) does not. This is just one of many differences that exist between the various ontology standards. As a consequence, some of the elements in this chapter depend on the chosen ontology standard, and might vary for other ontology standards. The Ontology Extraction Methodology that is presented in Wouters, Dillon, et al. (2002) and Wouters et al. (2004a, 2004b) addresses this problem by providing flexible support for multiple standards. This is possible by taking a high level approach, which can be seen in this chapter by the specification of a conceptual view, and an extraction methodology that considers the conceptual level. All the benefits from various ontology standards can be incorporated by extending the methodology with optimization schemes (which are on a lower level, i.e., standard or even language-specific level).

Views for the Semantic Web (SW-View)

The emergence of Semantic Web (SW) and the related technologies promise to make the Web a meaningful experience. Yet, high level modeling, design, and querying techniques still prove to be challenging tasks under the SW paradigm. Unlike relational database views, in SW, data semantics are usually defined at a higher level of abstraction. Therefore, a SW-view formalism should have their definitions captured at a higher level of abstraction (Volz et al., 2003a, 2003b; Wouters et al., 2004b) and provide some mechanisms to be able to execute over heterogeneous data and schemas without loss of view definitions semantics.

Views in general can be considered as a special kind of transformation. Figure 1 shows a generic partitioning of any transformation into the 3 E's: Extraction, Elaboration, and Extension. Extraction can informally be defined as taking a part of the original without any modifications. Elaboration is providing such an extracted part with additional levels of detail (also referred to as interpolation). Finally, Extension can be considered as the addition of completely new elements (i.e., they were not present in any shape or form in the original). Although this is an informal partitioning, it nonetheless agrees with the common vision in many areas that use extraction.

As stated, views are a certain type of transformation. When considering Figure 1, views are purely situated in the areas of Extraction and Elaboration. Throughout the remainder of this chapter, Extension will not be considered anymore. In addition, a SW-view formalism should be able to deal with not just one but

Copyright © 2006, Idea Group Inc. Copying or distributing in print or electronic forms without written permission of Idea Group Inc. is prohibited.

multiple data-encoding language standards and schemas (such as XML, RDF, OWL, etc.), as enterprise content may have not one, but multiple data-coding standards and ontology bases. Another issue that deserves investigation is the modeling techniques of views for SW. Though expressive, SW related technologies suffer from proven visual modeling techniques (Cruz, Decker, Euzenat, & McGuinness, 2002). This is because object-oriented modeling languages (such as UML) provide insufficient modeling constructs for utilizing semistructured (such as XML, RDF, OWL) schema-based data descriptions and constraints, while XML/RDF schema lack the ability to provide higher levels of abstraction (such as conceptual models) that are easily understood by humans. But many researchers have proposed OMG's UML (OMG-UML™, 2003b) (and OCL)-based solutions (Cruz et al., 2002; Gašević, Djuric, Devedzic, & Damjanovic, 2004a, 2004b; Wongthamtham, Chang, Dillon, Davis, & Jayaratna, 2003; Wouters et al., 2002; Wouters et al., 2004b), with added extensions to model semistructured data.

In this chapter, we propose a view formalism with conceptual and logical extensions for the SW (SW-view). Initially such view formalism was proposed for XML data models by Rajugan et al. (2003) (shown in Figure 2) with clear distinctions between the three levels of abstraction, namely: (a) conceptual, (b) logical (or schematic), and (c) document (or instance). Here it is adopted for the SW paradigm. In work with XML, the authors provide clear distinctions between conceptual, logical, and document levels views: as in the case of data engineering, there exists a need to clearly distinguish these levels of abstractions. But in the case of ontology views, though there exists a clear distinction between

Figure 1. 3 Es of views

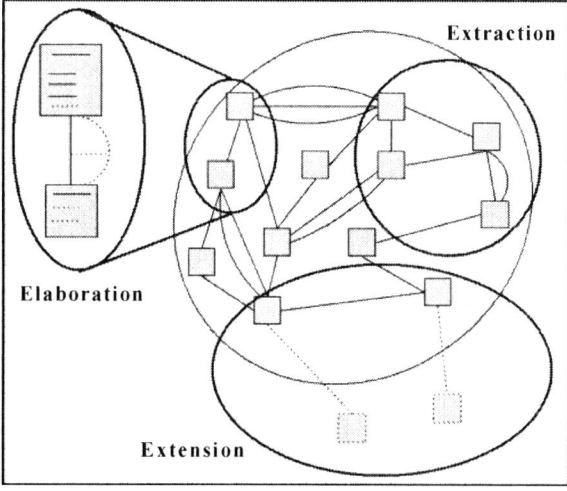

Copyright © 2006, Idea Group Inc. Copying or distributing in print or electronic forms without written permission of Idea Group Inc. is prohibited.

conceptual and logical models/schemas, the line between the logical (or schema) level and document (or instance) level tends to overlap due to the nature of ontologies, where concepts, relationships, and values may present mixed sorts, such as schemas and values.

Therefore, in the SW-view formalism, we provide a clear distinction between conceptual and logical views, but depending on the application, we allow an overlap between logical views and document views (thus it is shown as a dashed line in Figure 2). This is one of the main differences between the XML view formalism and the SW-views. To our knowledge, other than our work, there exist no research directions that explore the conceptual and logical view formalism for the Semantic Web paradigm. This notation of SW-view formalism has explicit constraints and an extended set of expressive conceptual operators to support Ontology Extraction Methodology (Wouters et al., 2002; Wouters et al., 2004a, 2004b).

Conceptual Views

The conceptual views are views that are defined at the conceptual level with conceptual level semantics using a higher-level modeling language such as UML (OMG-UML™, 2003a). To understand the SW-view and its application in

Figure 2. SW-view formalism and levels of abstraction

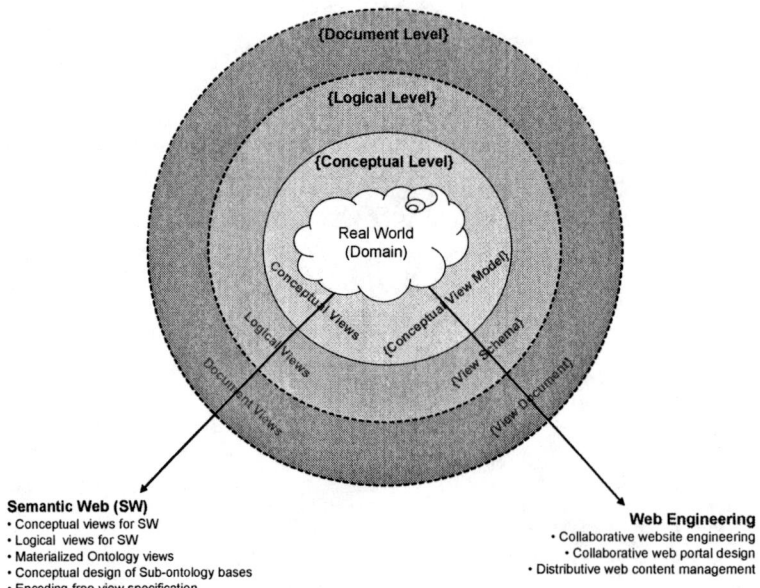

Copyright © 2006, Idea Group Inc. Copying or distributing in print or electronic forms without written permission of Idea Group Inc. is prohibited.

constructing ontology views, it is imperative to understand its concept and its properties. First, an informal definition of the view concept is given followed by a formal definition that serves the purpose of highlighting the view model properties and the modeling issues associated with such a high-level construct.

Definition 1: A conceptual view (Rajugan et al., 2003) is the one which is defined at the conceptual level with higher level of abstraction and semantics.

To utilize the SW-view model in applications, it is imperative that, one must first understand some of its unique properties and characteristics. In this section, we first provide some of the SW-view formal conceptual semantics followed by the derivation of the conceptual view definition. It should be noted here that, though more elaborated definitions are possible depending on the application domain, here we provide a simplified generic SW-view definition that can be easily applied to ontology extraction. Following the conceptual view definition are the sections that address some of the unique characteristics of the conceptual view formalism, including conceptual view hierarchy, conceptual operators (Rajugan, et al., 2005b), some of the modeling issues associated with the conceptual views, and the descriptive constraint model.

Conceptual Objects (CO): CO refers to model elements (objects, their properties, constraints, and relationships) and their semantic interrelationships (such as composition, ordering, association, sequence, all, etc.) captured at the conceptual level, using a well-defined modeling language such as UML (Feng, Chang et al., 2003; Rajugan, Chang, Dillon, & Feng, 2005), or XSemantic Nets (Feng et al., 2002) or E-ERD (Elmasri & Navathe, 2004), etc. A CO can be either of type simple content ($s_{content}$) or complex content ($c_{content}$) depending on its internal structure (Feng et al., 2002; Feng, Chang et al., 2003; Feng, Dillon, Weigand, & Chang, 2003). For example, CO that uses primitive types (such as integer, character, etc.) as their internal structure corresponds to $s_{content}$ and CO that uses composite objects to represent their internal structure corresponds to $c_{content}$.

Conceptual Schema (CS): We refer conceptual schema as the metamodel (or language) that allow us to define, model, and constrain COs. For example, the conceptual schema for a valid UML model is the MOF (combined with its associated MOF metamodel elements such as stereotypes and data dictionaries). Also, the UML metamodel provides the namespace of such schemas.

Copyright © 2006, Idea Group Inc. Copying or distributing in print or electronic forms without written permission of Idea Group Inc. is prohibited.

Like XML Schema, where the instance will be an XML document, here, an instance of the conceptual schema will be a *well-defined*, *valid* conceptual model (in this case in UML) or other conceptual schemas (i.e., metamodel such as MOF), which can be either visual (such as UML class diagrams) or textual (in the case of UML/XMI models).

Logical/Schema Objects (LO): When CO are transformed or mapped into the logical/schema level (such rules and mapping formalism described in works such as Feng, Chang et al. [2003]; Gašević et al., [2004b]), the resulting objects are called LO. These objects are represented in textual (such as a schema language, OWL) or other formal notations that support schema objects (such as a graph).

Postulate 1: A *context* (ε) is an item (or collection of items) or a concept that is of interest for the organization as a whole. It is more than a measure (Golfarelli, Maio, & Rizzi, 1998; Trujillo, Palomar, Gomez, & Song, 2001) and is a meaningful collection of model elements (classes, attributes, constraints, and relationships) at the conceptual level, which can satisfy one or more organizational perspective/(s) in a given domain. Simply said, it is a collection of concepts, attributes, and relationships that are of interest in construction of other ontology/(ies).

For example, in our bank case study example, "staff", "accounts", and "customers" can be some of the (simplified) examples of a context.

Postulate 2: A *perspective* (∂) is a viewpoint of an item (or a collection of items) that makes sense to one or more stakeholders of the organization or an organizational unit, at given point in time. That is, one viewpoint of a context at a given point in time.

For example, in our bank case study "BankStaff Who Look After VIP-Customers" or "Accounts Owned by Customers Who has NO-Income" are couple of perspectives in the context of "staff" and "accounts" respectively.

Definition 2: A conceptual view ($\mathcal{V}_{CO}$) (Rajugan et al., 2003) is a view, defined over a collection of valid *model elements, at the conceptual level. That is, it is a perspective for a given context at a given point in time.*

Let X be a collection of COs. Let $\mathfrak{R}$ be the rule set, constraints, and syntaxes that makes X a **valid** collection of CO (according to a metamodeling language such

Copyright © 2006, Idea Group Inc. Copying or distributing in print or electronic forms without written permission of Idea Group Inc. is prohibited.

as MOF or UML or XSemantic nets). Therefore it can be shown that, a *valid conceptual collection set X* is a function of $\Re$, shown as:

$$X = \Re(X).$$

A *valid conceptual view* V_{CO} of the *valid* CO set *collection X* is defined as the *perspective* ∂ constructed over a *context* ζ by the conceptual *construct* λ. The resulting conceptual view belongs to the *domain* $\mathcal{D}(V_{CO})$, (where $\mathcal{D}(V_{CO}) = \mathcal{D}_{CO}(\varepsilon)$) with *schema* $S_{CO}(V_{CO})$, (where $S_{CO}(V_{CO}) = S_{CO}(\partial)$). The conceptual view is said to be valid if it is a valid instance of the view schema $S_{CO}(\partial)$. Let V be a function of a view, therefore *conceptual view* V_{CO}:

$$V_{co} = V(\varepsilon, \partial, \lambda, X)$$

where the *view name* of V_{CO} is provided by the perspective ∂; the *domain* and the *namespace* for V_{CO} is provided by the context ε in the *valid* CO collection set of X; the view construction is provide by the conceptual construct λ (i.e., *conceptual operators* that construct the view over a given context); the *valid* collection set X set provides the data for the view V_{CO} instantiation; the *view schema* $S_{CO}(V_{CO})$ that constrains and validates the view instances of the view V_{CO}; and the domain $\mathcal{D}(V_{CO})$ provides the domain for the view V_{CO}.

For example, in our case study, a conceptual view "VIP-Customers" (*perspective*) can be constructed for a *context* of "customers", for those bank customers who has more than seven banking accounts (perspective/view *constraint$_i$*), each with an average (perspective/view *constraint*) balance of US$3,000 (perspective/view condition).

The Conceptual View Hierarchy

In the OO paradigm, a class has both attributes (may be nested or set-valued) and methods. A class can also form complex hierarchies (inheritance, part-of, etc.) with other classes. Owing to these complexities, definition of views for the OO paradigm is not straightforward. In one of the early discussions on OO data models, Kim and Kelly (1995) argue that an OO data model should be considered as one kind of extended relational model. Naturally, this statement reflected on most discussions of OO views (Abiteboul & Bonner, 1991; Chang, 1996; Kim & Kelly, 1995). In general, the concept of *virtual class* is considered as the OO equivalent of the relational view formalism (virtual relation).

Copyright © 2006, Idea Group Inc. Copying or distributing in print or electronic forms without written permission of Idea Group Inc. is prohibited.

Abiteboul and Bonner (1991) argue that *virtual classes* and stored classes are interchangeable in a class hierarchy. They also stated that, virtual classes are populated by creating new objects (imaginary objects) from existing objects and other classes (virtual objects).

But Kim and Kelly and Chang argue that, in contrast to relational models, class hierarchy and view (virtual) class hierarchy should be kept separate. It is inappropriate to include the view virtual class in the domain inheritance hierarchies because: (1) A view can be derived from an existing class by having fewer attributes and more methods. It would be inappropriate to treat it as a subclass unless one allow for the notion of selective inheritance of attributes. (2) Two views could be derived from the same subclass with different groups of instances. However the instances from one view definition could be overlapping with the other and nondisjoint. (3) View definitions, while useful for examining data, might give rise to classes that may not be semantically meaningful to users (Bertino, 1992). (4) Effects of schema changes on classes are automatically propagated to all subclasses. If a view is considered as a subclass, this could create problems (Kim & Kelly, 1995) in requiring the changes to be propagated to the view as it might be appropriate or inappropriate. (5) An inappropriate placement of the view in the inheritance hierarchy, could lead to the violation of the semantics because of the extent of overlapping with an existing class (Chang, 1996; Kim & Kelly, 1995).

In continuing the above discussion, to avoid confusion, we need to clarify the issue of the relationship between *stored* semistructured (XML, RDF, OWL, etc.) documents and the view documents. As in relational and OO systems, semistructured documents share some relational and many OO features. Naturally, new *view documents* may form new document hierarchies (inheritance, aggregation, nested, etc.), may extend the existing namespace of the stored XML namespace/(s) and may also be used to provide dynamic windows to one or more stored heterogenous data domains. Views may also be used to provide imaginary schema changes (such new simple/complex tags, new document hierarchy, restructuring, etc.). But, keeping in line with the arguments presented for OO views in Chang and Kim and Kelly, we believe that the stored documents hierarchy and the view document hierarchies should be kept separate. Many of the points made by Kim and Kelly, Chang, and Dillon and Tan (1993) for OO views should apply to our conceptual view hierarchy.

At a given instant, users can add new stored documents, modify/delete old stored documents, modify structures/schema of stored documents, create new schemas/hierarchies, or create new view definitions based on stored documents and/or existing views. But, at any given instant, users *cannot* create new *stored document/schema* based on existing view document(s)/schema(s).

Copyright © 2006, Idea Group Inc. Copying or distributing in print or electronic forms without written permission of Idea Group Inc. is prohibited.

The Conceptual View Construct

The conceptual constructor is a collection of binary and unary operators, that operates on CO (at the conceptual level) to produce result that is again a valid CO collection. The set of binary and unary operators provided here is a complete or basic set, or other operators, such as division operator (Elmasri & Navathe, 2004) and compression can be derived from this basic set of operators.

(I) *Conceptual Binary Operators*

The conceptual set operators are binary operators that take in two operands and produce a result set. The following algebraic operators are defined for manipulation of CO collection sets. A CO collection set can be represented in UML, XSemantic nets or other high-level modeling languages.

Let x,y be two valid CO collection sets (operands) that belongs to domains $\mathcal{D}_{co}(x) = dom(x)$ and $\mathcal{D}_{co}(y) = dom(y)$ respectively.

1. **Union Operator:** A Union operator $\bigcup_{(x,y)}$ of operands x,y produces a CO collection set $\mathcal{R}$, such that $\mathcal{R}$ is again a valid CO collection that includes all COs that are either in x or in y or in both x and y with no duplicates. This can be shown as:

$$\bigcup_{(x,y)} = \mathcal{R} = x \cup y = x \cup x' = y \cup y', \text{ where } dom(\mathcal{R}) = \mathcal{D}_{co}(x) \cup \mathcal{D}_{co}(y).$$

2. **Intersection Operator:** An *Intersection* operator $\bigcap_{(x,y)}$ of operands x,y produces a CO collection set $\mathcal{R}$, such that $\mathcal{R}$ is again a valid CO collection that includes all COs that are in both x and y.

$$\bigcap_{(x,y)} = \mathcal{R} = x \cap y, \text{ where } dom(\mathcal{R}) = \mathcal{D}_{co}(x) \cap \mathcal{D}_{co}(y).$$

Note: Since both Union and Intersection operators are *commutative* and *associative*, they can be applied to n-ary operands, i.e., for $\bigcup_{(x_i,x_j)}$ where $1 \leq i \leq n$; $1 \leq j \leq i$:

$$\bigcup_{(x_i,x_j)} = \mathcal{R} = x_1 \cup x_2 \cup ..... x_i \cup x_j .... = x_{n-1} \cup x_{n-2} \cup ..... x_j \cup x_i .... \text{ and}$$

$$\bigcup_{(x_i,x_j)} = \mathcal{R} = (x_1 \cup x_2) \cup x_3 ..... x_i \cup x_j .... = x_1 \cup (x_2 \cup x_3) ..... x_i \cup x_j ....$$

Copyright © 2006, Idea Group Inc. Copying or distributing in print or electronic forms without written permission of Idea Group Inc. is prohibited.

Similarity for $\bigcap_{(x_i, y_j)}$ where $1 \le i \le n$; $1 \le j \le i$:

$$\bigcap_{(x_i,x_j)} = \mathcal{R} = x_1 \cap x_2 \cap .....x_i \cap x_j.... = x_{n-1} \cap x_{n-2} \cap .....x_j \cap x_i.... \text{ and}$$

$$\bigcap_{(x_i,x_j)} = \mathcal{R} = (x_1 \cap x_2) \cap x_3.....x_i \cap x_j.... = x_1 \cap (x_2 \cap x_3).....x_i \cap x_j....$$

3. **Difference Operator:** A *Difference* operator $\overline{D_{(x,y)}}$ of operands x, y produces a CO collection set $\mathcal{R}$, such that $\mathcal{R}$ is again a valid CO collection that includes all COs that are in x but not in y.

$$\overline{D_{(x,y)}} = \mathcal{R} = x - y,$$

where $dom(\mathcal{R}) = \mathcal{D}_{CO}(x)$. Also note, the difference operator is NOT *commutative*.

4. **Cartesian Product Operator:** A *Cartesian product* operator $\times_{(x,y)}$ of operands x, y produces a CO collection set $\mathcal{R}$, such that $\mathcal{R}$ is again a valid CO collection that includes all COs of x and y, combined in combinatorial fashion.

$$\times_{(x,y)} = \mathcal{R} = x \times y, \text{ where } dom(\mathcal{R}) = \mathcal{D}_{co}(x) \times \mathcal{D}_{co}(y).$$

5. **Join Operator:** A *Join* operator can be shown in its general form as:

$$\triangleright\triangleleft_{(x,y)} = \mathcal{R} = x \triangleright\triangleleft_{[j_{condition}]} y,$$

where, optional join-condition provides meaningful merger of COs.

A join-condition $j_{condition}$ be of the form: (1) simple-condition: where the join-condition $j_{condition}$ is specified using CO simple content $s_{content}$ types, (2) complex-condition: where the join-condition $j_{condition}$ is specified using CO complex content $c_{content}$ types, and (3) pattern-condition: where the join-condition $j_{condition}$ is specified using a combination of one or more CO simple and complex content types in a hierarchy with additional constraints, such as ordering, etc.

Copyright © 2006, Idea Group Inc. Copying or distributing in print or electronic forms without written permission of Idea Group Inc. is prohibited.

(i) Natural Join

A natural join operator $\triangleright\triangleleft_{(x,y)}$ of operands x,y is a join operator with no join-condition specified, produces a CO collection set $\mathcal{R}$, such that $\mathcal{R}$ it is equivalent to a Cartesian product operator. This can be shown as:

$$\triangleright\triangleleft_{(x,y)} = R = x \triangleright\triangleleft y = \times_{(x,y)}.$$

(ii) Conditional Join

A join operator $\triangleright\triangleleft_{(x,y)}$ of operands x,y with explicit join-condition $j_{condition}$ specified, produces a CO collection set $\mathcal{R}$, such that $\mathcal{R}$ will have *only* the combination of CO collection set that satisfies the join condition. The join-condition $j_{condition}$ can only be of type: (1) simple-condition and (2) complex-condition. This join is comparable to the relational operator θ join. This can be shown as:

$$\triangleright\triangleleft_{(x,y)} = \mathcal{R} = x \triangleright\triangleleft_{(j_{condition1}[AND\ldots])} y.$$

(iii) Pattern Join

A join by pattern $\triangleright\triangleleft_{(x,y)}$ is a join by condition operator where the join-condition $j_{condition}$ is of type pattern-condition.

(II) *Conceptual Unary Operators*

We propose four unary conceptual operators to construct conceptual views without loss of CO semantics that are represented in the model. The four conceptual operators are projection, selection, rename, and restruct(ure).

1. **PROJECT Operator:** Given a valid CO collection set x, and a set of CO (either $s_{content}$ or $c_{content}$ or combination of both $s_{content}$ and $c_{content}$), the project operator $\prod_{(x)}$ will produce a CO collection set $\mathcal{R}$ where it has only the specified CO set with: (a) preserved node hierarchy, (b) preserved node order, and (c) preserved semantic relationships (if any). If need be, the projected CO set (in the case of hierarchical CO/(s) can be specified using the W3C XPath (W3C-XPath, 1999) standard. This can be shown as:

Copyright © 2006, Idea Group Inc. Copying or distributing in print or electronic forms without written permission of Idea Group Inc. is prohibited.

$$\Pi_{(x)} = \mathcal{R} = \Pi_{(CO_1, CO_2, \dots)}(x),$$

where the domain of $\mathcal{R}$ is $dom(\mathcal{R}) = \bigcup_{k=1}^{m} dom(CO_k)$.

2. **SELECT Operator:** Given a valid CO collection set x, the select operator $\sigma_{(x)}$ will produce a CO collection set $\mathcal{R}$, where it contains one or more matching CO (or collection) that satisfy the select-condition $s_{condition}$. In addition, the select-conditions can be combined using the AND, OR, NOT logical operators. This can be shown as:

$$\sigma(x) = \mathcal{R} = \sigma_{s_{condition}}(x).$$

Again, here, the select-condition $s_{condition}$ be of the form: (1) simple-condition: where the select-condition $s_{condition}$ is specified using CO simple content $s_{content}$ types and the select operator is called *value-based*, (2) complex-condition: where the select-condition $s_{condition}$ is specified using CO complex content $c_{content}$ types and the select operator is called *structure-based*, and (3) pattern-condition: where the select-condition $s_{condition}$ is specified using a combination of one or more CO simple and complex content types in a hierarchy with additional constraints, such as ordering, etc., where the select operator is called *structure-based*.

3. **RENAME Operator:** Given a valid CO collection set x, and a CO *src* (with old and new labels (l^{old}, l^{new})$\in L_{able}$), the rename operator $\rho_{(x)}$ will return x where the label of *src* is changed. A RENAME operation *cannot*; (a) alter *src* specific data types and (b) alter *src* specific contents, values or constraints. This can be shown as:

$$\rho_{(x)} = \mathcal{R} = \rho_{src(l^{old}, l^{new})}(x).$$

4. **RESTRUCT(ure) Operator:** Given a CO collection set x, and a CO, *src* (with a pair of positions, old and new (pos_1, pos_2)), where the positions can be either absolute or relative (in a CO hierarchy), the restructure operator $\delta_{(x)}$ will return $\mathcal{R}$, where the position of *src*(*src* can be either $s_{content}$ or $c_{content}$) is changed from pos_1 to pos_2. This can be shown as:

$$\delta_{(x)} = \mathcal{R} = \delta_{src(pos_1, pos_2)}(x).$$

Copyright © 2006, Idea Group Inc. Copying or distributing in print or electronic forms without written permission of Idea Group Inc. is prohibited.

But a restructure operation does not allow: (a) deletion of CO/(s) in the hierarchy, (b) alteration of CO structural relationships, constraints, names or cardinality, nor (c) alteration of CO data type or values.

Note: The operators presented above are referred to as extended or nonrestive *basic set*, as many secondary (e.g., DIVISION operator) and restrictive operators can be derived by combining one or more of these binary and unary operators.

Modeling Conceptual Views

In this chapter, to model conceptual views, we propose two OO modeling languages, namely OMG's UML (OMG-UML™, 2003a) (for modeling ontologies) and XSemantic nets. The only reason we use these notations in this chapter as the modeling standard for conceptual views is to demonstrate our concepts and applications and not to emphasis or promote these as the only modeling notation for conceptual views.

UML has established itself as the *defacto* modeling language of choice in OO conceptual modeling paradigm. UML provides a well-defined rich collection of tools to visually model a given domain into a needed level of abstraction. It can be said that UML helps to provide a well-defined blue print for a software system that is easily understood both by users and developers alike. UML also provides extensibility to the modeling language in the form of *stereotypes* which we utilise in defining our *conceptual views*. In the case of ontology engineering, UML provide classes (similar to concepts in ontology), attributes, and relationships that are used in defining ontology models (Wouters et al., 2004b) in this chapter.

Another reason we adopt UML is that its models are portable, in other words, many schemata transformation rules and mapping techniques exists for transforming UML models to: (a) XML Schema (Feng, Chang, et al., 2003), (b) Ontology Web Language (Gaševic et al., 2004b), (c) RDF, and (d) XMI. Therefore, for the purpose of this chapter, UML is visual modeling language of choice for OOCM and support abstraction from classical data models to ontology bases.

To model conceptual views in UML (OMG-UML™, 2003a), in addition to conceptual operators (constructor), we introduce a set of UML stereotypes (Rajugan, Chang, Dillon, & Feng, 2005) (view, OID, etc.) and constraints to visually model views. A stereotype is based on an existing base-model element or on a variant of the base-model element, to provide extensibility and model management for an existing, well-defined model. Since UML provides insufficient modeling constructs for conceptual views, XML schema description and

Copyright © 2006, Idea Group Inc. Copying or distributing in print or electronic forms without written permission of Idea Group Inc. is prohibited.

constraints (Feng et al., 2002; Feng, Chang et al., 2003; Rajugan, Chang, Dillon, & Feng, 2005), we provide a set of stereotypes and OCL to capture conceptual views in UML. Here, we use UML stereotypes to provide conceptual semantics to the view formalism, which is defined over a stored/domain data model (as shown in Figure 3). In addition, the constraint specification makes the view constraints more explicit and visible, where we use OMG's Object Constraint Language (OCL) (OMG-OCL, 2003; Warmer & Kleppe, 2003) based declarative constraint specification language. The following sections highlight some of the main stereotypes used to capture conceptual views in UML.

In the case of modeling conceptual views in XSemantic nets, it is straight forward (in comparison to UML) as it was proposed for semistructured data (namely XML). It consists of a rich set of constraints that can be represented, including all of class, object, attribute, relationship constraints, and special constraints such as ordering, disjunction, and class-attribute cardinality and dependency constraints.

Since XSemantic net is a directed graph, the model transformations between XSematic nets and the target schema/model is only a two-step schemata transformation. For example, models defined using this can be easily mapped to other schemas such as XML Schema, RDF, and OWL. The only difference between XSemantic net to XML schema transformation and the ontological language (such as OWL) is that the XML schemata transformation is graph-to-tree while the other is a directed-graph-to-graph transformation.

In the following sections, we show some of these view specific stereotypes and constraint specifications for conceptual views using UML/OCL, as they are well-understood without describing them in detail. A detailed discussion on XSemantic net model, constraints, and transformation can be found in Feng et al.

Constraint Specification

The constraint specification we used here is declarative; that is, it is simple, OCL-based, and helps to explain our view model constraints more explicitly in UML. This is shown in Figure 3, where a conceptual view is constructed from a stored domain class hierarchy and with OCL constraints (model, relationship, and view constraints).

In data modeling, specifications often involve constraints. In the case of views, it is usually specified by the data language in which they are defined in. For example, in the relational model, views are defined using SQL and a limited set of constraints can be defined using SQL (Date, 2003; Elmasri & Navathe, 2004), namely: (1) presentation-specific (such as display headings, column width, pattern, order, etc.), (2) range and string patterns for aggregate fields, (3) input

Copyright © 2006, Idea Group Inc. Copying or distributing in print or electronic forms without written permission of Idea Group Inc. is prohibited.

formats for updatable views, and (4) other DBMS specific (such view materialization, table block, size, caching options, etc.).

In object-relational and OO models, views had similar constraints but they are more extensive and explicit due to the data model. The views here are constructed and specified by DBMS specific (such as OQL [Cattell et al., 2000]) and/or external languages (such as C++, Java or O_2C [Abiteboul & Bonner, 1991]). It is a similar situation in views for a semistructured data paradigm, where a rich set of view constrains are defined using languages such as OQL-based LOREL (Abiteboul, Quass, McHugh, Widom, & Wiener, 1997). Today, in the case of ontology engineering (and in ontology views), this is still holds true, where constraints are specified using data definition languages, and not specified at the schema level. In doing so, the constraints are implicit and mostly accessible only at runtime of the system and not at the modeling and/or design time.

But the work by authors of Chen et al. (2002) provides some form of higher-level view constraints (under ORA-SS model) for XML views, while the work in Volz et al. (2003b) provides some form of logical level view constraints to be defined in views for in SW/RDF paradigm. Here, for our view formalism, we look into using UML/OCL as our view constraint specification language. Also, our work should not be confused with work such as Balsters (2003), where authors use OCL to "model" (not to specify) relation views (in contrast to ontology views), which utilizes OCL from a data-modeling point of view.

As our conceptual view mechanism is defined at a higher-level of abstraction, we can provide an explicit view constraint specification model, as most high-level OOCM languages (such as UML, XSemantic nets, E-ER) provide some form constraint specification.

In UML, the Object Constraint Language (OCL) (OMG-OCL, 2003), which is now a part of the UML 2.0 standard, can support unambiguous constraints specifications for UML models including specification of ontology model elements. In our conceptual view model, we incorporate OCL (in addition to built-in UML constraint features) as our view constraint specification language to explicitly state view constraints. It should be noted that we do not use OCL to define views, rather we use it to state additional constraints. OCL supports defining *derived* classes (OMG-OCL, 2003; Warmer & Kleppe, 2003), which is close to a view concept (Balsters, 2003).

To define our conceptual views, we show view classes visually, with the <<view>> stereotypes and the relationship between the stored class and the view as <<construct>> stereotype (Rajugan, Chang, Dillon, & Feng, 2005). Therefore, we do not require non-visual OCL view specification, but can be used to show some of the derivations rule for the attributes and/or operations to make the view definition more explicit and precise. It also supports specifying derived values and attributes in already existing views (and stored classes) (Rajugan, Chang, et al., 2005c).

Copyright © 2006, Idea Group Inc. Copying or distributing in print or electronic forms without written permission of Idea Group Inc. is prohibited.

In addition, further constraints can be defined for conceptual views including: (1) domain constraints (range of values, min, max, pattern, etc.), (2) constructional contents (set, sequence, bag, ordered-set), (3) ordering, (4) explicit homogenous composition/heterogeneous compositions, (5) adhesion and/or dependencies, (6) exclusive disjunction, and many more. Specifying these constraints using OCL expression in conceptual views are similar to that of stored domain objects.

(I) Constructor, <<construct>>

To show the relationship between a conceptual view and the stored class/(es) from which it is constructed, we use a directed-dashed line with <<construct>> keyword (shown above the line, Figure 3). This is to avoid confusion with the built-in UML dependency relationship and other stereotypes. As shown in Figure 3, where a conceptual view is constructed from a stored class hierarchy, the relationship is shown as <<construct>>. If a conceptual view is constructed over an existing conceptual view (view of a view), the same relationship is used show the hierarchy (the base conceptual view and the new conceptual view).

(II) Object Identifier, <<OID>>

In an OO system, an object has a unique system-wide identifier that is independent of the values of its attribute/(s), called *Object Identifier* or OID (Dillon & Tan, 1993). When created, an object will be referred to using its system assigned OID during its entire existence. In DBMS systems, OIDs can be either *logical* or *physical* depending on its nature.

In many OO conceptual models and diagrams, though the concept of OID is assumed to be an implicit concept (unlike primary keys in E/ER), in our work, with *conceptual views*, we have a need to explicitly state the OIDs and should be available to visualize at that highest level of abstraction. Therefore, here we provide a means of using OIDs for the purpose of IDs, similar to that of primary/ foreign key constraints available in E/ER models. We argue that just utilizing OID (a unique concept to OO systems) in our conceptual model provides additional semantics, such as providing Id/keys, referential, and integrity constraints that are visually lacking in many OO conceptual modeling techniques.

To visually model OID in a UML class diagram, we define a stereotype <<OID>>, shown in Figure 3, as an attribute type. Together with attribute name and optional type definition, the OID stereotype <<OID>> can be used in UML to indicate that the attribute that is an OID. Later in the implementation of the system, these OID can be mapped to XML Schema Specific **ID/KEY** and UNIQUE constraints (Rajugan, Chang, Dillon, & Feng, 2005).

Copyright © 2006, Idea Group Inc. Copying or distributing in print or electronic forms without written permission of Idea Group Inc. is prohibited.

Figure 3. An example conceptual view in UML (with constraints)

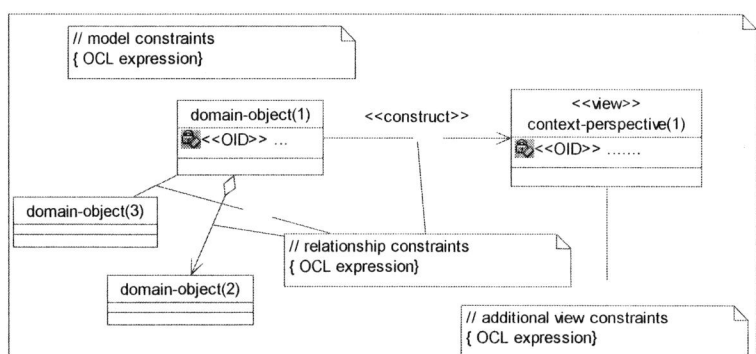

Figure 4. UML stereotype for an ordered composition

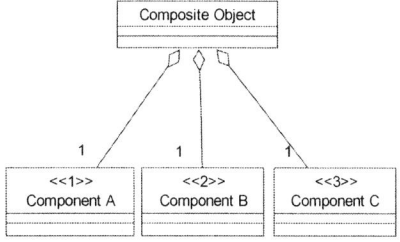

(III) Ordered/Unordered Composition

In the real-world, composite objects are in an aggregation with one or more sub-objects, and they also can be in a predefined order. For example in XML Schema construct such as with <xs:sequence>, we regularly observe that the tag <xs:sequence> signifies that the embedded elements are not only a simple assortment of components but have a specific ordering. This signifies an important OO concept, *ordered composition*.

Simply said, to capture ordering, we add a UML stereotype that allows capturing of the ordered composition utilizing stereotypes to specify the objects' order of occurrence such as <<1>>, <<2>>, <<3>>, ... ,<<n>>. This is shown in Figure 4. In related work (Nassis, R.Rajugan, Dillon, & Rahayu, 2005), authors have extensively discussed defining such ordered composition and mapping it to XML Schema. Due to page limitation we do not include that detailed discussion here.

Copyright © 2006, Idea Group Inc. Copying or distributing in print or electronic forms without written permission of Idea Group Inc. is prohibited.

Logical View and Document Views

Logical views are views that are defined at the logical or schematic level of abstraction. The logical views are intermediated views, that is, each logical view is derived from the corresponding conceptual view definition at the conceptual level and has equivalent document view definition (or construction) at the document level. Views at this level provide the schema (or template) for validation of the resulting view instance (from a conceptual view definition), at the document/instance level.

Conversely, in the case of document views, a document view can be used for many purposes (such as database views, materialised SW-views, etc.) and may be instantiated from simple projection of selected document tags to provide dynamic window into complex heterogenous documents. Based on how these documents are constructed, they can be classified into three categories (Rajugan et al., 2003): namely, Derived Imaginary documents, Constructed Imaginary documents, and Triggered Imaginary documents. In the following sections we highlight derivation and usage of such document views in the context of OEM.

As we stated earlier, unlike view models for data-oriented systems (e.g., relational, XML), in the SW-view model, the distinction between the conceptual and logical levels is clearly stated, but not between the logical and document levels. Here, in SW-views, the conceptual views are mapped to logical and document level views using the MOVE (Wouters et al., 2004a) system rather than as separate layer/(s). This implementation model is discussed in detail later in the later sections.

Mapping Conceptual Operators to Document View Query Expressions

In the case of SW-view model, one of the major differences between the XML-view and the SW-view is the transformation of the conceptual operators. In the case of XML-views, at the document level, the conceptual operators (from the conceptual view definition) are mapped to a native XML query language specific syntax (e.g. XQuery or SQL 2003) (Rajugan, Chang, et al., 2005b). Here, the conceptual operators (from the SW-view definition) are mapped to restricted ontology operators, which in turn mapped to ontology extraction algorithms. The following sections describe these issues in detail.

Copyright © 2006, Idea Group Inc. Copying or distributing in print or electronic forms without written permission of Idea Group Inc. is prohibited.

Semantic Web View for Ontology Extraction

The previous sections discussed the background of a view, the difference in interpretation when applied to databases or Semantic Web. Furthermore, a formal conceptualization of a Semantic Web view was presented. This section will apply the SW-view to the area of Ontology Extraction. This requires an introduction to some specific characteristics of an ontology, and subsequently, a restriction to the SW-view. This restricted SW-view is then used as the result of an extraction process.

Our previous work specified a high-level formal way to perform ontology extraction, called the Ontology Extraction Methodology (Wouters et al., 2002; Wouters et al., 2004a, 2004b). We will informally present the parts of the OEM that are needed to comprehend the use case given in the earlier section.

The most important part of this section is the description of the actual extraction process, or how the view is constructed. More formally, the manner in which the specification of the query to construct the view is obtained is quintessential here. In a later section, a fully automatic approach to the extraction process is proposed. This is then applied to a practical use case, not only showing low-level interpretations of the formal Semantic Web view specification, but also indicating the potential of an automated extraction process (that uses the SW-view).

Views for Ontology

The notion of materialization needs to be reconsidered for ontologies. One of the main differences between an ontology and a database is the fact that (according to some ontology standards, such as OWL) an ontology may include the whole interconnected concepts, attributes and relationships definitions and instances, which are normally strictly separated into "schema" and "instances" in a database system. The firm separation between schema and data both at the definition and storage levels is the very nature of a database system. While it is clear that a view is defined on the data or instances level in traditional database systems, this notion is no longer always practical in ontology as the traditional notion of "instances" is often integrated within the ontology as independent concepts. Relating this back to databases, it can be seen that the explicit mentioning of the actual schema is not used there.

The ontology, even though it may contain instance data, is primarily a metadata structure. A view has its own database schema (i.e., table definition), and this is always "materialized", as it is stored on the database server. For ontologies,

Copyright © 2006, Idea Group Inc. Copying or distributing in print or electronic forms without written permission of Idea Group Inc. is prohibited.

this is a very important fact, as dynamic mapping can be established without the need for a materialized metalevel (or ontology). As discussed further, the term materialized serves to make explicit the resulting schema (or metalevel/ontology).

Restricting Semantic Web View

Considering Figure 1, it was already discussed that a view (also the Semantic Web view) is situated in the areas of Extraction and Elaboration. The OEM (Wouters et al., 2002; Wouters et al., 2004a, 2004b) further restricts the view that is used to only be situated in the Extraction area. In other words, elaboration is not allowed in ontology extraction (as the name already suggests). To reflect this, a new definition is introduced, that restricts the SW-view. Such a view is called a Strict SW-view, or Ontology View.

Definition 3: Informal Definition for Ontology View

A Strict Semantic Web View (or Ontology View) is a materialized Semantic Web View that is derived from an ontology (called the base ontology). The derivation can consist of any (combination) of the following operations: synonymous rename, selection, and compression.

Intuitively, the allowed operations, including the restrictions on the result, can be interpreted as follows:

- A synonymous rename of an element indicates that the same element is still present in the Ontology View, but in a (syntactically) alternative form. Semantically, however, the same or less information is present. Note that this is a restricted version of the "rename" unary operator presented earlier.

- A selection of an element indicates that that element is still present in the Ontology View, in an unaltered form. These selected elements amount to the selection of a CO, which is an allowed unary operator.

- A compression of elements indicates that those elements are replaced by a single element in the Ontology View. The element itself can be a new element, but it will not provide additional semantic information (compared to the base ontology). The compression operator is a complex operator, constituted of the concatenation of multiple unary operations applied in sequence.

Copyright © 2006, Idea Group Inc. Copying or distributing in print or electronic forms without written permission of Idea Group Inc. is prohibited.

As shown, this super set of operators can be supported by one or more basic conceptual operators described in the earlier sections. From here on we will use this higher level set of operators to specify the extraction process. Note that the result of the extraction process is always a valid conceptual view, as they operators used can be broken down in the allowed atomic operators. However, many conceptual views can be constructed that would not constitute a valid Ontology View (hence the name "Strict" Semantic Web View).

Materialized vs. Virtual

It can be plainly seen from the definition that the intent and setup of the Ontology View is the same as those of the Database View. Note that in the definition the term "materialized" is used. As discussed earlier, depending on which ontology standard is followed, what is considered instance data is sometimes an integral part of the ontology. So, having a "virtual" view is not possible, as the ontology view definition also has the instance data intertwined with the metadata. In the case of ontology standards that do differentiate between metadata and instance data, the term "materialized" remains important, as it then indicates that the metalevel is materialized. In databases, there is no real distinction in terminology to indicate what level is considered. The metalevel of a view (the table or view definition) is always materialized. To clearly indicate the same for ontologies (also considering metalevel), the term "materialized" is still used. However, this does not necessarily mean that the instance data is stored with the ontology view as well.

Automated Ontology Extraction: An Overview

Previous sections have introduced operators that can be used to construct conceptual views. However, the process of constructing such a conceptual view is not detailed. Implicitly, it is typically considered a manual process, in other words, a knowledgable engineer decides what the elements of the view should be and constructs it to conform to them using the allowed operators. The OEM focuses on large-scale ontologies, and as such, a manual construction of the allowed view (using only the restricted operators defined above) would be a tedious task. The remainder of this chapter discusses how the Ontology View, which was a strict version of the formal conceptual view, can be constructed automatically using the OEM. The semantics of ontologies can be used to arrive at a qualitative automated extraction process. A similar use of intrinsic ontology features can be found in the heuristic rules of PROMTDiff, which has a high

Copyright © 2006, Idea Group Inc. Copying or distributing in print or electronic forms without written permission of Idea Group Inc. is prohibited.

success rate in predicting the complex changes that have occurred between two ontologies (Noy & Musen, 2002).

Although there are many important parts to the Ontology Extraction Methodology, for the purpose of this chapter we focus on the automation feature. The OEM provides a framework for extending itself through Optimization Schemes. More information on how this automation fits in the OEM is provided further, when important steps of the Materialized Ontology View Extractor (MOVE) system are presented.

Informally, Optimization Schemes are logically grouped sets of rules and algorithms to enforce those rules. The logical grouping is done in such a manner that they address a certain interpretation of an aspect of quality. In the OEM, quality has been broken down into many aspects, and for each aspect, certain interpretations exist. For instance, the size of the resulting ontology view after extraction may be an aspect that a user finds important. The interpretation the user may give is that the smaller the result, the better (given it provides all the information the user needs). Every Optimization Scheme (OS) addresses such an interpretation of an aspect of quality, and for this example the OS may be called "Total Simplicity Optimization Scheme". Here it is important that the user does not have to construct the Ontology View, but that the (query for the) view is generated automatically, through the rules and algorithms of the TSOS.

The rules and algorithms for the TSOS are created by an Ontology Engineer, and from then on, can be used by anyone, through selection of the OS. An example rule (informally stated) for the TSOS is:

If no interconnected graph can be constructed from the concepts and relationships of the extracted Ontology View up to this point adapt the query to include more relationships, so that at least a minimal spanning tree can be obtained. (Wouters et al., 2004b)

For this rule, there is an algorithm (constructed by the Ontology Engineer) that enforces this rule when applied to a partially extracted Ontology View. More information on how logical groupings of rules and algorithms are formed, and examples of applications of such OS algorithms can be found in Wouters et al. (2002) and Wouters et al. (2004b).

Additionally, to fine-tune the extraction, a user can specify through labeling what they consider important in the base ontology. Accordingly, "selected" and "deselected" labels are given to these elements, and the OS can take that into account.

As a single OS is almost never sufficient to meet all the user requirements, any combination of available OSs is allowed. All these elements combined provide

Copyright © 2006, Idea Group Inc. Copying or distributing in print or electronic forms without written permission of Idea Group Inc. is prohibited.

the user with a lot of flexibility to obtain a high-quality extracted Ontology View. The use case demonstrates in practice how the automated Ontology Extraction is used to seamlessly provide different users with different views of a base ontology, without spending a lot of time constructing the query for the view. Both the optimization schemes and the ontology labeling are shown in the Materialized Ontology View Extractor system architecture.

MOVE Workings

For the practical use case that is presented in the next sections, the Materialized Ontology View Extractor (Wouters et al., 2004a) system was used. The MOVE allows the user to provide high level requirements about the type of resulting view they would like, and the system automatically generates the requested Ontology View. The main benefit of MOVE for this purpose is that it can be used and operated by nonIT users. It does not require an ontology engineer anymore to manually construct a view and ensure its quality.

The general architecture of MOVE is given in Figure 5. Brief explanations about the workings of this system are given here. For a more detailed discussion we refer to Wouters et al. (2004a, 2004b).

As the inner workings of MOVE are not the focus here, but rather how the system is practically used, the steps (black, numbered circles in Figure 5) that are external to the system are discussed. More specifically, steps 1, 2, 3, and 7 are described.

- **Step 1:** Ontology Standard Import. The first step in the process is the importing into the system of the ontology standard file. MOVE is not standard dependent, and by reconfiguring itself to a given ontology standard file it is compatible with every standard, even supporting all the features exclusive to that standard (unlike many other tools that claim compatibility, but only deliver compatibility on the common features).

- **Step 2:** Ontology Import. The original ontology (also called base ontology) on which the Ontology View is based. Naturally, this ontology has to be in the ontology standard specified by the ontology standard import file, such as OWL (W3C-OWL, 2002) or the IOC (Wouters et al., 2004a).

- **Step 3:** User Requirements Import. The user provides the system with the necessary information for it to be able to arrive at the required high-quality extracted materialized ontology view (discussed next).

- **Step 7:** Materialized Ontology View Export. The final results are transformed in the MOVE system to an ontology that is in the same format as

Copyright © 2006, Idea Group Inc. Copying or distributing in print or electronic forms without written permission of Idea Group Inc. is prohibited.

Figure 5. MOVE high level architecture

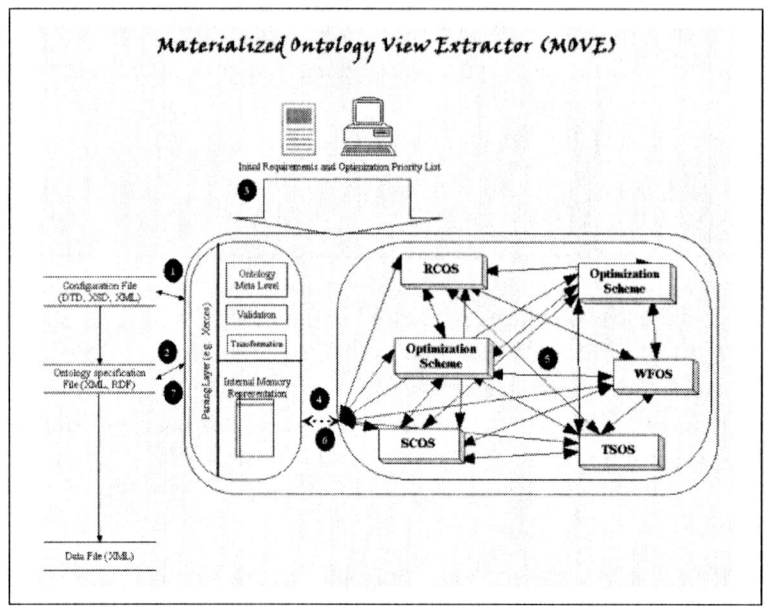

the imported ontology. This means that a Ontology View is saved to a file, but that it can be used as an independent ontology (so to a user it is transparent that it was actually extracted).

The third step in this process requires more explanation, as it is the primary way in which the user (nonIT) interacts with the system. As discussed previously, there are two ways in which information is provided to the system.

First, there is the specification of key points of interest and disinterest (called the ontology labeling). In a very user-friendly manner, the user indicates which elements of the ontology they find interesting, and which ones they do not. Note that this does not have to be done for every element, just some key ones.

The second way the user conveys requirements to the MOVE is by specifying a list of optimization schemes, which are used to configure the MOVE for this particular extraction process. As discussed previously, these optimization schemes are the quintessential part of the OEM.

Copyright © 2006, Idea Group Inc. Copying or distributing in print or electronic forms without written permission of Idea Group Inc. is prohibited.

Bank Use Case

The previous sections provided some theory on how the database view can be reinterpreted to suit the Semantic Web needs. The Strict SW-view, or Ontology View, was introduced together with its function in the ontology extraction methodology. This section discussed a practical scenario that uses the Semantic Web view and the ontology extraction ontology. First, the use case is introduced, followed by a solution using Ontology Views.

Use Case: Company Ontology

Motivated by the promises held by ontologies, a bank took all the different knowledge bases and databases of its departments, and merged them all into a single ontology, that would cater for everyone's needs. Although successful in this attempt, the bank is now confronted with a new problem: The departments all have a more complex structure to deal with, and a lot of what is present in this ontology is not of interest to a particular department. Although there are many advantages as well, these new problems have prevented a real gain from the conversion.

For the remainder of this use case, we only consider two departments of the bank: (1) the personnel department and (2) the product department. Figure 6 introduces a part of the ontology the bank is using now (depicted in Protege, using Ontoviz tab[1]) which is only relevant to one of the two departments. For the purpose of this chapter, only a part of the ontology is presented.

Considering only the personnel department, it is now illustrated that there is too much complexity in the Bank Ontology. Figure 7 shows a Web interface to the old knowledge base that was used by this department, so before anything was merged into a single ontology. Only relevant info to this department is shown here.

Figure 8 shows the same Web interface adapted to work with the ontology. The complexity has increased, and a lot of insignificant information (again, to this department) is present. Note also that these are merely simplified versions, and that they only consider two departments. The full ontology increases complexity significantly more.

It is clear that although old problems have been solved, new problems have been introduced, foremostly the increased complexity for users. The next section shows how Ontology Views can be used to provide a solution to this problem, while maintaining all the advantages of the Bank Ontology.

Copyright © 2006, Idea Group Inc. Copying or distributing in print or electronic forms without written permission of Idea Group Inc. is prohibited.

Figure 6. Extract of bank ontology

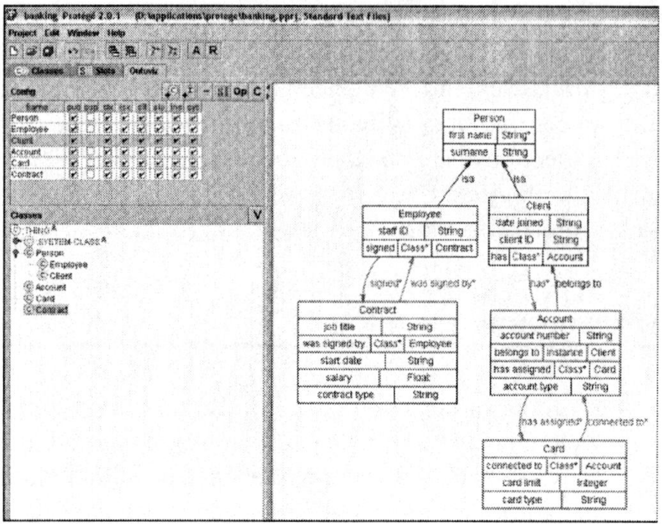

MOVE Toward a Solution

From the previous section it follows that it does not suffice to merge all the structures into a single ontology, and assume that all users will be happy using this ontology (even if they all agree on its contents). The fact that an ontology is a shared agreement does not infer that everyone wants to use all of it. This section will show how MOVE was used to provide the personnel department a new ontology (or more precisely an Ontology View of the Bank Ontology) that is strongly connected to the Bank Ontology (like views still are connected to the original tables, and use their information), but for them appears to be only slightly (or not at all) different from the system they were used to. First the extraction process is discussed, and as a result the same adapted Web interface is shown to indicate the similarity with the Web interface to the legacy system (see Figure 7).

The first step in arriving at a suitable Ontology View is for a domain expert (someone from the personnel department) to indicate some key point of interest and disinterest. This is done using straight forward selection and deselection from a list of elements (shown in Figure 9).

Secondly, the user indicates they want to use certain optimization scheme. In our example, an optimization scheme had been constructed, based on the Total Simplicity Optimization Scheme (TSOS) (Bhatt et al., 2004; Wouters et al., 2002; Wouters et al., 2004b). This optimization scheme (called Medium Downsize

Copyright © 2006, Idea Group Inc. Copying or distributing in print or electronic forms without written permission of Idea Group Inc. is prohibited.

Figure 7. Web interface of legacy system (personnel department)

Client Details Page

(no filters applied)

Chester Goldstein	Surname	Goldstein	
	First name	Chester	+
	Is an Employee		
	Staff ID	S532212	
	Contract type	A3	
	Salary	50.000	
	Job Title	Teller	
	Started Working	01-01-2001	

Figure 8.

Client Details Page

(no filters applied)

Chester Goldstein	Surname	Goldstein		
	First name	Chester	+	
	Client number	ID030772351		
	Is an Employee			
	Staff ID	S532212		
	Contract type	A3		
	Salary	50.000		
	Job Title	Teller		
	Has Account	123-123456-78		
		789-18236011	Has card connected	Bankcontact
			Has card connected	VISA
	Started Working	01-01-2001		
	Joined Bank	03-07-1972		

Figure 9. Selection and deselection of elements

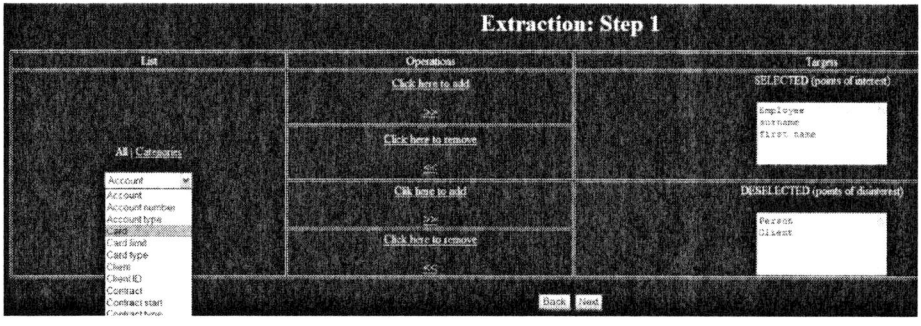

Copyright © 2006, Idea Group Inc. Copying or distributing in print or electronic forms without written permission of Idea Group Inc. is prohibited.

Optimization Scheme [MDOS]) provided a more balanced result, which seemed to be adequate for these departments (or rather; for this size of base ontology and size of desirable Ontology View). Although not every company can use this custom optimization scheme, it still is kept sufficiently generic to be applicable in many situations (especially with regards to requirements of the same size ratio between base ontology and Ontology View). Note that in real situations, the extraction from MOVE is not the final result, but some manual pruning can still be done. This has to be done by an ontology engineer, but is considerably less work than the traditional entire manual extraction.

In this specific case, the ontology file fed to MOVE was in RDF, more exactly the format given by the save as function in Protege. Also the Ontology View is saved in this file format, and Figure 10 shows the Ontology View as it is imported into Protege again.

The same Web interface adaptation that was constructed to work with the base ontology can be reused to work with the Ontology View as well, as this is in the same standard. The resulting Web interface only minimally differs from the Web interface to the legacy system. Although there might be a slight increase in complexity, this is negligible, especially when comparing it to the original increase in complexity from the use of the Bank Ontology. Note that the same optimization schemes can also be used for other departments to arrive at an equivalent result. Note that it was already indicated that sometimes additional manual pruning is required, but that at least the majority of the ontology is already correct, saving

Figure 10. Ontology view representation (Protege/OntoViz view)

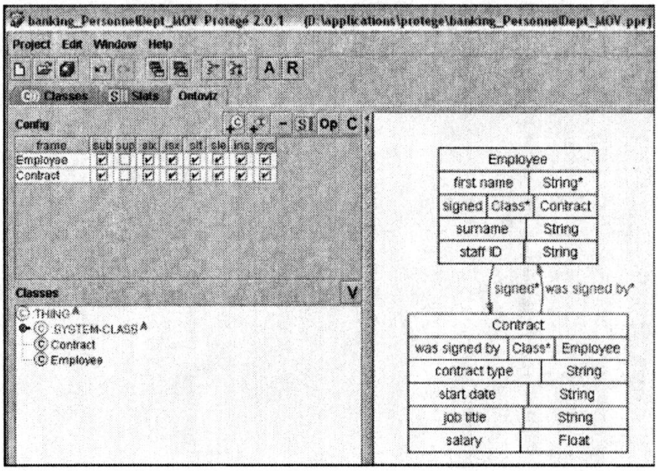

Copyright © 2006, Idea Group Inc. Copying or distributing in print or electronic forms without written permission of Idea Group Inc. is prohibited.

the ontology engineer time and cost. Furthermore, the automated extraction of a Ontology View will enable different users in the same department to apply a different pruning or to make a different selection of elements. Therefore, they can customize the results to their own needs with no (or minimal) knowledge of ontologies.

Conclusion and Future Work

Views have proven to be very useful in DB area, and this chapter examined the potential usefulness for Semantic Web and ontologies of a similar concept, here referred to as Semantic Web View. It was shown that the same benefits from DB area can be obtained for Semantic Web and ontologies.

In order to do this, first a formal approach was taken to provide the definitions of the Semantic Web View (SW-view). The different layers of abstraction for modeling were discussed, and the work in this chapter mainly is situated in the highest level of abstraction. Both the formalization for the SW-view and the Ontology Extraction Methodology (OEM) take this high-level approach. However, to show some of the applications and practical usability of this high-level specifications, an example of conversion of conceptual SW-view to logical view was given. Furthermore, the OEM demonstrated by example one of its applications, using a prototype called the Materialized Ontology View Extractor (MOVE).

The practical Bank Use Case example used all the high-level formalizations seamlessly, and combines them into a user-friendly, straightforward application on the Semantic Web.

Our future work includes the mechanism to maintain consistency between the Ontology Views with the main ontology. This is important considering the fact that an ontology will dynamically evolve with time. We are going to evaluate the possibilities of adopting the concept of view update in relational or object-oriented databases. Techniques using the concept of incremental updates will be developed for this purpose, where it will determine what new instances, or concepts, have to be added, deleted, or modified in a materialized view when a change is applied to the base ontology.

Copyright © 2006, Idea Group Inc. Copying or distributing in print or electronic forms without written permission of Idea Group Inc. is prohibited.

References

Aberer, K., et al. (2004, June 17-19). *Emergent semantics systems.* First International IFIP Conference on Semantics of a Networked World (ICSNW 2004) - Revised Selected Papers, Paris.

Abiteboul, S. (1999). On views and XML. *Proceedings of the 18th PODS'99.*

Abiteboul, S., & Bonner, A. (1991). Objects and views. *Proceedings of the International Conference on ACM SIGMOD.*

Abiteboul, S., et al. (1997). *Views for semistructured data.* Workshop on Mgmt. of Semistructured Data.

Abiteboul, S., Quass, J., McHugh, J., Widom, J., & Wiener, J. (1997). The Lorel query language for semistructured data. *International Journal on Digital Libraries, 1*(1), 68-88.

Balsters, H. (2003, September). *Modelling database views with derived classes in the UML/OCL-framework.* The Unified Modeling Language: Modeling Languages and Applications (UML'03).

Berners-Lee, T. (1998). *Semantic Web road map.* Retrieved from http://www.w3c.org/DesignIssues/Semantic.html

Bertino, E. (1992). *A view mechanism for object-oriented databases.* The 3rd International Conference on Extending Database Technology (EDBT '92), Vienna, Austria.

Bhatt, M., et al. (2004, March 29-31). *A distributed approach to sub-ontology extraction.* The 18th International Conference on Advanced Information Networking and Applications (AINA'04), Volume 1, Fukuoka, Japan.

Cattell, R. G. G., et al. (Eds.). (2000). *The object data standard: ODMG 3.0.* Morgan Kaufmann.

Chan, S., Dillon, T. S., & Siu, A. (2002). Applying a mediator architecture employing XML to retailing inventory control. *The Journal of Systems and Software, 60*, 239-248.

Chang, E. J. (1996). *Object oriented user interface design and usability evaluation.* Published Doctor of Philosophy (PhD), La Trobe University, Melbourne, Australia.

Chen, Y. B., Ling, T. W., & Lee, M. L. (2002). Designing valid XML views. *Proceedings of the 21st International Conference on Conceptual Modeling (ER'02),* Tampere, Finland.

Cluet, S., Veltri, P., & Vodislav, D. (2001). Views in a large scale XML repository. *Proceedings of the 27th VLDB Conference (VLDB'01),* Roma, Italy.

Copyright © 2006, Idea Group Inc. Copying or distributing in print or electronic forms without written permission of Idea Group Inc. is prohibited.

Cruz, I., Decker, S., Euzenat, J., & McGuinness, D. (Eds.). (2002). *The emerging Semantic Web: Selected papers from the first Semantic Web Working Symposium.* Tokyo: IOS Press.

Date, C. J. (2003). *An introduction to database systems* (8th ed.). New York: Pearson/Addison Wesley.

Dillon, T. S., & Tan, P. L. (1993). *Object-oriented conceptual modeling.* Australia: Prentice Hall.

Do, H. H., & Rahm, E. (2004). Flexible integration of molecular-biological annotation data: The genmapper approach. *Proceedings of the 9th International Conference on Extending Database Technology (EDBT'04),* Greece.

Elmasri, R., & Navathe, S. (2004). *Fundamentals of database systems* (4th ed.). New York: Pearson/Addison Wesley.

Feng, L., Chang, E., & Dillon, T. S. (2002). A semantic network-based design methodology for XML documents. *ACM Transactions on Information Systems (TOIS), 20*(4), 390-421.

Feng, L., Chang, E., & Dillon, T. S. (2003). Schemata transformation of object-oriented conceptual models to XML. *International Journal of Computer Systems Science & Engineering, 18*(1), 45-60.

Feng, L., et al. (2003, September 1-5). *An XML-enabled association rule framework.* The 14th International Conference on Database and Expert Systems Applications (DEXA'03), Prague, Czech Republic.

Gaševic, D., et al. (2004a). *Approaching OWL and MDA through technological spaces.* The 3rd Workshop in Software Model Engineering (WiSME 2004), Lisbon, Portugal.

Gaševic, D., et al. (2004b). Converting UML to OWL ontologies. *Proceedings of the 13th International WWW Conference,* New York.

Golfarelli, M., Maio, D., & Rizzi, S. (1998). The dimensional fact model: A conceptual model for data warehouses. *International Journal of Cooperative Information Systems, 7*(2-3), 215-247.

Graham, I., et al. (2001). *Object-oriented methods: Principles & practice* (3rd ed.). Harlow: Addison-Wesley.

HyOntUse. (2003). *User oriented hybrid ontology development environments.* Retrieved from http://www.cs.man.ac.uk/mig/projects/current/hyontuse/

KAON. (2004). *KAON project.* Retrieved from http://kaon.semanticweb.org/Members/rvo/Folder.2002-08-22.1409/Module.2002-08-22.1426/view

Kim, W., & Kelly, W. (1995). On view support in object-oriented database systems. In *Modern database systems* (pp. 108-129). Addison-Wesley.

Copyright © 2006, Idea Group Inc. Copying or distributing in print or electronic forms without written permission of Idea Group Inc. is prohibited.

Liefke, H., & Davidson, S. (2000, September 4-6). View maintenance for hierarchical semistructured. *Proceedings of the Second International Conference DaWak '00*, London.

Lucie-Xyleme. (2001, July 16-18). *Xyleme: A dynamic warehouse for XML data of the Web.* International Database Engineering & Applications Symposium (IDEAS'01), Grenoble, France.

Ludaescher, B., Papakonstantinou, Y., Velikhov, P., & Vianu, V. (1999). *View definition and DTD inference for XML.* Post-ICDT Workshop on Query Processing for Semistructured Data and Non-Standard Data Formats.

Mohania, M. K., Karlapalem, K., & Kambayashi, Y. (1999). *Data warehouse design and maintenance through view normalization.* The 10th International Conference on DEXA '99, Florence, Italy.

Nassis, V., Rajugan, R., Dillon, T. S., & Rahayu, W. (2005). Conceptual and systematic design approach for XML document warehouses. *International Journal of Data Warehousing and Mining, 1*(3).

Noy, N., & Musen, M. (2002). Promptdiff: A fixed-point algorithm for comparing ontology versions. *Proceedings of the 18th National Conference on Artificial Intelligence.*

OMG-OCL. (2003). *UML 2.0 OCL.* Retrieved from http://www.omg.org/cgi-bin/doc?ptc/2003

OMG-UML™. (2003a). *UML 2.0 final adopted specification.* Retrieved from http://www.uml.org/#UML2.0

OMG-UML™. (2003b). *Unified Modeling Language™ (UML) Version 1.5 Specification.* OMG.

Rajugan, R. Chang, E., Dillon, T. S., & Ling, F. (2003). *XML Views: Part 1.* The 14th International Conference on Database and Expert Systems Applications (DEXA'03), Prague, Czech Republic.

Rajugan, R., et al. (2005). *XML views, Part III: Modeling XML conceptual views using UML.* The 7th Int. Conf. on Enterprise Information Systems (ICEIS'05), Miami, FL.

Rajugan, R., Chang, E., Dillon, T. S., & Ling, F. (2005a, September 21-23). *Engineering XML solutions using views.* The 5th International Conference on Computer and Information Technology (CIT'05), Shanghai, China.

Rajugan, R., et al. (2005b, September 21-23). *A layered view model for XML repositories & XML data warehouses.* The 5th International Conference on Computer and Information Technology (CIT '05), Shanghai, China.

Rajugan, R., Chang, E., Dillon, T. S., & Ling, F. (2005c, October 24-28). *A three-layered XML view model: A practical approach.* The 24th International Conference on Conceptual Modeling (ER'05), Klagenfurt, Austria.

Copyright © 2006, Idea Group Inc. Copying or distributing in print or electronic forms without written permission of Idea Group Inc. is prohibited.

Roussopoulos, et al. (2001). *The opsis project: Materialized views for data warehouses and the Web.* Advances in Informatics (Revised selected papers) 8th Panhellenic Conference on Informatics (PCI '01).

Spyns, P., Meersman, R., & Mustafa, J. (2002). *Data modeling versus ontology engineering.* SIGMOD.

Trujillo, J., et al. (2001, December). Designing data warehouses with OO conceptual models. *IEEE Computer Society, "Computer"*, 66-75.

Volz, R., Oberle, D., & Studer, R. (2003a, July 16-18). *Implementing views for light-weight Web ontologies.* Seventh International Conference on IDEAS'03, Hong Kong, SAR.

Volz, R., Oberle, D., & Studer, R. (2003b). Views for light-weight Web ontologies. *Proceedings of the ACM Symposium on Applied Computing (SAC '03).*

W3C-OWL. (2002). *OWL Web Ontology Language 1.0 reference.*

W3C-RDF. (2004, February 4). *Resource Description Framework (RDF).* Retrieved from http://www.w3.org/RDF/

W3C-SW. (2005a). Retrieved from http://www.w3.org/2001/sw/

W3C-SW. (2005b). *Semantic Web.* Retrieved from http://www.w3.org/2001/sw

W3C-XML. (2004, February 4). *Extensible Markup Language (XML) 1.0.* Retrieved from http://www.w3.org/XML

W3C-XPath. (1999, November). *XML Path Language (XPath) Version 1.0.*

Warmer, J. B., & Kleppe, A. G. (2003). *The object constraint language: Getting your models ready for MDA* (2nd ed.). Boston: Addison-Wesley.

Wongthamtham, P., et al. (2003, December 2-5). *Ontology based solution for software development.* International Conference on Software Engineering and Applications (ICSSEA '03), Paris.

Wouters, C., et al. (2002, September 2-6). *A practical walkthrough of the ontology derivation rules.* The 13th International Conference on DEXA'02, Aix-en-Provence, France.

Wouters, C., et al. (2004a, March 17-19). *Ontologies on the MOVE.* The 9th International Conference on Database Systems for Advanced Applications (DASFAA '04), Jeju Island, Korea.

Wouters, C., Dillon, T. S., Rahayu, J. W., Chang, E., & Meersman, R. (2004b). A practical approach to the derivation of a materialized ontology view. In D. Taniar, & W. Rahayu (Eds.), *Web information systems.* Hershey, PA: Idea Group Publishing.

Copyright © 2006, Idea Group Inc. Copying or distributing in print or electronic forms without written permission of Idea Group Inc. is prohibited.

Zhuge, Y., & Garcia-Molina, H. (1998). Graph structured views and incremental maintenance. *Proceeding of the 14th IEEE Conference on Data Engineering (ICDE '98)*.

Endnote

[1]   http://protege.stanford.edu/

Copyright © 2006, Idea Group Inc. Copying or distributing in print or electronic forms without written permission of Idea Group Inc. is prohibited.

Chapter II

Representation of Web Application Patterns in OWL

Pankaj Kamthan, Concordia University, Canada

Hsueh-Ieng Pai, Concordia University, Canada

Abstract

Patterns are distilled forms of knowledge from past experience and expertise in solving recurring problems in a domain. The Semantic Web provides an environment where the knowledge inherent in patterns can be adequately represented to be broadly accessible and be reasoned with. This chapter describes the process of creating OWAP, an ontology in the language OWL for Web Application Patterns. The problems faced in each phase and steps taken to resolve them are given. The significance and limitations of tools during OWAP design, implementation, and testing are outlined. The lessons learned in engineering OWAP are cast as an aggregated list of guidelines. Finally, some directions for future enhancements of OWAP are pointed out.

Copyright © 2006, Idea Group Inc. Copying or distributing in print or electronic forms without written permission of Idea Group Inc. is prohibited.

Introduction

Patterns are abstractions of knowledge gained by experts during their experience in solving problems that occur repeatedly in a domain (Alexander, 1979). Over the last decade, patterns have found applications in a variety of domains of interest and emerged as an indispensable tool in the hands of computer scientists and software engineers. In recent years, patterns have been introduced in a variety of domains such as use cases (Adolph et al., 2002), software design (Gamma et al., 1995), human-computer interaction (Borchers, 2001), electronic business (Adams et al., 2001), and configuration management (Berczuk & Appleton, 2003), to name a few in computing.

Formally, a pattern is a solution to a recurring problem in a given context (Alexander, 1979). Application of one pattern often leads to new context(s) and thus to the need for new pattern(s). For example, the application of a personalization pattern can give rise to privacy issues (new contexts), and thus require further application of new pattern(s). Therefore, the interest is in the *collective* of patterns that works cooperatively to solve a larger problem. These pattern collections need to be suitably represented so as to be widely accessible to their user community. The Semantic Web has recently emerged as an extension of the current Web that adds technological infrastructure for better knowledge representation, interpretation, and reasoning (Hendler, Lassila, & Berners-Lee, 2001). This chapter proposes the use of the Semantic Web as a new vehicle for communication of patterns. In doing so, we present our experience in engineering a large-scale ontology of Web Application Patterns (OWAP), focusing both on the process and on the product. It is our hope that this case study will benefit those interested in a similar undertaking.

The outline of the chapter is as follows: Next, the chapter gives the necessary background and outlines related work. This is followed with aspects involved in planning for OWAP and a discussion on preliminary tasks carried out for OWAP analysis. A detailed description of OWAP design is given and the following section provides detail of the implementation of OWAP. Next, the chapter outlines the testing and evaluation of OWAP and then briefly lists some useful inferences that can be drawn from OWAP and discusses the lessons learned during OWAP engineering project in the hope that they will be useful for those who pursue similar endeavors. Future trends, including possible research avenues, are discussed, followed by concluding remarks.

Copyright © 2006, Idea Group Inc. Copying or distributing in print or electronic forms without written permission of Idea Group Inc. is prohibited.

Background

In the last few years, authors have begun to express their pattern collections in HyperText Markup Language (HTML) and make them publicly available on the Web (Borchers, 2001). However, the focus is mainly on the presentation rather than on the organization of these pattern collections. This makes precise access and efficient retrieval of desirable patterns rather difficult and thereby inhibits their broad use. Furthermore, finding relevant patterns by manually traversing the links has limited utility as the pattern collections grow.

The Extensible Markup Language (XML) (Bray et al., 2004) is a metalanguage that provides directions for expressing the syntax of descriptive markup languages. The syntactical structure of patterns can be expressed in XML (Pai, 2002; Kamthan & Pai, 2006) and stored in XML Database Management Systems (DBMS) that can be accessed via the Web in a more precise fashion than HTML. However, an XML DBMS suffers from several limitations. For instance, it has no specific mechanism for differentiating between synonyms or homonyms, it is not able to extract implicit knowledge (such as hidden dependencies), and it can only provide limited inference and reasoning capabilities, if at all. The patterns in a collective can be related in complex ways, and these relationship types can not be satisfactorily modeled using XML alone. Thus, XML does not have the capabilities to adequately represent a pattern collective and as a consequence, there may exist some patterns that are important for a pattern user, but are not retrieved. The core of the problem lies in the need for an adequate representation of the semantical *knowledge* inherent in collections of patterns.

In order to represent the knowledge inherent in a domain, we need to classify the domain into concepts and relations between them. The declarative knowledge of a domain is often modeled using an *ontology*. Formally, an ontology is defined as "an explicit formal specification of a conceptualization" that consists of a set of concepts in a domain and relations among them (Gruber, 1993). This conceptualization provides an abstract, simplified view of the world that one wishes to represent for some purpose. By explicitly defining the relationships and constraints among the concepts, the *semantics* of a concept is constrained by restricting the number of possible interpretations of the concept. As shown later, the problems mentioned previously with the use of XML DBMS for patterns can be circumvented with a *suitable* ontological representation.

The need for an ontological representation of patterns has been put forth in recent years (Devedzic, 2002; Kamthan & Pai, 2006). However, initiatives towards actual ontological representation of patterns, particularly in modern ontology specification languages, are in their infancy. In their simplest incarnation, patterns could be organized as an informal taxonomy such as in the case of

Copyright © 2006, Idea Group Inc. Copying or distributing in print or electronic forms without written permission of Idea Group Inc. is prohibited.

Web Site Traversal Patterns (Gillenson, Sherrell, & Chen, 2000). However, directory-style hierarchies provide limited possibilities for automated reasoning and only very simple inferences can be drawn from them. Therefore, for an ontology to be broadly useful, some level of formalization (via mathematical logic, for example) is necessary.

In the last five years, a number of initiatives for ontology specification languages for the Semantic Web, with varying degrees of formality and target user communities, have been proposed (Gómez-Pérez, Fernández-López, & Corcho, 2004). These include Simple HTML Ontology Extensions (SHOE), Ontology Inference Layer (OIL), DARPA Agent Mark Up Language (DAML), and DAML+OIL, to name a few. The OWL Web Ontology Language (Dean & Schreiber, 2004) is the successor of these efforts, and is the language adopted for OWAP. Indeed, an ontology for interaction design patterns is proposed in Henninger (2002), but the focus is on the implementation in DAML rather than on the process, details are sketchy, and no specifics of possible inferences are given.

Among the domains for which patterns have been developed, this chapter focuses on the development of an ontology for Web Application Patterns (WAP) for a special reason. For the sake of this chapter, a Web Application is a Web site that is created programmatically and (dynamically) provides opportunities for interaction to a user. The purpose of WAP is to improve the development and use of Web Applications that may reside in the realm of the Semantic Web; conversely, the Semantic Web ontology language can be utilized to advance the communicability of WAP. Thus, controlled use of the Semantic Web technologies can contribute to the improvement of the information architecture of the Semantic Web.

OWAP Planning

Prior to the actual development of the ontology, it is important to carry out infrastructure and resource planning. This includes putting forth a philosophical basis, addressing feasibility issues, selecting a suitable process to be followed, and choosing an appropriate ontology representation language and corresponding tools for development. This section describes how some of these preliminary tasks were done for OWAP.

Basic OWAP Principles

For long-term sustainability of the ontology, certain software and knowledge engineering principles guided the development and formed the basis of OWAP.

Copyright © 2006, Idea Group Inc. Copying or distributing in print or electronic forms without written permission of Idea Group Inc. is prohibited.

They, listed in broadly categorized form, are:

- **User-Centeredness.** An ontology is of little use if it does not meet the needs of its target stakeholders. Therefore, OWAP must be designed with the user and user goals in mind. To that regard, OWAP is sensitive to the terminology of the domain and documented appropriately.
- **Evolvability.** Since change is inevitable, OWAP must be designed to be evolvable. The ideas behind the Agile Software Development (Highsmith, 2002) initiatives, including flexibility for iterative development, and the classical principles of separation of concerns, modularity, high cohesion, and low coupling provide the motivation in this respect.

OWAP Development Process

From the perspective of scale, required resources, and human involvement, there is parity between engineering ontology and engineering software. The success of iterative, incremental, and adaptive software process models has led to customizations of these models to ontological engineering context (Gómez-Pérez et al., 2004).

We adopted a *heterogeneous* approach to engineering OWAP. In particular, aspects of use case modeling (the Unified Process); principles of pair modeling (Kamthan, 2005); principles of testing frequently and refactoring (Extreme Programming); and the practice of proper documentation (Waterfall Model), were adapted and followed accordingly.

Ontology Specification Language for OWAP

OWL was chosen as the underlying language for OWAP due to its certain unique characteristics. OWL is designed to be compatible with the architecture of the Web (Jacobs & Walsh, 2004) in general, and the Semantic Web in particular. It benefits from using XML as its serialization syntax, has foundations in well understood declarative semantics, has the ability to be distributed across many systems, and is in agreement with the Web standards for accessibility and internationalization.

Specifically, OWAP was implemented using OWL DL, one of the three sub-languages of OWL, which provides the right balance for ontological representation of patterns due to its semantical foundations in expressive Description Logics (DL) (Baader et al., 2003) and available tool support.

Copyright © 2006, Idea Group Inc. Copying or distributing in print or electronic forms without written permission of Idea Group Inc. is prohibited.

Deployment of Tools in OWAP

It was deemed important that the tools used for OWAP be nonproprietary, readily available, mature, and well-documented. Access to software reviews and publicly available documentation was useful in this regard. The primary ontology authoring environment was Protégé-2000 (Noy et al., 2001). Protégé-2000 has established itself as the de facto Integrated Development Environment in the arena of knowledge representation in general and, over the years, provided support for many of the ontology specification languages for the Semantic Web, including OWL. The main testing tool used was Racer (Haarslev & Möller, 2001). Racer is a complete and fairly stable reasoner with support for OWL DL that, unlike its competitors, provides both TBox and ABox reasoning (the intentional and extensional knowledge in a DL knowledgebase, respectively).

Reuse Potential

In order to reduce duplication of work and foster possibility for reuse, a preliminary search for an ontology in the domain of patterns in general or WAP in particular was carried out. However, the results did not reveal any potential candidates, and therefore OWAP did not make any ontology reuse.

Having outlined the steps in OWAP planning, we now turn our attention to the OWAP analysis.

OWAP Analysis

This section discusses the first phase in the actual ontology development to determine the goals and scope of OWAP, and acquire the domain knowledge (Noy & McGuinness, 2001). The results of these efforts were formally documented as requirements.

OWAP Goals and Scope

The knowledge inherent in a domain such as WAP is open-ended and the "boundaries" of an ontology that captures that knowledge are determined by its goals and scope. The main goal of OWAP is to help engineers who plan to use WAP by retrieving the desired patterns with precision and with efficiency. In

Copyright © 2006, Idea Group Inc. Copying or distributing in print or electronic forms without written permission of Idea Group Inc. is prohibited.

addition, OWAP is intended to help pattern-authors better organize their patterns so that they can be easily located by users.

In order to determine the scope of OWAP, certain *competency questions* (Noy & McGuinness, 2001) from a pattern user's point of view were asked. Some examples are:

- If I want to create an E-Commerce Web Site, what kind of Web Pages do I need to include in it and what type of information/functionality should each Web Page include?

- What are the different ways of adding search functionality to an existing Web Site?

- In what situation (context) do I need to include a Privacy Policy Page in my Web Site?

- What patterns are related to Travel Site pattern?

- What are the known uses of the Hot List pattern?

- What Page(s) should or could a Product List page link to?

- What kind of action(s) should be triggered when a user presses the Login button?

These questions helped in defining the scope of OWAP. By virtue of necessity, it was not the intention of OWAP to cover everything in the domain of WAP. For instance, OWAP only addressed structural patterns; time-dependent knowledge such as behavioral patterns is difficult to deal with ontologically. OWAP did not deal with details on aesthetic aspects of Web Applications, or with the problem of how certain functionality can be implemented using a particular programming language. In addition, OWAP did not include any anti-patterns.

Knowledge Acquisition for OWAP

To develop OWAP, it was necessary to acquire credible knowledge from two different domains, that of patterns and of Web Applications. To acquire knowledge in the domain of patterns, classical references that have built the foundation of the subject (Alexander, 1979; Gamma et al., 1995) were used. This helped in obtaining the defining characteristics, including necessary terminology and constraints involved, in the domain of patterns that must be taken into account in the design of OWAP.

To acquire knowledge in the domain of Web Applications, two complementary approaches were taken: the use of documented WAP resources (Van Duyne,

Copyright © 2006, Idea Group Inc. Copying or distributing in print or electronic forms without written permission of Idea Group Inc. is prohibited.

Landay, & Hong, 2003; Weiss, 2003) and examination of carefully selected representative Web sites. The former approach provided a collection of patterns and helped identify the terms used in labeling them as well as the relations and constraints among those terms. The latter approach gave an illustration of instances of these patterns in actual real-world use.

Once the analysis of problem domain was complete, the next phase focused on the solution domain and the design of the ontology.

OWAP Design

We now turn to details involved in the design of OWAP, which was the most time-consuming and effort-intensive phase. This section describes the design of major concepts and properties in OWAP. The limitations of space do not permit us to discuss each and every concept and respective properties.

Design of OWAP Concepts

The relatively abstract nature of the domain of patterns, compared to other more tangible domains (such as food or wine [Smith, Welty, & McGuinness, 2004]) for which ontologies have been developed, posed several challenges in conceptual modeling. Furthermore, constructing the class hierarchy required considerable time and attention due to the lack of an available reusable hierarchy for WAP.

OWAP has a total of 76 concepts. The concept hierarchies are described in text as well as visually using the Unified Modeling Language (UML) Class Diagrams (Booch, Jacobson, & Rumbaugh, 2005). The use of UML has been made only for the purpose of visual compactness and to separate design-level consider-ations from that of the implementation. A UML profile for OWL has been proposed in Kendall (2002), but it is based on a dated version of OWL. An initiative to provide a standard mapping between UML and ontology languages such as OWL is in progress under the auspices of the Object Management Group (OMG). We also note that the expressions in the UML diagrams presented here could be stated more precisely using the Object Constraint Language (OCL) (Warmer & Kleppe, 2003).

The next two subsections outline the main class hierarchies defined in OWAP and the supporting concepts used to define the main concepts.

Copyright © 2006, Idea Group Inc. Copying or distributing in print or electronic forms without written permission of Idea Group Inc. is prohibited.

Figure 1. Physical component hierarchy

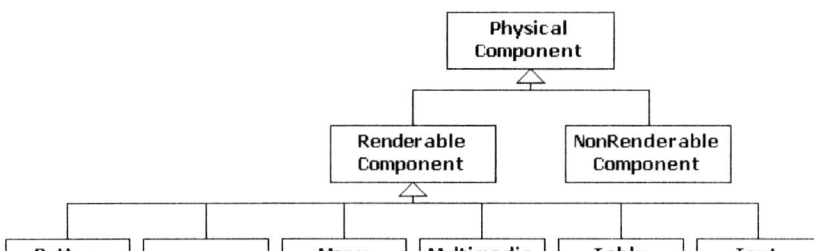

Main Concepts

There are two main class hierarchies in OWAP: one of the *physical components* that a Web Application can be composed of, and the other of the WAP, that is the *logical components* that a Web Application can be composed of.

The concept Physical Component consists of components that can physically exist in a Web Application. It includes Renderable Component, that is, components that are rendered to the user (such as a table, button, and so on) and NonRenderable Component, that is, components that usually are not rendered to the user (such as metadata, style rules, and so on). The top three levels of the Physical Component hierarchy are shown in Figure 1.

The concepts related to Physical Component were designed using a bottom-up approach. All the components that can physically exist in a Web Application were first identified and then based on their similarity and nature were grouped together into a hierarchical form.

Due to limitations of space, we describe only one of the sub-concepts under Renderable Component hierarchy, namely the Menu Component, in detail. This component is chosen to exemplify the difficulties encountered in its design. The Menu Component groups together concepts related to menu. In particular, Menu Item is a Menu Component, and a Menu is a Menu Item (see Figure 2). The reason to design Menu as a subclass of Menu Item is to enable complex menu patterns such as the Fly-Out Menu. As shown in Figure 3, "How to Buy" is a Menu Item of the main menu (left), but it is a Menu itself (that is, the sub-menu on the right).

The second main class hierarchy for OWAP is the Web Application Pattern. In contrast to the Physical Component, the Web Application Pattern also contains the logical components that a Web Application can be composed of. It is designed

Copyright © 2006, Idea Group Inc. Copying or distributing in print or electronic forms without written permission of Idea Group Inc. is prohibited.

Figure 2. Menu component design

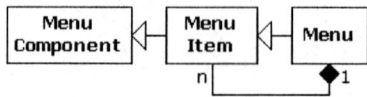

Figure 3. Example of Fly-Out menu

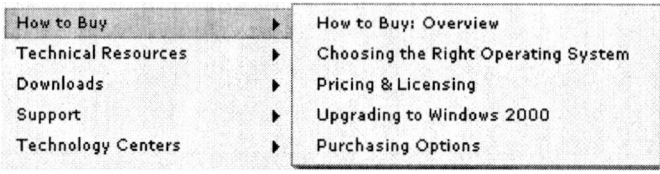

Figure 4. Web application pattern hierarchy

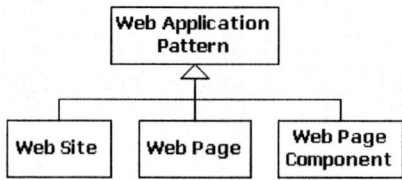

using a top-down approach and used as a placeholder for different types of WAP, including Web Site, Web Page, and Web Page Component (see Figure 4).

The Web Site can be composed of another Web Site (in case of a Web Portal), one or more Web Page (among which there must be a Home Page), and a Menu (see Figure 5). The reason for including Menu at the Web Site level rather than at Web Page level is to avoid redundancy. It was realized that each Web Site generally has its own type of Menu, and this Menu generally appears on every Web Page of such Web Site. So, instead of including the Menu on each possible Web Page for each type of Web Site, a particular Menu is included as part of each type of Web Site.

The Web Page is composed of Web Page Component and Physical Component (see Figure 6). Each Web Page must have a Header and Footer. Besides that, it may optionally contain Metadata, Style Sheet, and Grammar. Furthermore, depending

Copyright © 2006, Idea Group Inc. Copying or distributing in print or electronic forms without written permission of Idea Group Inc. is prohibited.

Figure 5. Web site design

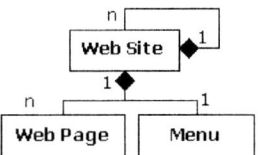

Figure 6. Web page design

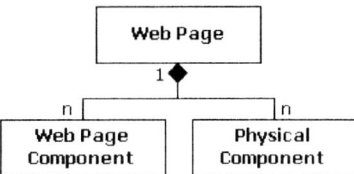

on the nature of each Web Page, other components (such as Search Area) can be included. Also, each Web Page can link to Web Site, another Web Page, and/or Multimedia Component (such as an MP3 file).

The Web Page Component is composed of other Web Page Components and/or Physical Component (see Figure 7). However, there must also be at least one Renderable Component to form a Web Page Component. The reason to introduce the concept Web Page Component is to minimize redundancy and to facilitate reuse. There are some components that are often used as part of different Web Pages. For example, it is common to see a Search text field along with a Search button to enable users to search some information on the Web Site. Web Page Component was introduced to represent such reusable components. For instance, instead of including Search text field and Search button on every page that contains the search facility, a Search Area (which is an individual of Web Page Component) is used.

Supporting Concepts

Besides the main class hierarchies, there are a few supporting classes in OWAP. One supporting concept worthy of mention in detail here is Boolean. The reason for introducing Boolean as a concept rather than as a datatype property is that current OWL reasoners (such as Racer) do not yet support the implementation of Boolean datatype properties. So, if two concept definitions differ from each

Copyright © 2006, Idea Group Inc. Copying or distributing in print or electronic forms without written permission of Idea Group Inc. is prohibited.

Figure 7. Web page component

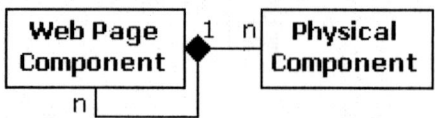

other only by the value of a certain datatype property, and if one wants to specify that those two concepts are disjoint, an "inconsistency" error will be returned from Racer. This problem is solved by using Boolean as a concept instead of datatype property. Furthermore, the performance for checking ABox consistency is greatly improved. When Boolean datatype property was used, it took Racer a few hours to check the consistency of ABox for OWAP. However, when Boolean concept is used instead, checking ABox consistency using Racer took only a few minutes. Thus, in this case the OWAP design was driven by implementation needs.

Design of OWAP Properties

This section outlines the design of properties along with the rationale behind those design decisions. There are 30 object properties and four datatype properties in OWAP. Earlier in the design phase, there were fewer object properties and more datatype properties. However, as mentioned previously, the use of Boolean datatype properties leads to certain undesirable effects. So, in a later iteration, all the Boolean datatype properties (such as isEditable, allowMultipleSelection, and so on) were recast as functional object properties with range equal to the concept Boolean.

Object Properties

Among the object properties, the most important ones are isComposedOf, isPartOf, mayComposeOfForLogicalComponent, hasKnownUse, and isRelatedTo.

The isComposedOf is a super-property of isComposedOfForPhysicalComponent and of isComposedOfForLogicalComponent, each of which is a transitive property. The reason for distinguishing between isComposedOf for physical components and for logical components is to provide a more strict/rigorous definition of the concepts and the individuals. In theory, a physical component can be composed of only physical components. On the other hand, a logical component (that is, a pattern) can be composed of both logical and physical components. So,

Copyright © 2006, Idea Group Inc. Copying or distributing in print or electronic forms without written permission of Idea Group Inc. is prohibited.

if only one kind of isComposedOf is used, the range of this property must be the union of physical and logical components. This opens the possibility for defining an instance of a physical component (such as a Table) to be composed of logical component (such as a Search Page). This obviously is not possible.

The isPartOf property is the super-property of isPartOfPhysicalComponent and isPartOfLogicalComponent. These three properties are the inverse properties of isComposedOf, isComposedOfForPhysicalComponent, and isComposed OfForLogicalComponent.

The mayComposeOfForLogicalComponent property is introduced to overcome the uncertainty problem related to patterns. In the world of WAP, many patterns are defined in terms of the components that a given pattern must have, and the components that this pattern can optionally have. For example, a Product Comparison Page must contain product comparison information. However, there may be some other optional components included in it such as Search Area, Hot List, and so on. The introduction of the mayComposeOfForLogicalComponent property helps deal with these optional components. A similar solution was applied to deal with the uncertainty in Web page linking: the isLinkedTo property was used to describe the required linking between the Web pages, and the mayLinkTo property is used for optional linking.

The hasKnownUse property is used to express the "known use" relationship for WebApplicationPattern. Unlike other pattern-related properties (such as solveProblem), hasKnownUse is defined as an object property instead of String datatype property. This is because the known uses for a pattern must be at least three, but a datatype property allows at most one value per instance. The inverse for hasKnownUse is also defined (called usePattern) so that the user can find out what kind of patterns does a particular Web site use.

The isRelatedTo property (due to the fact it is symmetric) is used to express all the Web Application Patterns that are related to a given Web Application Pattern.

Datatype Properties

All the datatype properties in OWAP are of the String type. The datatype properties solveProblem, isUsedWhen, and hasSolution are used to describe the problem, context, and solution for a Web Application Pattern, while hasTitle is used to describe the title of each Web Page. The reason behind the use of String datatype properties (instead of making them as object properties) is because XML in general and OWL in particular does not allow white space in instance names. However, in the domain of WAP, the problem, context, and solution are often described in narrative (sentence) form. For example, the Search Page pattern solves the problem "user needs to find an item." If object property is used

Copyright © 2006, Idea Group Inc. Copying or distributing in print or electronic forms without written permission of Idea Group Inc. is prohibited.

to describe the "problem" of this pattern, one has to use underscore or some other technique to connect the words of the sentence together to form the instance name, such as "user_needs_to_find_an_item". This is awkward to read, and makes the parsing of the query result difficult. Hence, datatype property is used to include white space within the string value in this case.

OWAP Implementation

OWAP was implemented using OWL DL as it provides maximum expressiveness along with guarantees of computational completeness and decidability of reasoning systems. OWAP concepts, object properties, and datatype properties discussed in the design, along with hundreds of individuals, were expressed in OWL DL. The individuals in OWAP were included for the purpose of testing only and not for completeness. An OWL DL ontology could be represented in a variety of syntaxes: XML presentation syntax, OWL abstract syntax, RDF/XML, and Notation 3 (N3), of which only RDF/XML is considered to be normative and can be exported by Protégé-2000. Since an RDF/XML serialization of OWAP is prohibitively verbose, we have not included it here due to considerations of space.

We now outline the tools deployed in developing OWAP and some problems encountered during the implementation.

Tools Deployed in Implementing OWAP

OWAP was implemented primarily using Protégé-2000. Although most of the OWL markup was generated automatically using Protégé-2000, the ordinary text editor was also occasionally used to modify the markup. For instance, since Protégé-2000 did not allow moving a sub-property upwards to become a top-level property, text editor was used so that the markup fragment <rdfs:subPropertyOf>...</rdfs:subPropertyOf> can be removed from the OWL file.

Problems Encountered During Implementing OWAP

There were certain problems encountered during the implementation of OWAP and provisions were made to circumvent them where possible.

First, since isComposedOf property and its related properties are transitive, no number restriction can be applied to it (else, it ceases to be legal OWL DL). As

Copyright © 2006, Idea Group Inc. Copying or distributing in print or electronic forms without written permission of Idea Group Inc. is prohibited.

a result, it was not possible to make assertions like "CheckBoxGroup is composed of at least two CheckBoxes".

The second set of major problems encountered during the implementation was related to the OWL editing environment. The copies of Protégé-2000 used were unstable and resistant to modifications. They often failed to update the interface after changes were made to the concept definition. For a concept that already had individuals, modifying the concept definition or property definition in the OWL file led to the Protégé-2000 freezing or crashing, or corrupting the file. Protégé-2000 often showed Anonymous classes when an existing OWL project file (in its native format) was opened. Since Protégé-2000 did not support the Undo feature, the steps of deletion and modification were irreversible. Protégé-2000 often did not preserve the structure of markup in an existing OWL file: it either imposed its own structure, including addition of XML namespace Uniform Resource Identifier (URI) that were not needed, or scrambled the original structure. This was inconvenient when the files had to be edited using other means.

Testing and Evaluation of OWAP

Testing formed an integral part of OWAP during development. Tests were carried out on a regular basis to ensure that the concepts and individuals of OWAP were both syntactically and semantically correct. In particular, validity and consistency checking were done frequently during the development process, after addition of almost every concept and every two or three individuals.

We now discuss the tools deployed for testing OWAP, problems encountered during testing, and evaluation of quality of OWAP with respect to its goals and requirements.

Tools Deployed for Testing OWAP

Most of OWAP testing was carried out using Racer. Racer was used through different means: via command-line, using the graphical interface in Racer Interactive Client Environment (RICE), using the graphical interface in Protégé-2000, and using the graphical interface in OntoXPL (Haarslev, Lu, & Shiri, 2004). Protégé-2000 was used for basic syntactic checking of OWL files, for checking the consistency of TBox and ABox, and to view the inferred class hierarchy. Although Protégé-2000 provides the query facility, it was deemed too primitive for testing OWAP completely. Running Racer both from command-line

Copyright © 2006, Idea Group Inc. Copying or distributing in print or electronic forms without written permission of Idea Group Inc. is prohibited.

and from RICE enabled the use of the full power of Racer. Hence, they were used for testing most of the time. OntoXPL provides a tree view and basic query facility to explore the ontology, and was periodically used for visual inspection and testing of OWAP.

Problems Encountered During Testing OWAP

This section describes certain problems that were encountered and tackled during the testing of OWAP.

As the number of individuals increased, so did the time to carry out any kind of query/inference related to transitive properties. (When the property is not transitive, the response time was within a few minutes, including when inverse property was used in the query.)

Racer does not implement the entire OWL DL. For example, Racer does not explicitly implement the OWL oneOf class constructor. Since hasValue is not supported by Racer, whenever hasValue is used, such value becomes a new root concept. There is a similar problem when enumeration is used. This led to adjustments in design and implementation.

Evaluation of OWAP Quality

To be broadly useful, an ontology must strive for high quality (Noy & McGuinness, 2001). The quality of OWAP was evaluated based on the following criteria:

- **Completeness (Domain Goal[s]).** OWAP completely describes the domain that it claims to model. Some sample queries that answer the competency questions are provided in the Appendix.

- **Correctness.** OWAP captures the domain knowledge accurately and in an unambiguous manner.

- **Decidability.** OWAP is strictly based on OWL DL, and is therefore decidable.

- **Maintainability (Extensibility/Modifiability).** The domain underlying the ontology may evolve. OWAP was designed to reflect that.

- **Minimal Redundancy.** OWAP was designed to minimize redundancy and steps were taken to ensure that.

- **Rich Axiomatization.** OWAP defines complex relationships between the concepts, which are detailed enough so that nontrivial, meaningful infer-

Copyright © 2006, Idea Group Inc. Copying or distributing in print or electronic forms without written permission of Idea Group Inc. is prohibited.

ences can be made. In addition, it was designed in such a way that answers to queries can be made as precisely and as detailed as possible.

- **Efficiency.** OWAP is able to answer most of the queries within a few seconds to a few minutes. However, when the number of individuals becomes large, a query that involves the use of transitive property takes many hours to formulate a response.

These quality attributes were, for most part, checked informally and via inspections. It would be useful to devise a quality model and a means for automatic formal verification of these.

OWAP Usage and Experience

An ontology is only as useful as the quality of inferences that can be drawn from it. Therefore, the interest is in pattern ontology with logical constraints that can be reasoned with, and from which complex and "interesting" inferences can be derived.

Inferences from OWAP

One of the motivations for using DL-based ontologies in the study of patterns is the interesting inferences that can be drawn based on *concept* match that would otherwise not have been possible with a traditional representation of a pattern language as a simple taxonomy. The quality of retrieval related to patterns is largely concerned with two attributes: accuracy and efficiency.

The Appendix presents examples of some possible inferences that can be made from OWAP. They are based on responses to the competency questions that were posed during OWAP analysis.

Lessons and Guidelines

This section shares both progressive and cautionary lessons that we learned during the OWAP development process, including that on ontology design for change, proper use of OWL constructs, and optimal use of authoring and reasoning tools. We hope that they can serve as guidelines to those who plan to develop ontologies for similar domain contexts.

Copyright © 2006, Idea Group Inc. Copying or distributing in print or electronic forms without written permission of Idea Group Inc. is prohibited.

Ontologies versus Databases. There are certain advantages of using ontology over the use of relational database, which is a current practice to store patterns. In OWL, it is possible to define number restrictions. For example, a necessary condition for a pattern is that it obeys the "Rule of Three": this can be modeled in OWL as "hasKnownUse property must be greater than or equal to 3". This constraint can not be expressed in a typical relational database. In OWL, it is possible to define transitive property, such as isComposedOf, and then infer that all the components that a certain individual is composed of using, for example, one simple Racer command. This is not possible with a standard SQL statement. One can enumerate all the possible individuals of a class in OWL, but in database, it is not possible to enumerate all the values of a table. For instance, it is possible to say that the only possible values for concept Boolean are True and False, and nothing else. It is possible to specify classes in OWL based on the existence of particular property values. So, an individual will be a member of such a class whenever at least one of its property values is equal to the hasValue constraint. For example, in OWAP, all Web sites' isComposedOf property must have at least one value that is equal to HomePage. In other words, among the pages that a certain Web site is composed of, there must be a HomePage. This kind of constraint cannot be specified in an ordinary database.

Necessity of Logic Prerequisites. Although not as much emphasized in the OWL specification as it probably should be, knowledge of DL is essential for a detailed understanding of the ontology during development including testing, subsequent evolution, and for deriving meaningful inferences from a reasoner.

Property Characteristics. Extra care is needed in assigning property characteristics in OWL. An important consideration is that inheritance does not always apply to property characteristics: if a property P has a characteristic C, then it is not necessary that the sub-properties of P also have C. For example, the functional property is inherited but symmetry is not.

Property Restrictions. The scope of property restrictions (such as (owl:allValuesFrom, owl:someValuesFrom, and owl:hasValue) need to be kept into consideration. They restrict a property only locally, not globally.

OWL DL versus DL. There are certain differences between the traditional DL and to the extent it manifests itself in OWL DL constructs that may not be immediately evident from the OWL specification. DL follows the Unique Name Assumption (UNA): two individuals with different names are considered to be different. This is not the case in OWL as UNA is not reasonably possible on the

Copyright © 2006, Idea Group Inc. Copying or distributing in print or electronic forms without written permission of Idea Group Inc. is prohibited.

Web. Thus, two individuals with different names are still considered to be the same. OWL does, however, provide constructs (owl:differentFrom and owl:AllDifferent) to label them as different. However, frequent explicit declarations of disjointness of individuals will evidently increase the size of the ontology. The traditional definitions of conjunction and disjunction in DL give us only necessary conditions. However, in OWL DL, owl:oneOf, owl:equivalentClass, owl:intersectionOf, owl:unionOf and owl:complementOf constructs allow us to define both necessary and sufficient conditions (logical equivalence) for establishing class membership.

Inherent Limits of OWL. OWAP provided an opportunity to stretch the boundary of OWL to the limit. The current definition of OWL is not designed to deal with representation of uncertainty in knowledge. Therefore, such cases may have to be dealt with using special means. OWL DL lacks the ability to express uncertain concepts and properties. For example, as seen in OWAP design, OWL DL (and, in fact, DL itself) does not provide sufficient constructs to allow one to express that some individuals are optionally related to each other through a certain property (such as a certain Web page mayLinkTo another Web page only occasionally). As a related limitation, OWL DL does not allow the expression of nonbinary (that is, n-ary) relations (which is how the certainty values for properties are expressed in many application domains). Although some approaches have been proposed to express n-ary relations in OWL DL (Noy & Rector, 2004), these proposals do not provide a clean/elegant solution. The issue of uncertainty has also been addressed using other approaches, including directly embedding uncertainty information in the knowledge base (Heinsohn, 1994; Jaeger, 1994; Giugno & Lukasiewicz, 2002) and using Bayesian networks (Koller, Levy, & Pfeffer, 1997; Ding, Peng, & Pan, 2004), but the reasoning support in tools for such extensions is lacking. OWL DL does not allow transitive properties to have number restriction. Hence, some of the semantics of OWAP cannot be fully expressed using OWL DL.

Ontology Design Principles Reiterated. In the development of OWAP, we "rediscovered" two known ontology design principles whose understanding only improved after an actual experience: (a) There is no one correct way to model a domain and the "best" solution almost always depends on the underlying goal and the anticipated extensions, and (b) ontology development is necessarily an iterative process that will likely continue through the entire lifecycle of the ontology.

Use of Authoring and Testing Tools. It is a common practice for XML users to use text editors for marking up documents from scratch. There is no substitute

Copyright © 2006, Idea Group Inc. Copying or distributing in print or electronic forms without written permission of Idea Group Inc. is prohibited.

for authoring at the source level, but it seems that an exception to this may need to be made (by using graphical interface-based editors, at least partially) for ontologies based on OWL. This is because the markup corresponding to, for example, even a single concept can be prohibitively verbose and error-prone particularly due to the number of relationships involved. The errors tend not to be localized and reasoners are not forgiving. Furthermore, the size of nontrivial ontologies can grow rapidly. Also, an ontology editor will likely be able to communicate with a reasoner within its environment, which makes the testing process a bit more convenient and efficient. The justified choice of authoring tool and reasoner are paramount in this regard.

Future Trends

The Semantic Web can serve as a powerful medium for communication of engineering artifacts, and ontological representation is the driving force behind it. However, to bring this endeavor to its full potential, it is imperative to consider how these artifact representations are reflected upon with respect to *both* technical and social layers of the Semantic Web architecture (Jacobs & Walsh, 2004). To this regard, we must engineer ontologies with a strong emphasis towards semiotic quality, taking the end user into consideration. This inevitably also leads to the need for a quality model suitable for ontologies in a Semantic Web framework. Since tools play a crucial role in large-scale ontological engineering, we also demand more promise from the current ontology authoring tools and reasoners in order for them to be trustworthy participants towards quality improvement.

There are a few research areas that emanate from this work that we now discuss.

OWAP was designed to match the domain as closely as possible, to be extensible, and with minimum redundancy. Therefore, it can be extended in the future to include programming and aesthetic details of WAP.

OWAP is only one way of engineering an ontology for WAP. As the development of ontologies for the Semantic Web grows, the need for best practices or patterns for ontology design (Rector et al., 2004) arises. Once the patterns for ontology design mature and actually form a coherent collection, it would be of interest to revisit and analyze OWAP with respect to them.

Patterns are only one part of a systematic approach to engineering Web Applications. They do not exist in isolation and are closely related to other

Copyright © 2006, Idea Group Inc. Copying or distributing in print or electronic forms without written permission of Idea Group Inc. is prohibited.

domains such as quality characteristics, refactoring methods, and collection of metrics for Web Applications. It is equally important that these entities of knowledge be incorporated in the growing ontology universe as well. It is this collective of ontologies, working together in synchrony throughout the lifecycle of a Web Application, that a Web Engineer can truly benefit from.

Although budgetary and time constraints were in place and kept in check, no formal cost (effort, schedule) estimation (Boehm et al., 2001) or feasibility analysis (Guinta & Praizler, 1993) were carried out. It would be of interest to monitor the dynamics of these factors as they impact each other (in response to, for example, the variations in ontology size or changes in personnel) in an ontological development process or thereafter during maintenance. We anticipate these as directions for future research.

Conclusion

Patterns are reusable form of knowledge that can benefit by participating in the Semantic Web, and OWAP is one step in that direction. Ontologies such as OWAP when viewed as organizational assets of shareable knowledge require a disciplined approach towards development. To that regard, a goal-oriented, user-centered view towards large-scale ontological engineering that satisfactorily addresses quality concerns is highly desirable. This requires appropriate planning and more attention on aspects in early phases of development. Since there are often several competing concerns, a feasibility analysis as part of decision making process may be necessary.

For a successful ontological development, like any other large project that spans over several months, adequate training is essential. This includes (roughly in order of significance) a comprehensive understanding of the domain, sensitivity to user needs, knowledge of features and limitations of the ontology specification language, and adequate skills in using processing tools. It is our hope that the lessons and guidelines provided as a distillation of our experience will be beneficial to those who plan to develop ontologies for similar domain contexts.

Availability of robust, reliable, and efficient tools is integral to large-scale ontological engineering. The existence of tools for ontology development simplified implementation and testing of OWAP but evidently also revealed some shortcomings. Therefore, care needs to be taken in selection of tools and caution must be exercised during their use.

Copyright © 2006, Idea Group Inc. Copying or distributing in print or electronic forms without written permission of Idea Group Inc. is prohibited.

Acknowledgments

The authors would like to thank Dr. Volker Haarslev (Concordia University, Montreal, Canada) for his expertise in Description Logics and insights into the Racer architecture, and the reviewers for their helpful feedback and suggestions for improvement.

References

Adams, J., Koushik, S., Vasudeva, G., & Galambos, G. (2001). *Patterns for e-business: A strategy for reuse.* IBM Press.

Adolph, S., Bramble, P., Cockburn, A., & Pols, A. (2002). *Patterns for effective use cases.* Addison-Wesley.

Alexander, C. (1979). *The timeless way of building.* Oxford University Press.

Baader, F., McGuinness, D., Nardi, D., & Schneider, P. P. (2003). *The description logic handbook: Theory, implementation and applications.* Cambridge University Press.

Berczuk, S., & Appleton, B. (2003). *Software configuration management patterns: Effective teamwork, practical integration.* Addison-Wesley.

Boehm, B. W., Abts, C., Brown, A. W., Chulani, S, Clark, B. K, Horowitz, E., et al. (2001). *Software cost estimation with COCOMO II.* Prentice Hall.

Booch, G., Jacobson, I., & Rumbaugh, J. (2005). *The Unified Modeling Language reference manual* (2nd ed.). Addison-Wesley.

Borchers, J. (2001). *A pattern approach to interaction design.* John Wiley & Sons.

Bray, T., Paoli, J., Sperberg-McQueen, C. M., Maler, E., & Yergeau, F. (2004). *Extensible Markup Language (XML) 1.0* (3rd ed.). W3C Recommendation. World Wide Web Consortium (W3C).

Dean, M., & Schreiber, G. (2004). *OWL Web Ontology Language Reference.* W3C Recommendation. World Wide Web Consortium (W3C).

Devedzic, V. (2002). Understanding ontological engineering. *Communications of the ACM, 45(4),* 136-144.

Ding, Z., Peng, Y., & Pan, R. (2004, November 15-18). A Bayesian approach to uncertainty modeling in OWL ontology. *Proceedings of the 2004*

Copyright © 2006, Idea Group Inc. Copying or distributing in print or electronic forms without written permission of Idea Group Inc. is prohibited.

International Conference on Advances in Intelligent Systems — Theory and Applications (AISTA2004), Luxembourg-Kirchberg, Luxembourg.

Gamma, E., Helm, R., Johnson, R., & Vlissides, J. (1995). *Design patterns: Elements of reusable object-oriented software*. Addison-Wesley.

Gillenson, M., Sherrell, D. L., & Chen, L. (2000). A taxonomy of Web site traversal patterns and structures. *Communications of the AIS, 3*(4).

Giugno, R., & Lukasiewicz, T. (2002). P-SHOQ(D): A probabilistic extension of SHOQ(D) for probabilistic ontologies in the Semantic Web. *Proceedings of the European Conference on Logics in Artificial Intelligence (JELIA 2002)*, Cosenza, Italy (pp. 86-97). Springer-Verlag.

Gómez-Pérez, A., Fernández-López, M., & Corcho, O. (2004). *Ontological engineering*. Springer Verlag.

Gruber, T. R. (1993). Toward principles for the design of ontologies used for knowledge sharing. In *Formal ontology in conceptual analysis and knowledge representation*. Kluwer Academic Publishers.

Guinta, L. R., & Praizler, N. C. (1993). *The QFD book: The team approach to solving problems and satisfying customers through quality function deployment*. Amacom Books.

Haarslev, V., Lu, Y., & Shiri, N. (2004). OntoXPL: Exploration of OWL ontologies. *Proceedings of the 2004 International Workshop on Description Logics (DL-2004)*, Whistler, Canada (pp. 60-69).

Haarslev, V., & Möller, R. (2001). Description of the Racer system and its applications. *Proceedings of the 2001 International Workshop on Description Logics (DL2001)*, Stanford, (pp. 132-141).

Heinsohn, J. (1994). Probabilistic description logics. *Proceedings of the 10th Annual Conference on Uncertainty in Artificial Intelligence (UAI '94)*, Seattle, WA (pp. 311-318). Morgan Kaufmann.

Hendler, J. Lassila, O., & Berners-Lee, T. (2001). The Semantic Web. *Scientific American, 284*(5), 34-43.

Henninger, S. (2002). Using the Semantic Web to construct an ontology-based repository for software patterns. *Proceedings of the 2002 Workshop on the State of the Art in Automated Software Engineering*, Irvine (pp. 18-22).

Highsmith, J. (2002). *Agile software development ecosystems*. Addison Wesley.

Jacobs, I., & Walsh, N. (2004). *Architecture of the World Wide Web, Volume One. W3C Recommendation*. World Wide Web Consortium (W3C).

Copyright © 2006, Idea Group Inc. Copying or distributing in print or electronic forms without written permission of Idea Group Inc. is prohibited.

Jaeger, M. (1994). Probabilistic reasoning in terminological logics. *Proceedings of the 4th International Conference (KR '94)*, Bonn, Germany (pp. 305-316). Morgan Kaufmann.

Kamthan, P. (2005, January 14-16). Pair modeling. *Proceedings of the Canadian University Software Engineering Conference (CUSEC 2005)*, Ottawa, Canada.

Kamthan, P., & Pai, H.-I. (2006). Knowledge representation in pattern management. In D. Schwartz (Ed.), *Encyclopedia of knowledge management*. Hershey, PA: Idea Group Reference.

Kendall, E. F. (2002, October 21-24). An introduction and UML profile for the Web Ontology Language (OWL). *Proceedings of the OMG's 3rd Workshop on UML for Enterprise Applications: Model Driven Solutions for the Enterprise*, San Francisco.

Koller, D., Levy, A., & Pfeffer, A. (1997). P-CLASSIC: A tractable probabilistic description logic. *Proceedings of the Fourteenth National Conference on Artificial Intelligence (AAAI-97)*, Providence (pp. 390-397).

Noy, N. F., & McGuinness, D. L. (2001, March). *Ontology development 101: A guide to creating your first ontology*. Stanford Knowledge Systems Laboratory Technical Report KSL-01-05 and Stanford Medical Informatics Technical Report SMI-2001-0880.

Noy, N. F., Sintek, M., Decker, S., Crubezy, M., Fergerson, R. W., & Musen, M. A. (2001). Creating Semantic Web contents with Protege-2000. *IEEE Intelligent Systems, 16*(2), 60-71.

Noy, N., & Rector, A. (2004). *Defining n-ary relations on the Semantic Web: Use with individuals*. W3C Working Draft. World Wide Web Consortium (W3C).

Pai, H. (2002). *Applications of Extensible Markup Language to mobile application patterns*. Master's Thesis, McGill University, Canada.

Rector, A., Schreiber, G., Noy, N. F., Knublauch, H., & Musen, M. A. (2004). Ontology design patterns and problems: Practical ontology engineering using Protege-OWL tutorial. *Proceedings of the Third International Semantic Web Conference (ISWC 2004)*, Hiroshima, Japan.

Smith, M. K., Welty, C., & McGuinness, D. L. (2004). *OWL Web Ontology Language Guide. W3C Recommendation*. World Wide Web Consortium (W3C).

Van Duyne, D. K., Landay, J., & Hong, J. I. (2003). *The design of sites: Patterns, principles, and processes for crafting a customer-centered Web experience*. Addison-Wesley.

Copyright © 2006, Idea Group Inc. Copying or distributing in print or electronic forms without written permission of Idea Group Inc. is prohibited.

Warmer, J., & Kleppe, A. (2003). *The Object Constraint Language: Precise modeling with UML* (2nd ed.). Addison-Wesley.

Weiss, M. (2003). Patterns for Web Applications. *Proceedings of the 10th Conference on Pattern Languages of Programs (PLoP 2003)*, Urbana.

Copyright © 2006, Idea Group Inc. Copying or distributing in print or electronic forms without written permission of Idea Group Inc. is prohibited.

Appendix

In this Appendix, some interesting inferences that can be made using OWAP are provided to show its usefulness. These inferences bring forth two important characteristics of patterns, namely, *guided justifiable use* (Examples 1, 2, 3, 5, 6 and 7), where a user is steered towards designing an application at hand, and *prospects for reuse* (Example 4), that highlight reuse potential of patterns as same pattern(s) could be returned repeatedly in response to different queries. The queries are formatted using the language supported by Racer, namely the Racer Query Language (RQL).

Example 1: What kind of components is an E-Commerce Web Site composed of?

Solution:

(individual-fillers |http://a.com/ontology#WebSite_ECommerce|
|http://a.com/ontology#isComposedOf|)

Example 2: I want to add search functionality to my existing Web site. What are the different ways to do so?

Solution:

Find all the individuals of WebApplicationPattern that contains the substring "search" in its solveProblem field. (This cannot be done with a single Racer command.)

Example 3: In what situation (context) do I need to include a Privacy Policy Page in my site?

Solution:

First, call the following command to find out the corresponding object name.
(individual-attribute-fillers |http://a.com/ontology#WebPage_PrivacyPolicy|
|http://a.com/ontology#isUsedWhen|)

Copyright © 2006, Idea Group Inc. Copying or distributing in print or electronic forms without written permission of Idea Group Inc. is prohibited.

Suppose the object name is (O123), use the following command to retrieve the value for this datatype property.

(told-value |O123|)

Example 4: What are the patterns that are related to Travel Web Site pattern?

Solution:

(individual-fillers |http://a.com/ontology# WebSite_Travel|
|http://a.com/ontology#isRelatedTo|)

Example 5: What are the known uses for Hot List pattern?

Solution:

(individual-fillers |http://a.com/ontology#WebPageComponent_HotList|
|http://a.com/ontology#hasKnownUse|)

Example 6: What page(s) should a Product List Page link to? What are other optional Pages that a Product List Page could link to?

Solution:

(individual-fillers |http://a.com/ontology#WebPage_ProductList| |http://a.com/
ontology#isLinkTo|)
(individual-fillers |http://a.com/ontology#WebPage_ProductList| |http://a.com/
ontology#mayLinkTo|)

Example 7: What kind of action(s) should be triggered when a user presses the Login button?

Solution:

(individual-fillers |http://a.com/ontology#Button_Login| |http://a.com/
ontology#hasAction|)

Copyright © 2006, Idea Group Inc. Copying or distributing in print or electronic forms without written permission of Idea Group Inc. is prohibited.

Chapter III

Contextual Ontology Modeling Language to Facilitate the Use of Enabling Semantic Web Technologies

Laura Caliusco, Universidad Tecnológica Nacional - FRSF, Argentina
César Maidana, Universidad Tecnológica Nacional - FRSF, Argentina
Maria R. Galli, INGAR-CONICET-UTN, Argentina
Omar Chiotti, INGAR-CONICET-UTN, Argentina

Abstract

A common approach to represent semantics on Semantic Web area is to use an ontology. However, there is an emerging approach that combines an ontology with its context definition. So, the misunderstanding can be avoided if the context is explicitly defined. The resulting structure is called contextual ontology. To process a contextual ontology at run time, it has to be expressed in a machine processable language. However, for the analysis and design phase of a Web domain, a more appropriate ontology modeling

Copyright © 2006, Idea Group Inc. Copying or distributing in print or electronic forms without written permission of Idea Group Inc. is prohibited.

language is needed. To this aim, this chapter presents a metamodel for modeling explicit and formal contextual ontologies that assists Web domain designers in modeling contextual ontologies. Furthermore, the relationship between XML specifications and ontologies in order to add formal and explicit semantics to Web domain designs is analyzed.

Introduction

In recent years, the Semantic Web has evolved to an important research and development topic. However, there is not a widespread agreement on what the Semantic Web is, what it is for, and how it may or should evolve.

According to Berners-Lee et al. (2001), the Semantic Web is about bringing "structure to the meaningful content of Web pages, creating an environment where software agents roaming from page to page can readily carry out sophisticated tasks for users." This vision states that software agents will be pervasive on the Web, carrying out a multitude of everyday tasks. They view the Semantic Web as an extension of the current Web, in which the information is given well-defined meaning (Uschold, 2001; Anwar et al., 2004).

Most people would say that this definition does not satisfy their vision of the Semantic Web. They consider that the Semantic Web is more than an extension of the current Web. Consequently, it is a vision of the *next-generation Web* which enables Web Applications to automatically collect Web contents from diverse sources, integrate and process information, and interoperate with other applications in order to execute sophisticated tasks for humans. For these purposes, however, it is necessary to develop appropriate information technologies.

So, all works done from a pragmatic point of view taking into account the current Web structure are an intermediate step for the full adoption of Semantic Web technologies and content description languages. Furthermore, this pragmatism can open different doors to make Semantic Web a reality.

On the one hand, for the current Web to evolve into the Semantic Web, tremendous effort has been made in defining and developing various supporting standards and technologies (Davies et al., 2002). The research community, industrial participants, and software vendors are working with the World Wide Web Consortium (W3C) to define specifications and enabling Semantic Web technologies (W3C, 2004). The two key Semantic Web technologies are: the revised Resource Description Framework (RDF) and the Web Ontology Language (OWL). However, while standardization can be often a major reason why adoption of a new technology succeeds, another requirement is easy to use

Copyright © 2006, Idea Group Inc. Copying or distributing in print or electronic forms without written permission of Idea Group Inc. is prohibited.

(Gannod & Timm, 2004). So, in order to facilitate the adoption of these enabling Semantic Web technologies, a bridge should be created. This could be achieve by using the Model Driven Architecture (MDA) initiatives.

On the other hand, it is well known that ontologies are an enabling technology for Semantic Web. A high number of different ontologies over the Web requires automatic and effective techniques for ontology matching in order to fulfill semantic interoperability. Taking into account that the meaning of ontology concepts depends basically on their context, an emerging approach is contextual ontologies.

In this chapter, we present a metamodel for modeling explicit and formal contextual ontologies that assists Web domain designers in modeling Web Semantics. First, we discuss the role of contextual ontologies on the Semantic Web. Then, we show how ontology modeling languages fit into the MDA initiative. Following, we present the elements of the metamodel. Then, we analyze the relationship between XML specifications and ontologies in order to add formal and explicit semantics to Web domain designs. Finally, conclusions and future works are presented.

Contextual Ontologies and the Semantic Web

Semantics is the heart of the Semantic Web. Semantics *means the study of meaning*. However, there is no agreement on how this applies to the term "Semantic Web". Uschold (2001) proposes the idea of real-world semantics as a form of capturing the essence of the main use of the term "semantics" in a Semantic Web field. This idea proposes to define the semantics of an "item", which might be a tag or a term, or possibly a complex expression in some language.

The major problem to be tackled in relation to semantics is the *semantic heterogeneity*. That is, how to integrate different domains that use the same item to refer to different concepts or different items to refer to the same concept. Moreover, an item could be described by using different granularity or different perspective.

A common approach to overcoming semantic heterogeneity in the Semantic Web is to define an ontology (Berners-Lee et al., 2001). However, there is another approach for solving problem of semantic interoperability: the creation of a context, which is a global representation that integrates local representation (Bouquet et al., 2003). In this section we show how these approaches could be

Copyright © 2006, Idea Group Inc. Copying or distributing in print or electronic forms without written permission of Idea Group Inc. is prohibited.

integrated in order to achieve semantic interoperability in the Semantic Web area.

Ontologies

Ontologies emerged as an alternative to represent knowledge in artificial intelligence. However, they have been used to support a great variety of tasks. So, it is possible to find several definitions of ontologies in the literature (Fensel, 2001). One of the most quoted definitions is: *an ontology is a formal, explicit specification of a shared conceptualization* (Gruber, 1993). In this context, it should be clear that an "explicit" object is a concrete, symbol-level object. But "conceptualization" is not clear and sometimes "conceptualization" is defined as an abstract model of some phenomenon in the world by having identified the relevant concepts of that phenomenon.

Our attitude in this chapter is quite pragmatical. We define an ontology as a set of concepts. Concepts imply a set of *terms* and *relations* between them. Furthermore, it is suitable to add *properties* and *axioms* to enrich the ontology. The set of properties defines the characteristics of *terms*. *Axioms* are properties of the relations (Caliusco et al., 2004). Following, we formalize this definition.

Definition 1.
An ontology O_i is a 4-tuple $<T_i, P_i, R_i, A_i>$ where:
i identifies the domain or source an ontology is associated with,
T_i is a set of terms t_j of O_i,
P_i is a set of properties of terms $t_j \in T_i$,
R_i is a set of relations between t_j and $t_x \in T_i$,
A_i is a set of axioms that characterizes each relation of R_i.

Ontology Mapping

The advent of the Semantic Web has dramatically increased the need for efficient and flexible mechanisms to provide semantic mapping among ontologies. Mapping may become necessary as Web communities usually have their own ontology and could use ontology mapping to facilitate data exchange (Oberle et al., 2005).

The problem of mapping different ontologies is a well-known problem in knowledge representation (Kalfoglou & Schorlemmer, 2003). However, the development of computationally economical techniques for semantic mapping

Copyright © 2006, Idea Group Inc. Copying or distributing in print or electronic forms without written permission of Idea Group Inc. is prohibited.

are required in order to make the Semantic Web a reality. One emerging approach is *contextual ontologies*.

Context

There is not a unique definition about what a context is. Different approaches use their own assumptions to define a context and use it for different purposes. A survey of these approaches can be found in Brézillon (1999) and Theodorakis and Spyratos (2002).

In this work, we propose to define a context by a collection of relevant assumptions that make a situation unique and composed by the real content. The assumptions are attributes and facts that define the context; and the real content is the ontology. Then, the context definition is formalized as:

Definition 2.

Let J be a set of indexes j, a context C_j ∀j ∈ J can be defined as a 3-tuple $<c, D_j, O_{i,j}>$, where:

c_j is the unique identifier of context j,

D_j is a set of assumptions about context j,

$O_{i,j}$ represents ontology i within context j.

Context Mapping

Context mapping allows us to state that a certain property holds between elements belong to different ontologies defined in different contexts (Bouquet et al., 2002). Then, context mapping is defined by *bridge rules* as linking rules between contexts. Following, we define context mappings and bridge rules.

Definition 3.

A context mapping $M_{s,t}$ can be defined as a 3-tuple $<c_s, c_t, BR>$ where: c_s identifies the context source, c_t identifies the context target and BR represents the set of bridges rules that map an element from source context to elements of the target context.

Mappings are directional, i.e., $M_{s,t}$ is not the inverse of $M_{t,s}$. A mapping $M_{s,t}$ might be empty (Bouquet et al., 2002). That means that there is not relation between both contexts.

Copyright © 2006, Idea Group Inc. Copying or distributing in print or electronic forms without written permission of Idea Group Inc. is prohibited.

Definition 4.

A bridge rule br can be defined as a 3-tupple $<e_s, e_t, R>$ where: e_s is an element from source context, e_t is an element from target context, and R is the relation between elements.

A bridge rule from context s to context t is a statement of one of the following forms:

$$c_s : e_i \xrightarrow{\equiv} c_t : e_j \tag{1}$$

$$c_s : e_i \xrightarrow{\perp} c_t : e_j \tag{2}$$

$$c_s : e_i \xrightarrow{*} c_t : e_j \tag{3}$$

$$c_s : e_i \xrightarrow{\subseteq} c_t : e_j \tag{4}$$

$$c_s : e_i \xrightarrow{\supseteq} c_t : e_j \tag{5}$$

where e_i and e_j are elements of context c_s and c_t respectively.

Rule (1) means that e_i is similar to e_j. For example, Forecasting:Item $\xrightarrow{\equiv}$ Scheduling:Item.

Rule (2) means that both elements are disjointed. For example, Forecasting:Forecast $\xrightarrow{\perp}$ Scheduling:Employee.

Rule (3) means that e_i and e_j are compatible elements. For example, Forecasting:Forecast $\xrightarrow{*}$ Scheduling:Schedule (a Schedule derives from a Forecast).

Rule (4) means that e_i is less general than e_j and rule (5) means that e_i is more general than e_j. For example, Forecasting:Bucket $\xrightarrow{\supseteq}$ Scheduduling:Date (bucket: valid forecast time period).

Copyright © 2006, Idea Group Inc. Copying or distributing in print or electronic forms without written permission of Idea Group Inc. is prohibited.

Contextual Ontology Modeling Languages and MDA

The objective of this section is to show how contextual ontology modeling languages fit into the MDA (Model Driven Architecture) initiative (Mellor, 2004). This initiative is a standard produced by the Object Management Group (OMG) and its goal is to separate the design of applications or business rules from the implementation platform. To achieve this goal, MDA uses abstract high level models based on a four-layer architecture, as shown in Figure 1. Furthermore, MDA depends on and makes use of several other OMG standards including the Unified Modeling Language (UML) and Meta-Object Facility (MOF).

UML has been proposed to model ontology since UML class diagram can be used to express concepts in terms of classes and relationships among them. Cranefield (2001) proposed an ontology representation formalism based on a subset of the UML together with its associated Object Constraint Language (OCL) for agent software communication. One advantage of using UML for ontology modeling is that it is easily understood by Web designers. But, UML itself does not satisfy needs for representation of ontology concepts that are borrowed from Descriptive Logic and that are included in ontology specification languages (Djuric et al., 2003), like the Web Ontology Language (OWL). Furthermore, there are many features which can be only expressed in an ontology language, like transitive and symmetric properties in OWL (Bockmans et al., 2004).

The Ontology Working Group is defining the Ontology Definition Metamodel (ODM) (ODM, 2003). This metamodel is a MOF compliant metamodel, shown in Figure 1 that allows a user to define ontology models using the same terminology and concepts as those defined in OWL. OWL is a semantic markup language for publishing and sharing ontologies on the World Wide Web (McGuiness & van Harmelen, 2003). The main disadvantage of ODM modeling language is that it has no elements for context modeling. The context definition is supported by using annotations that are not easily transformed into a machine processable language.

It is well known that a human being does not reason without context. So, a well-designed semantic model has to describe contextual facts and contextual interrelationship (Strang et al., 2003). In resume, the need for a dedicated contextual ontology modeling language stems from the observation that a contextual ontology cannot be sufficiently modeled with UML or ODM. This new modeling language resides on the M2 layer of the MDA initiative and derives from MOF. In the next section we describe the elements of the language metamodel.

Copyright © 2006, Idea Group Inc. Copying or distributing in print or electronic forms without written permission of Idea Group Inc. is prohibited.

Figure 1. Four-layer architecture

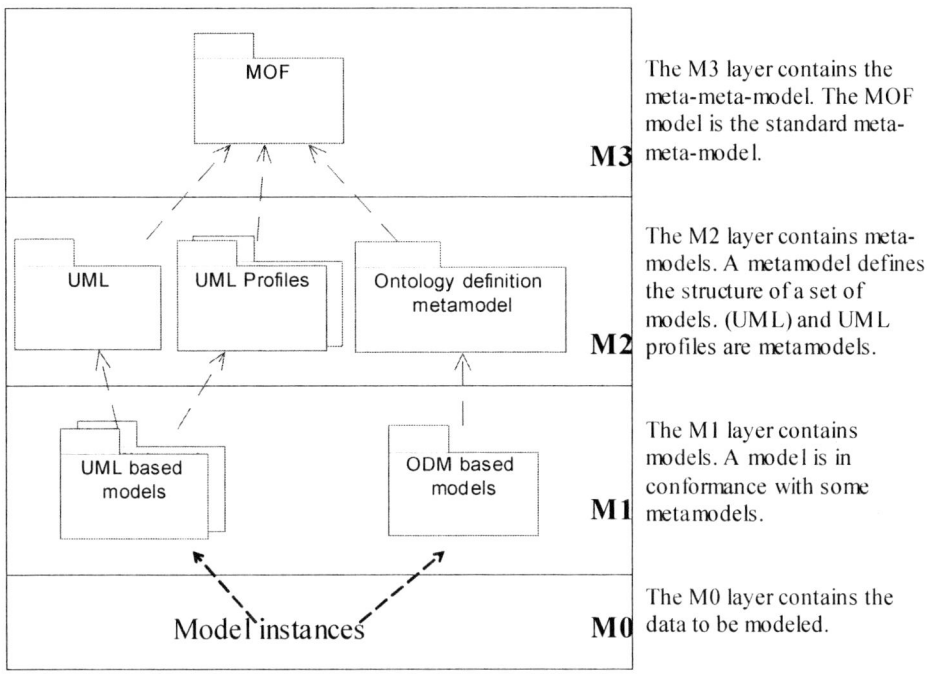

The M3 layer contains the meta-meta-model. The MOF model is the standard meta-meta-model.

The M2 layer contains meta-models. A metamodel defines the structure of a set of models. (UML) and UML profiles are metamodels.

The M1 layer contains models. A model is in conformance with some metamodels.

The M0 layer contains the data to be modeled.

Contextual Definition Metamodel

The goal of this section is to define a metamodel that assists Web designers in modeling contextual ontologies for Semantic Web. The typical role of a metamodel is to define the semantics for how an element in a model gets instantiated. In order to define the proposed metamodel we have imported some elements of the Core::Abstractions and Core::PrimitiveTypes Packages of the "UML 2.0: Infrastructure" specification (UML, 2003). The Core::Abstractions contains a set of metaclasses, most of which are abstract, to be specialized when defining new metamodels complying with MOF2. The Core::PrimitiveTypes simply contains a number of predefined types that are commonly used when metamodeling. Additionally, OCL constraints, which specific invariants that have to be fulfilled by all models that instantiate it, were defined in our metamodel.

Copyright © 2006, Idea Group Inc. Copying or distributing in print or electronic forms without written permission of Idea Group Inc. is prohibited.

Design Principles and Language Infrastructure

The main design principles of our metamodel are:

1.   Being easy to use in rapid development of contextual ontologies by ordinary persons,
2.   High independence degree of ontology specification languages, and
3.   Modularity.

In order to follow the modularity design principle, the metamodel constructs were grouped into packages according to the elements needed to define an ontology. The main package is the Kernel Package which imports the reused elements from Infrastructure::Core Package, as shown in Figure 2. All metamodel elements are derived from the Kernel elements. These packages are:

1.   **Kernel Package:** This package contains classes and associations that form the kernel of the metamodel, which are used by all other packages. This package imports and specializes elements from InfrastructureLibrary::Core.
2.   **DataType Package:** This package contains classes and associations that can be used to create data types and data values to be employed in defining an ontology.
3.   **Ontology Package:** This package contains classes and associations that can be used to define an ontology.
4.   **Terms and Properties Package:** This package contains classes and associations that can be used to model terms and their properties.
5.   **Relations Package:** This package contains classes and associations that can be used to model relations between terms belonging to an ontology.
6.   **Axioms Package:** This package contains classes and associations that can be used to describe axioms about relations.
7.   **Context Package:** This package contains classes and associations that can be used to describe a context.
8.   **Context Mapping Package:** This package contains classes and associations that can be used to define mappings between terms that belong to different contexts.

Copyright © 2006, Idea Group Inc. Copying or distributing in print or electronic forms without written permission of Idea Group Inc. is prohibited.

Figure 2. Packages defined in the contextual ontology modeling language

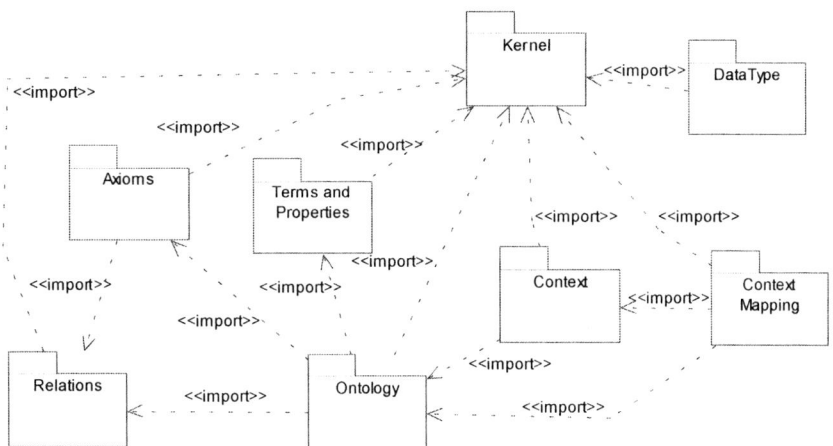

Ontology Package

Figure 3 represents the classes and associations of the Ontology Package. The main component is *Ontology* class that includes definition of concepts used to describe and represent a domain. This class is associated with the *OntologyElements* abstract metaclass, which groups the objects of an ontology metamodel. If an ontology is removed, so are its elements. The *imports* association represents the fact that an ontology could contain definitions whose meanings are defined in other ontologies. The association *prior_Version* identifies the referred ontology as a prior version of one ontology. Each ontology element could be described by a comment represented by the *Documentation* class.

Properties and Terms Package

Terms represent the set of concepts that an ontology designer wants to represent in an ontology. A term can be simple or complex. Simple terms have literal of some kind as their values. Complex terms are composed by simple terms.

Properties describe the features of a term. For example, allowed values, the number of values, and other features of values that a simple or complex term could take.

Copyright © 2006, Idea Group Inc. Copying or distributing in print or electronic forms without written permission of Idea Group Inc. is prohibited.

Figure 3. Elements defined in Ontology Package

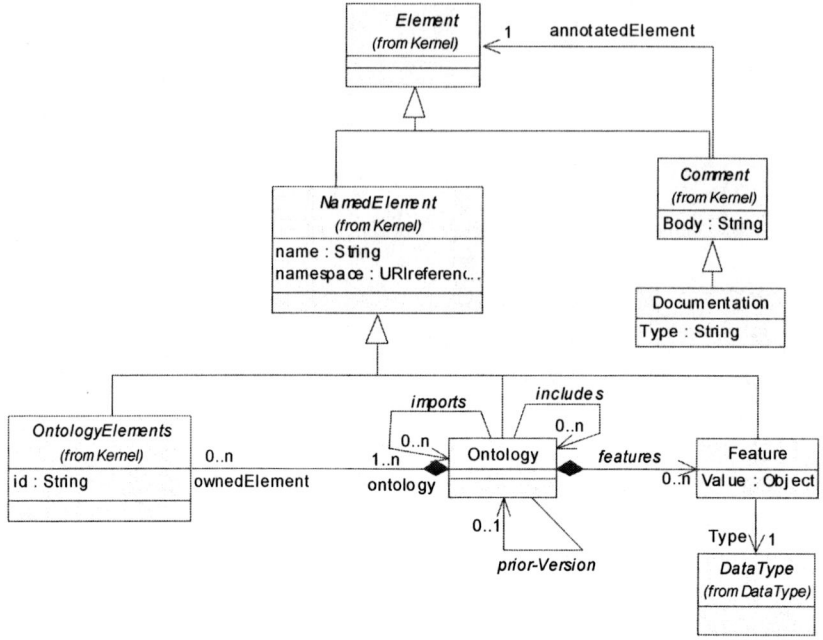

Figure 4. Elements defined in Properties and Terms Package

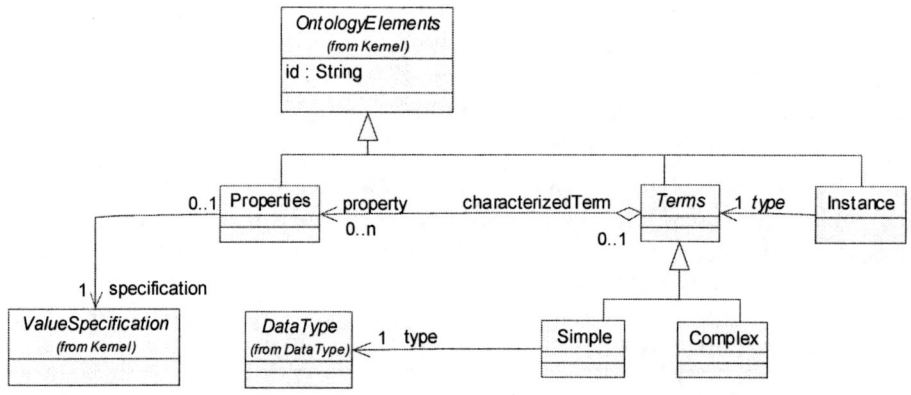

The metamodel that represents the relation between *Properties* and *Terms* is presented in Figure 4. In the proposed metamodel, the class *Properties* defines the features of a term so that if a term is removed, its properties have to be removed.

Copyright © 2006, Idea Group Inc. Copying or distributing in print or electronic forms without written permission of Idea Group Inc. is prohibited.

Relations Package

Relations can be divided into hierarchical relations (is-a and part-of), conceptual relations (synonym and antonym) and particular relations (defined by the ontology designer).

The relations metamodel is presented in Figure 5. *Terms* and *Relations* classes are associated via the *RelationEnd* class. An instance of *Relations* class has to be associated at least with two instances of *RelationEnd* class, this fact is indicated with the relations *source* and *target*. The *RelationEnd* contains the information about cardinality and the role of terms. Furthermore, this class has the *Navigable* attribute to represent the direction of the relation.

One important requirement for ontologies is the ability to structure the relations into hierarchies, i.e., to define sub-relations of a relation. Furthermore, it is suitable to define equivalent relations and inverse relations. *Subrelationof, inverseof,* and *equivalentto* relations model these characteristics.

Figure 5. Elements defined in the Relations Package

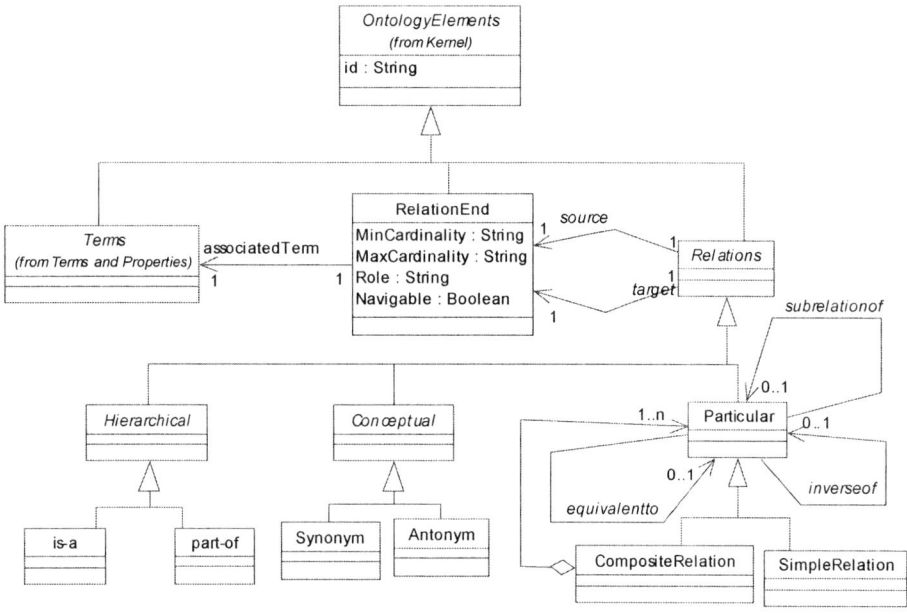

Copyright © 2006, Idea Group Inc. Copying or distributing in print or electronic forms without written permission of Idea Group Inc. is prohibited.

Axioms Package

Axioms are properties of relations and help to constrain the interpretation of concepts. Furthermore, they provide guidelines for automated reasoning. In the knowledge engineering area, axioms have been represented using logic languages. In the UML class diagram, axioms could be expressed by OCL (Staab & Mädche, 2000). For example, OCL constraints have to be used to declare a transitive property of a relation between terms. However, describing such constraints may involve writing moderately complex OCL expressions that are not immediately understood by human readers. In addition, there may be several different expressions encoding the same constraint. An interesting issue is to represent axioms as objects (ODM, 2003).

Axioms can be classified into two subsets: the set of axioms for relational algebra and the set of particular axioms, in other words, axioms defined by users. We include symmetric, asymmetric, reflexive, transitive, and functional axioms as relational algebra axioms.

In addition, the metamodel allows us to represent temporal axioms by the *Period* class. This class represents the period of time in which the axiom is valid.

Figure 6 represents the metamodel for modeling axioms and their associations with the class *Relations*.

Figure 6. Elements defined in the Axioms Package

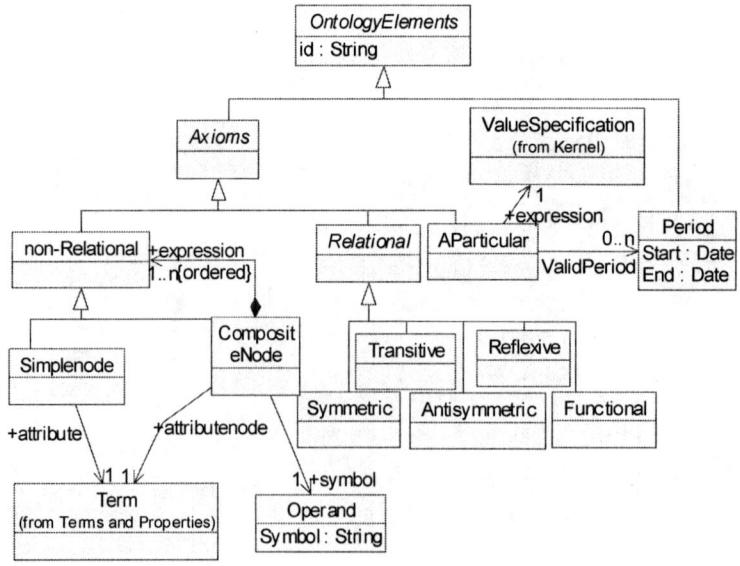

Copyright © 2006, Idea Group Inc. Copying or distributing in print or electronic forms without written permission of Idea Group Inc. is prohibited.

Context Package

We associate the *Context* class with an *Ontology* class and with one or more *Assumptions* classes, shown in the class diagram of Figure 7. In Bouquet et al. (2002) the following components are defined as assumptions: the owner of the context, the group in which the context has been developed, the security information, and information on how a context was generated. But, we preferred to define the class in general in order to allow a user to define their own *Assumptions*. In addition, a context could be derived from another context and this is modeled by the *derivedfrom* association.

Furthermore, a context could be simple or complex. That is, a context could be formed by other contexts. For example, a Web site of an enterprise that commercializes information technology could be viewed as a context. This context could be composed by "hardware" and "software" contexts.

Context Mapping Package

Figure 8 represents classes and associations to model the context mapping, bridge rules, and domain space. The *DomainSpace* class is associated with one or more *Context* and *Mapping* classes. A *Rule* class associated with the *BrigdeRule* class could be one of the five rules defined above.

Figure 7. Elements defined in the Context Package

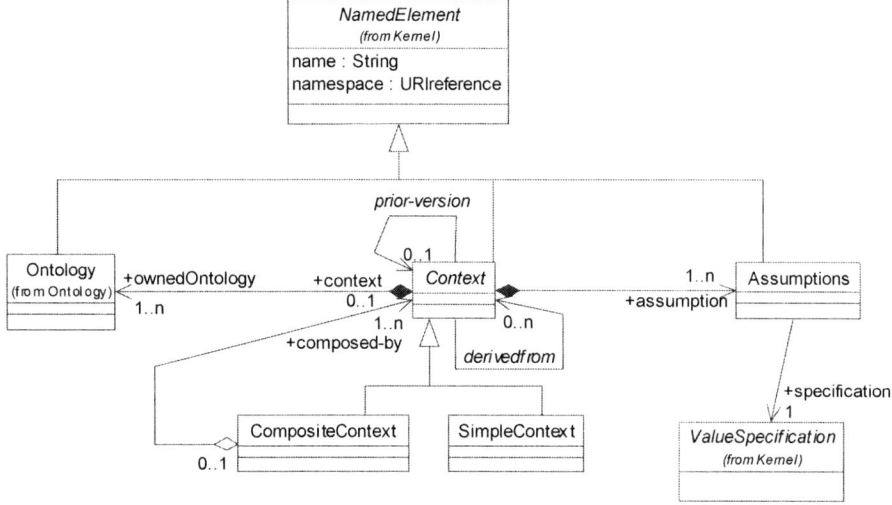

Copyright © 2006, Idea Group Inc. Copying or distributing in print or electronic forms without written permission of Idea Group Inc. is prohibited.

Figure 8. Elements defined in the Context Mapping Package

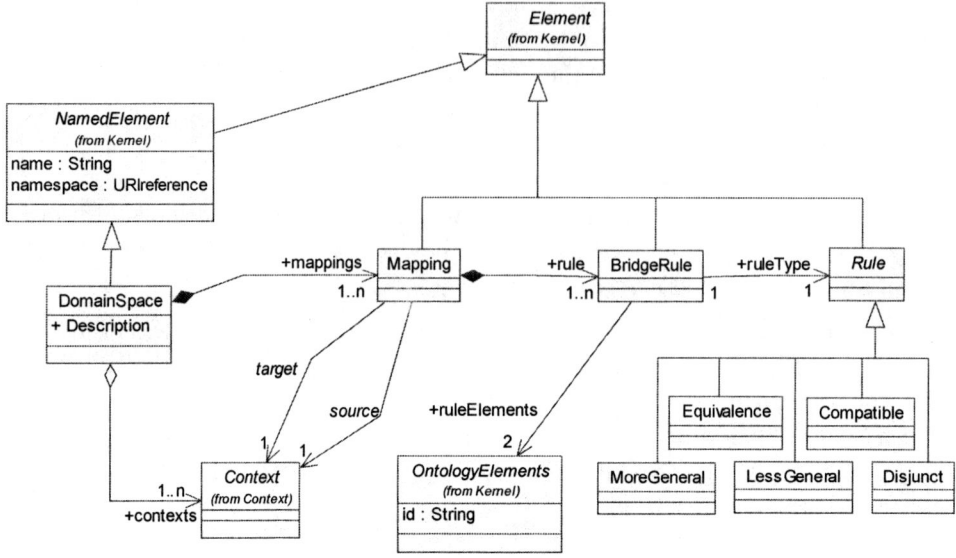

Finally, we have to define the container of all the above defined elements. This container is the domain space concept. That is, the contexts and the rules that relate the elements between context and context mapping are contained in a domain.

A Study Case

The goal of this section is to show the use of the metamodel above presented and its advantages in the design phase of a Web page. Let's suppose that enterprise A uses the catalog definition shown in Figure 9 to publish its products in the Web. The catalog was defined using XML schema (Bray et al., 2000).

This catalog is maintained by a person who is the Web designer of the enterprise. Without a graphical tool, the Web designer has to define the XML document and the associated contextual ontology document by creating a text document. This is a time-consuming task because the syntax of a contextual ontology language such as C-OWL, is not natural for human beings. Furthermore, it is well known that a graphical tool is a good support when creating complex Web. In addition,

Copyright © 2006, Idea Group Inc. Copying or distributing in print or electronic forms without written permission of Idea Group Inc. is prohibited.

Figure 9.   Catolog XML schema

```
1.   <xs:simpleType name = "Description">
2.   <xs:restriction base = "xsd:string">
3.   <xs:maxLength value = "40"/>
4.   </xs:restriction>
5.   </xs:simpleType>
6.   <xs:element name = "Item" type = "Items"/>
7.   <xs:annotation>
8.   <xs:documentation> Catalog Item </xs:documentation>
9.   </xs:annotation>
10.  </xs:element>
11.  <xs:complexType name = "ItemLine">
12.  <xs:sequence>
13.  <xs:element ref="Item"/>
14.  </xs:sequence>
15.  </xs:complexType>
16.  <xs:complexType name="CatalogLine"/>
17.  <xs:extension base="ItemLine">
18.  <xs:sequence>
19.  <xs:element name="LineNumber" Type="xs:nonNegativeInteger">
20.  </xs:sequence>
21.  </xs:extension>
22.  </xs:complexType>
23.  <xs:complexType name="Catalog"/>
24.  <xs:sequence>
25.  <xs:element name="ProductDescription" type="Description"/>
26.  <xs:element name="Line" type="CatalogLine"/>
27.  </xs:sequence>
28.  </xs:complexType>
29.  <xs:element name="CatalogDocument" type = "Catalog"/>
```

an ontology expert is not necessary if graphical tools for designing semantic models are used. So, if the design process is to become effortless, a graphical tool is needed. This graphical tool has to:

1.   Support the XML document definition.
2.   Generate a contextual ontology model from XML documents using the previously defined metamodel.
3.   Support the contextual ontology definition.
4.   Automatically generate the contextual ontology in a machine-processable language.

In order to generate a contextual ontology model from XML documents, we have defined a set of rules. The rules are the following:

Copyright © 2006, Idea Group Inc. Copying or distributing in print or electronic forms without written permission of Idea Group Inc. is prohibited.

1. The *xs:complexType* element has to be modeled by the class *Complex*. For example, from the document defined in Figure 9, line 11, line 16, and line 23, we can model **ItemLine**, **CatalogLine**, and **Catalog** terms as complex terms.

2. The *xs:simpleType* element has to be modeled by the class *Simple*. For example, from the document presented in Figure 9, line 1, we can model **Description** element as a simple term.

3. Then the elements defined by *xs:element* tag could be simple or complex depending on its type definition. That is, if they are defined as a base xs type, they are simple terms. For example, on line 19 the **LineNumber** element is defined as *xs:nonNegativeInteger*. So, this element has to be modeled as a simple term. Then, on line 29, the **CatalogDocument** element is defined as *Catalog*. Since *Catalog* is a complex term then *CatalogDocument* is a complex term, too. On the other hand, **ProductDescription** (line 25) is a simple term because *Description* is a simple one.

4. Furthermore, the *xs:restriction* element has to be modeled by *Properties* class. For example, on line 3 the **Description** element is restricted by using *<xs:maxLength value = "40"/>* definition, thus this characteristic has to be modeled as a property of the **Description** term.

5. Finally, the *xs:annotation* element has to be modeled by *Documentation* class.

In addition, the relationships between terms can be derived from an XML document as follows:

1. The *xs:extension* element represents the *is-a* relationship between terms. For example, in Figure 9 line 17, the *<xs:extend base = "ItemLine">* definition states that the previously defined element (**CatalogLine**) *is-a* **ItemLine**.

2. When an element is defined as another element, the relationship between those elements is the relation *is-a*. For example, the definition *<xs:element name="ProductDescription" type="Description"/>* states the **ProductDescription** *is-a* **Description**.

3. The combination of *complexType* and *element* primitives represents the *part-of* relationship between terms. For example, all elements defined between the *complexType* primitives that define the **Catalog** term are related to it by the *part-of* relation. That is, **ProductDescription** and **Line** are *part-of* **Catalog**.

Copyright © 2006, Idea Group Inc. Copying or distributing in print or electronic forms without written permission of Idea Group Inc. is prohibited.

Figure 10 presents the contextual ontology model generated from the XML documents defined in Figure 9. To automatically generate the contextual ontology model, the above defined rules have been used.

Conclusion and Future Works

It is widely accepted that the Web will reach its full potential only when it becomes a place where data can be shared and processed by automated tools as well as by people. The Semantic Web is a vision of the next-generation Web which enables Web Applications to automatically collect Web contents from diverse sources, integrate and process information, and interoperate with other applications in order to execute sophisticated tasks for humans.

Figure 10. Contextual ontology model

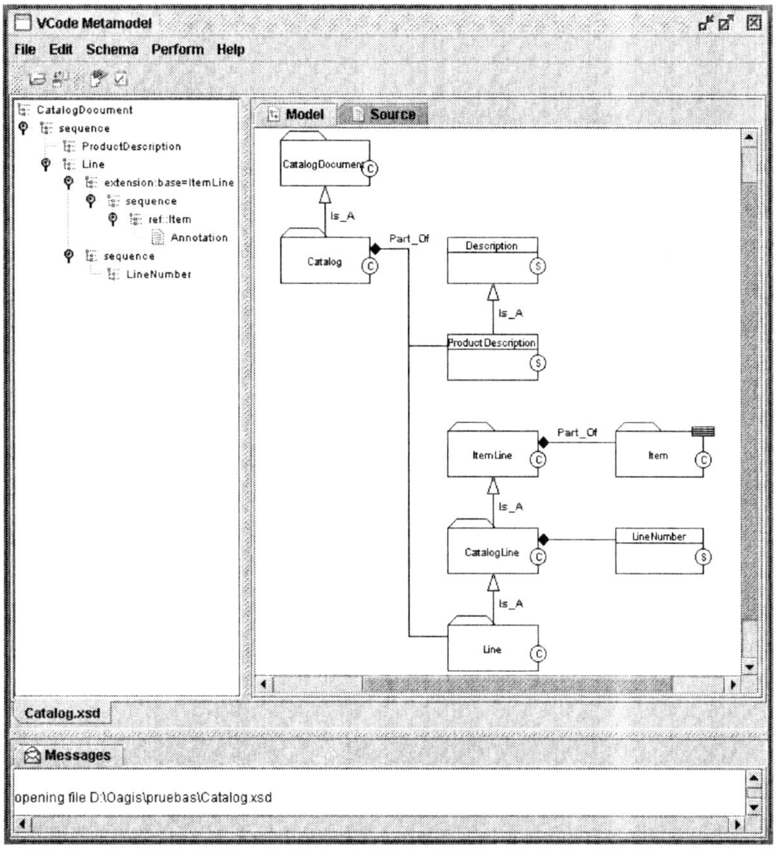

Copyright © 2006, Idea Group Inc. Copying or distributing in print or electronic forms without written permission of Idea Group Inc. is prohibited.

Representation of semantics is a key factor for Semantic Web success and will bring lots of important applications. Ontologies are developed to provide a machine-processable semantics of information sources on the Web. However, we think that a more suitable representation of semantics is contextual ontology.

In order to fill the gap between people involved in the Web design and contextual ontology specification languages, we have defined a Contextual Ontology Definition Metamodel. This metamodel allows Web designers to explicitly model contextual ontologies. Ontologies and contexts have to be integrated in order to solve the semantic heterogeneity problem and this is the main characteristic of the proposed metamodel. On the one hand, ontologies define concepts and relationships between them. On the other hand, contexts are useful tools to model concepts since the latter are true or false according to their contexts.

Furthermore, we have presented a graphical tool for supporting the contextual ontology definition from a XML document. The advantage of having such a tool is that it provides a user-friendly interface for contextual ontology creation and maintenance.

Future work is aimed at defining the rules to automatically generate a machine-processable contextual ontology from the metamodel presented in this chapter.

References

Anwar, M., Ding, W., Fang, S., &Wang, L. (2004). *The Semantic Web. Semantics for data on the Web*. Technical Report. University of Houston, Texas.

Berners-Lee, T., Hendler, J., & Lassila, O. (2001) The Semantic Web. *Scientific American, 7*-15.

Bouquet, P., Dona, A., Serafini, L., & Zanobini, S. (2002, July). Contextualized local ontologies specification via CTXML. *Proceedings of AAAI-02 Workshop on Meaning Negotiation (MeaN-02),* Edmonton, Canada.

Bouquet, P., Giunchiglia, F., van Hamerlen, F., Serafini, L., & Stuckenshmidt (2003). C-OWL: Contextualizing ontologies. *Proceedings of the 2nd International Semantic Web Conference* (pp. 164-179).

Bouquet, P., Magnini, B., Serafini, L., & Zanobini, S. (2003). A SAT-based algorithm for context matching. *Proceedings of IV International and Interdisciplinary Conference on Modeling and Using Context (CONTEXT 2003),* Stanford University (pp. 23-25).

Copyright © 2006, Idea Group Inc. Copying or distributing in print or electronic forms without written permission of Idea Group Inc. is prohibited.

Bray, T., Paoli, J., Sperberg-McQueen, C. M., & Maler, E. (2000). *Extensible Markup Language (XML) 1.0* (2nd ed.). Retrieved from http://www.w3.org/TR/1998/REC-xml-19980210

Brézillon, P. (1999). Context in problem solving: A survey. *The Knowledge Engineering Review, 14*(1), 1-34.

Brockmans, S., Volz, R., Eberhart, A., & Löffler, P. (2004). Visual modeling of OWL DL ontologies using UML. In S. A. McIlraith, et al. (Eds.), *Proceedings of the 3rd International Semantic Web Conference* (pp. 198-213). Springer.

Caliusco, M. L., Galli, M. R., & Chiotti, O. (2003). Ontology and XML-based specifications for collaborative B2B relationships. *Proceedings of III Jornadas Iberoamericanas de Ingeniería de Software e Ingeniería de Conocimiento (JIISIC)*.

Caliusco, M. L., Maidana, C., Chiotti, O., & Galli, M. R. (2004a). A semantic definition metamodel. *Proceedings of XXX Conferencia Latinoamerica de Informatica,* Arequipa, Peru.

Caliusco, M. L., Maidana, C., Chiotti, O., & Galli, M. R. (2004b). Propuesta de un sistema multiagente para asistir al modelado de documentos de negocio. *Proceeding of Argentine Symposium on Information Systems (ASIS 2004)*.

Corcho, O., & Gómez-Pérez, A. (2002). Ontology languages for the Semantic Web. *IEEE Intelligent Systems, 17*(1), 54-60.

Cranefield, S., Haustein, S., & Purvis, M. (2001). UML as an ontology modeling language. *Proceedings of the Workshop on Ontologies in Agent Systems, 5th International Conference on Autonomous Agents*.

Davies, J., Fensel, D., & van Harmelen, F. (2002). *Towards the Semantic Web: Ontology-driven knowledge management.* John Wiley.

Djuric, D., Gasevic, D., & Devedzic, V. (2003). A MDA-based approach to the ontology definition metamodel. *Proceedings of the 4th Workshop on Computational Intelligence and Information Technologies,* Faculty of Electronics, Niš, Serbia.

Fensel, D. (2001) *Ontologies: A silver bullet for knowledge management and electronic commerce.* Springer.

Gannod, G., & Timm, J. (2004). An MDA-based approach for facilitating adoption of Semantic Web service technology. *Proceedings of the IEEE EDOC Workshop on Model-Driven Semantic Web (MDSW'04)*.

Copyright © 2006, Idea Group Inc. Copying or distributing in print or electronic forms without written permission of Idea Group Inc. is prohibited.

Giunchiglia, F., & Bouquet, P. (1997). Introduction to contextual reasoning. An artificial intelligence perspective. In B. Kokinov (Ed.), *Perspectives on cognitive science, Vol. 3*. Bulgaria: NBU Press.

Grüber, T.R. (1993). A translation approach to portable ontology specification. *Knowledge Acquisition, 5*, 199-220.

Guha, R., McCool, & Fikes, R. (n.d.). *Contexts for the Semantic Web*. The 3rd International Semantic Web Conference (ISWC2004).

Kalfoglou, Y., & Schorlemmer, M. (2003, January). Ontology mapping: The state of the art. *The Knowledge Engineering Review, 18*(1), 1.

Kogut, P., Cranefield, S., Hart, L., Dutra, M., Baclawski, K., Kokar, M., & Smith, J. (2002). UML for ontology development. *Knowledge Engineering Review Journal Special Issue on Ontologies in Agent Systems, 17*(1), 61-64.

McGuinness, D., & van Harmelen, F. (2003, December). *OWL Ontology Web Language: Overview*. W3C. Retrieved from http://www.w3.org/TR/owl-features/

Mellor, S., Scott, K., Uhl, A., & Weise, D. (2004). *MDA distelled — Principles of model driven architecture*. Addison-Wesley.

Oberle, D., Staab, S., Studer, R., & Volz, R. (2005). Supporting application development in the Semantic Web. *ACM Transactions on Internet Technology (TOIT)* (forthcoming).

Ontology Definition Metamodel (ODM). (2003, August). Initial Submission to OMG.

Staab, S., & Mädche, A. (2000). Axioms are objects, too — Ontology engineering beyond the modeling of concepts and relations. In V. R. Benjamins, A. Gómez-Pérez, & N. Guarino (Eds.). *Proceedings of the ECAI 2000 Workshop on Ontologies and Problem-Solving Methods*. Berlin.

Strang, T., Linnho-Popien, C., & Frank, K. (2003). CoOL: A context ontology language to enable contextual interoperability. In J. B. Stefani, I. Dameure, & D. Hagimont (Eds.), *LNCS 2893: Proceedings of the 4th IFIP WG 6.1 International Conference on Distributed Applications and Interoperable Systems (DAIS2003), Volume 2893 of Lecture Notes in Computer Science (LNCS)* (pp. 236-247). Paris: Springer Verlag.

Theodorakis, M., & Spyratos, N. (2002). Context in artificial intelligent and information modeling. *Proceedings of the 2nd Hellenic Conference on Artificial Intelligence (SETN-02)*, Thessaloniki.

Copyright © 2006, Idea Group Inc. Copying or distributing in print or electronic forms without written permission of Idea Group Inc. is prohibited.

UML 2.0 Infrastructure — Final Adopted Specification. (2003, September).

Uschold, M. (2001). *Where is the semantics in the Semantic Web*? Workshop on Ontologies in Agent Systems (OAS) at the 5th International Conference on Autonomous Agents.

W3C. (2004). *World Wide Web Consortium: Semantic Web Activity Statement.* Retrieved from http://www.w3.org/2001/sw/Activity

Copyright © 2006, Idea Group Inc. Copying or distributing in print or electronic forms without written permission of Idea Group Inc. is prohibited.

Section II

Ontologies and
Enterprise Systems

Copyright © 2006, Idea Group Inc. Copying or distributing in print or electronic forms without written permission of Idea Group Inc. is prohibited.

Chapter IV

Ontology Management for Large-Scale Enterprise Systems

Juhnyoung Lee, IBM T.J. Watson Research Center, USA

Richard Goodwin, IBM T.J. Watson Research Center, USA

Rama Akkiraju, IBM T.J. Watson Research Center, USA

Abstract

Semantic markup languages such as RDF (Resource Description Framework) (RDF, 1999) and OWL (Web Ontology Language) (OWL, 2004) are increasingly being used to externalize metadata or ontologies about data, software and services in a declarative form. Such externalized descriptions in ontological format are used for purposes ranging from search and retrieval to information integration and to service composition (RDF Projects, 2004; OWL Tools, 2004). Ontologies could significantly reduce the costs of deploying, integrating, and maintaining enterprise systems. The barrier to more widespread use of ontologies for such applications is the lack of support in the currently available middleware stacks used in enterprise computing. This chapter presents our work on developing an

Copyright © 2006, Idea Group Inc. Copying or distributing in print or electronic forms without written permission of Idea Group Inc. is prohibited.

enterprise-scale ontology management system that will provide APIs and query languages, and scalability and performance that enterprise applications demand. We describe the design and implementation of the management system that programmatically supports the ontology needs of enterprise applications in a similar way to a database management system supporting the data needs of applications. In addition, we present a novel approach to representing ontologies in relational database tables to address the scalability and performance issues. The state of the art ontology management systems are either memory-based or use ad-hoc solutions for persisting data, and so provide limited scalability.

Introduction

In recent years, there has been a surge of interest in using ontological information for communicating knowledge among software systems. Particularly, the effort has been lead by the Semantic Web initiative by W3C (Semantic Web, 2001). As a result, an increasing range of software systems need to engage in a variety of ontology management tasks, including the creation, storage, search, query, reuse, maintenance, and integration of ontological information. Recently, there have been efforts to externalize such ontology management burden from individual software systems and put them together in middleware known as an ontology management system. An ontology management system provides a mechanism to deal with ontological information at an appropriate level of abstraction. By using programming interfaces and query languages the ontology management system provides, application programs can manipulate and query ontologies without the need to know their details or to reimplement the semantics of standard ontology languages. Such a setting is analogous to the way a database management system allows applications to deal with data as tables and provides a query engine that can understand and optimize SQL queries.

In this chapter, we describe the design and implementation of the SnoBase ontology management system (SnoBase, 2003), which is being developed at IBM T.J. Watson Research Center. While there are a number of projects for building ontology support tools (Chaudhri et al., 1989; Das et al., 2004; Missikoff, 2003; Oberle et al., 2004; Jena, 2003; pOWL, 2004; Protégé, 2005), the SnoBase project is different from others in its objective of developing an industry-strength ontology management system. The design of the SnoBase system focuses on providing reliability, scalability, and performance for enterprise computing, and also providing functionality robust and sufficient for different levels of practitioners of ontological information.

Copyright © 2006, Idea Group Inc. Copying or distributing in print or electronic forms without written permission of Idea Group Inc. is prohibited.

To make the system fit well into the current software development environment and reduce rather than increase the burden on software architects, programmers and administrators, we synthesize concepts familiar to software developers with ideas from the Semantic Web and ontology communities. The SnoBase system programmatically supports the ontology needs of applications in a similar way a database management system supports the data needs of applications. For programmers, SnoBase provides a Java API referred to as Java Ontology Base Connector (JOBC), which is the ontological equivalent of Java Data Base Connector (JDBC) (J2EE JDBC, 2002). JOBC provides a simple-to-use but powerful mechanism for application programmers to utilize ontologies without dealing with the details of ontological information. In addition, the SnoBase ontology management system supports a number of query languages. At present, SnoBase supports a variant of OWL Query Language (OWL-QL) (Fikes et al., 2003) and RDF Query Language (RDQL) (RDQL, 2004) as ontological equivalents of SQL (Structured Query Language) of relational database systems.

In addition, to address the scalability and performance issues of the state of the art ontology management systems that are either memory-based or use ad-hoc solutions for persisting data, SnoBase provides a novel approach to representing ontologies directly in database tables and providing logical reasoning by using plain SQL triggers. The provision of inference in an ontology management system requires the description and enforcement of semantics of semantic markup language constructs as rules. It turns out that most useful part of the semantics of W3C Semantic Web markup languages (such as RDF and OWL) can be effectively expressed as plain SQL triggers. The execution of the triggers automatically enforces the semantics of the ontology markup language constructs. With this database-based reasoning approach, an ontology management system gains the scalability, reliability, and query performance of the mature relational database system without extra cost. This chapter describes the schematic architecture and design considerations of the ontology management system.

The rest of this chapter is structured as follows: Next, we describe several enterprise application contexts in which an ontology management system can be useful. The chapter then provides technical background information on ontology using an example and discusses technical challenges on its management support. We then explain how we address the challenges and provide a schematic overview of the SnoBase ontology management system. Additionally, we briefly describe each component of the SnoBase system. This is followed by an explanation on ontology query languages which the SnoBase system supports. We proceed with a description of the design of JOBC API with examples and an approach to representing ontologies directly in database tables and providing logical reasoning capabilities by using plain SQL triggers. The chapter concludes

Copyright © 2006, Idea Group Inc. Copying or distributing in print or electronic forms without written permission of Idea Group Inc. is prohibited.

with a summary which summarizes the previous work on the ontology management problem, and discusses how the presented work is different from them.

Use Cases in Enterprise Computing

Ontology is similar to a dictionary, taxonomy, or glossary, but with structure and formalism that enables computers to process its content. It consists of a set of concepts, axioms, and relations, and represents an area of knowledge. Unlike taxonomy or glossary, ontology allows to model arbitrary relations among concepts, also model logical properties and semantics of the relations such as symmetricity, transitivity, and inverse, and logically reason about the relations. Ontology is often specified in a declarative form by using semantic markup languages such as RDF and OWL. It provides a number of potential benefits in processing knowledge, including the separation of domain knowledge from operational knowledge, sharing of common understanding of subjects among humans and computer programs, and the reuse of domain knowledge.

This section describes several enterprise application contexts in which an ontology management system can be useful. In addition to these examples, ontology management can be beneficial to any system dealing with multiple domain concepts that are interrelated and needs to use the concepts to describe the behavior or capabilities of its programs. Some other examples would include information retrieval and search systems for semantic-based search capabilities, video retrieval systems to annotate media with metadata, and collaboration management systems to provide a common understanding to collaboration contexts and annotate them.

Business Process Integration with Web Services

Web services facilitate the creation of business process solutions in an efficient, standard way. However, the process integration with Web services requires the automation of discovery and composition of Web services. Ontologies are applied to resolve two basic issues with the Web service-based business process integration: the discovery of Web services based on the capabilities and properties of published services, and the composition of business processes based on the business requirements of submitted requests. Application providers can annotate their Web services using semantic markup languages and make these services available for business partners via business service registries such as UDDI. Suppose that there is a semantic matching service available to help

Copyright © 2006, Idea Group Inc. Copying or distributing in print or electronic forms without written permission of Idea Group Inc. is prohibited.

match service requests with services offered by providers. Such a matching service would need to refer to ontologies in which domain knowledge describing the requirements or capabilities of services is defined. Then, it would be able to infer degrees of match between the requested and available services. An ontology management system that can manage the underlying ontology representation models and reasoning will enable the matching service to perform such semantic-based matches (Akkiraju, 2004).

Collaboration Using Corporate Social Networks

In any large organization such as a university, a company and a government, employees have to collaborate with a wide variety of people in order to perform different kinds of tasks. They collaborate with people, not only in their own group or department, but also with people in other departments, particularly, service departments such as Human Resources, Legal, Finance, IT, Purchasing, Facilities, Public Relations, etc. Large organizations often have specific people assigned for these tasks that an employee can contact. However, this contact information is often not available explicitly in the organization database. Therefore, it becomes difficult for a person, especially a new employee, to discover his/her assigned contact for accomplishing a certain kind of task. A possible approach to alleviating this problem is to use the concept of "social networks" based on the "friend" relation among people (Kerbs, 2002). The informal social networks of an organization setting can be further extended with richer ontological information on the types of relations. For example, each employee in the corporate social network has links to various kinds of contacts (e.g., HR, Legal, IT, etc.). The system then explores the corporate social network to find the right contact for a certain task. It was reported that one of the most popular applications of RDF is the Friend of a Friend (FOAF) project, which is about creating a Web of machine-readable homepages describing people, links between them and things they create and do (Lee & Goodwin, 2005; RDF Projects, 2004).

Model-Driven Business Transformation

An approach to business transformation is to employ business models such as process-oriented models or business component models to identify opportunities for saving costs or improve business processes (Zhu et al., 2004). The model-driven approach requires a model representation of business entities such as business processes, components, competencies, activities, resources, metrics, KPIs (Key Performance Indicators), etc., and their relations. Semantic models

Copyright © 2006, Idea Group Inc. Copying or distributing in print or electronic forms without written permission of Idea Group Inc. is prohibited.

using ontology markup languages provide useful representation of business models because they are not limited in representing different types of relations among business entities. Also, the automatic reasoning capability of semantic models provides an effective method for analyzing business models for identifying cost-saving or process improvement opportunities. For example, business performance metrics are naturally fit well with business activities and traditionally represented that way. By using this relation between business activities and metrics, and also the relation between business components and business activities represented in a semantic model, a business analyst can infer relations between business components and metrics. This relation can provides business insights into how the corporate can improve its performance metrics by addressing issues with the business components associated with the selected set of metrics. Then, by identifying, again in the semantic model, IT systems (a type of resources) associated with the business components, the analysts may be able to suggest recommendations about IT system management to improve performance metrics.

Technical Background

Figure 1 illustrates a simple example of ontology, which represents knowledge of a university domain (Horrocks, 2000). The ontology is displayed as a tree, as shown in a popular ontology editor, Protégé. In this simple example, all the relations among nodes are subClassOf. Note that ontology, in general, represents a graph, not a tree, because it allows arbitrary relations among nodes, and also allows a node to inherit more than one parents. In this example, PhDStudent is a subclass of two classes: Researcher and GraduateStudents.

As described earlier, an ontology is often specified in a declarative form by using semantic markup languages. The Semantic Web initiative of W3C is a vision for the future of the Web in which information is given explicit meaning, making it easier for machines to automatically process and integrate information available on the Web. The initiative utilizes XML and XML Schema's ability to define customized tagging schemes and RDF's flexible approach to representing data. RDF provides a simple semantics for data, and the data models can be represented in an XML syntax. In addition, the initiative introduces an ontology language, OWL, which can formally describe the meaning of terms. OWL is part of the growing stack of W3C recommendations related to the Semantic Web. Figure 2 displays a portion of the OWL representation of the university ontology introduced in Figure 1. It shows a number of classes related to people in the domain and their subClassOf relations by using OWL and RDF constructs.

Copyright © 2006, Idea Group Inc. Copying or distributing in print or electronic forms without written permission of Idea Group Inc. is prohibited.

Figure 1. University ontology

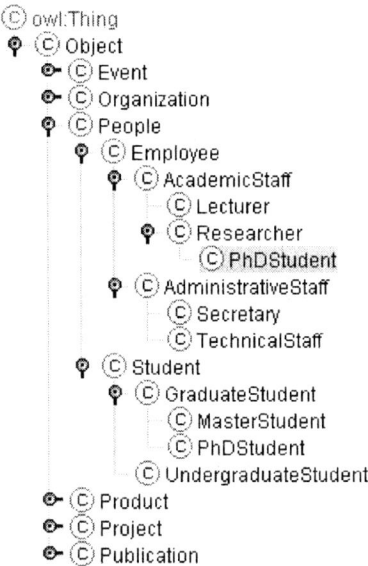

An ontology is expected to allow machines to perform useful reasoning tasks on documents and system it annotates with its concepts and relations. An inference engine provides this reasoning functionality. It extracts all the facts from the given ontology, and asserts them into its working memory. Also, it is equipped with a knowledge base which stores a set of action rules for interpreting constructs of RDF and OWL.

Before asserting the facts initially specified in OWL into the working memory, the inference engine often utilizes an OWL parser and translates them into a language, Notation 3 or N3 (N3, 2001), which is simplified, but basically equivalent to RDF and OWL in computational power. In RDF and OWL, information is simply a collection of statements, each with a subject, verb, and object, and nothing else. In N3, facts are written as triples, verb (subject, object). Figure 3 shows a portion of facts extracted from the university ontology, translated into N3, and numbered. They will be asserted into the working memory of inference engine for reasoning. Note that this set of facts is basically the same set shown in Figure 2, only in a different notation, i.e., N3.

An essential component of ontology management systems is the inference engine, which provides a mechanism for interpreting the semantics of an ontology language, represented as a set of language specific rules. The rules are

Copyright © 2006, Idea Group Inc. Copying or distributing in print or electronic forms without written permission of Idea Group Inc. is prohibited.

Figure 2. University ontology representation in OWL

```
<?xml version="1.0"?>
<rdf:RDF
  xmlns="http://a.com/ontology#"
  xmlns:rdf="http://www.w3.org/1999/02/22-rdf-syntax-ns#"
  xmlns:rdfs="http://www.w3.org/2000/01/rdf-schema#"
  xmlns:owl="http://www.w3.org/2002/07/owl#"
  xml:base="http://a.com/ontology">
<owl:Ontology rdf:about=""/>
<owl:Class rdf:ID="People">
  <rdfs:subClassOf>
    <owl:Class rdf:ID="Object"/>
  </rdfs:subClassOf>
</owl:Class>
<owl:Class rdf:ID="Employee">
  <rdfs:subClassOf rdf:resource="#People"/>
</owl:Class>
<owl:Class rdf:ID="AcademicStaff">
  <rdfs:subClassOf rdf:resource="#Employee"/>
</owl:Class>
<owl:Class rdf:ID="Researcher">
  <rdfs:subClassOf rdf:resource="#AcademicStaff"/>
</owl:Class>
<owl:Class rdf:ID="Lecturer">
  <rdfs:subClassOf rdf:resource="#AcademicStaff"/>
</owl:Class>
<owl:Class rdf:ID="PhDStudent">
  <rdfs:subClassOf rdf:resource="#GraduateStudent"/>
  <rdfs:subClassOf rdf:resource="#Researcher"/>
</owl:Class>

...
```

used to answer queries, when the requested fact is not immediately available, but must be inferred from available facts. For example, if the application requests the childrenOf an individual, but the working memory only contains parentOf relations, the inference engine can use the inverse property statements about childrenOf and parentOf to identify the correct response. For the inference engine component, most ontology management systems depend on pattern matching algorithms such as Rete (Forgy, 1982) originated from earlier work on the rule-based systems (Brownston et al., 1985). These systems comprises a working memory comprised of a set of facts representing the current status of the system, a knowledge base which stores a set of condition action rules, and a rule interpreter which applies the rules to the working memory. The pattern matching algorithms find all rules that are eligible to be fired by matching antecedent of rules to facts in working memory. If rules have variables, matching requires unification. The Rete algorithm does it efficiently because it avoids searching all rules every cycle by storing changes.

The inference engines of most ontology management systems currently available are based on pattern matching algorithms developed for expert systems in the

Copyright © 2006, Idea Group Inc. Copying or distributing in print or electronic forms without written permission of Idea Group Inc. is prohibited.

Figure 3. Asserted facts of the university ontology in N3 notation (part)

```
1       http://www.w3.org/1999/02/22-rdf-syntax-ns#type(
        http://rgoodwin.watson.ibm.com/snoscape/university.owl#Object,
        http://www.w3.org/2002/07/owl#Class)

2       http://www.w3.org/1999/02/22-rdf-syntax-ns#type(
        http://rgoodwin.watson.ibm.com/snoscape/university.owl#People,
        http://www.w3.org/2002/07/owl#Class)

3       http://www.w3.org/2000/01/rdf-schema#subClassOf(
        http://rgoodwin.watson.ibm.com/snoscape/university.owl#People,
        http://rgoodwin.watson.ibm.com/snoscape/university.owl#Object)

4       http://www.w3.org/1999/02/22-rdf-syntax-ns#type(
        http://rgoodwin.watson.ibm.com/snoscape/university.owl#Employee,
        http://www.w3.org/2002/07/owl#Class)

5       http://www.w3.org/2000/01/rdf-schema#subClassOf(
        http://rgoodwin.watson.ibm.com/snoscape/university.owl#Employee,
        http://rgoodwin.watson.ibm.com/snoscape/university.owl#People)

6       http://www.w3.org/1999/02/22-rdf-syntax-ns#type(
        http://rgoodwin.watson.ibm.com/snoscape/university.owl#AcademicStaff,
        http://www.w3.org/2002/07/owl#Class)

7       http://www.w3.org/2000/01/rdf-schema#subClassOf(
        http://rgoodwin.watson.ibm.com/snoscape/university.owl#AcademicStaff,
        http://rgoodwin.watson.ibm.com/snoscape/university.owl#Employee)

8       http://www.w3.org/1999/02/22-rdf-syntax-ns#type(
        http://rgoodwin.watson.ibm.com/snoscape/university.owl#Researcher,
        http://www.w3.org/2002/07/owl#Class)

9       http://www.w3.org/2000/01/rdf-schema#subClassOf(
        http://rgoodwin.watson.ibm.com/snoscape/university.owl#Researcher,
        http://rgoodwin.watson.ibm.com/snoscape/university.owl#AcademicStaff)

10      http://www.w3.org/1999/02/22-rdf-syntax-ns#type(
        http://rgoodwin.watson.ibm.com/snoscape/university.owl#PhDStudent,
        http://www.w3.org/2002/07/owl#Class)

11      http://www.w3.org/2000/01/rdf-schema#subClassOf(
        http://rgoodwin.watson.ibm.com/snoscape/university.owl#PhDStudent,
        http://rgoodwin.watson.ibm.com/snoscape/university.owl#Researcher)

...
```

1970s and 1980s. Their model for working memory and knowledge base is memory-based or uses ad hoc solutions for managing data. While this model is adequate for dealing with class hierarchies in small to medium scales, it does not scale for applications that involve large-scale ontologies and large amounts of instance data. We will explain how we address this scalability problem with the inference engines of ontology management systems in section 6. First, we will describe the architecture and APIs of the SnoBase ontology management system.

Copyright © 2006, Idea Group Inc. Copying or distributing in print or electronic forms without written permission of Idea Group Inc. is prohibited.

SnoBase Ontology Management System

Ontologies are becoming increasingly prevalent and important in a wide range of enterprise applications. They are used to support parametric searches, enhanced navigation and browsing, to integrate heterogeneous data sources and applications, to configure software, products, and services, and to qualitatively analyze business processes and components. In addition, applications such as information and (Web) service discovery and composition, and autonomous agents that are built on top of the Semantic Web require extensive use of ontologies.

One of challenges in the design of an industry-strength ontology management system is the versatility of Application Programming Interfaces and query languages for supporting such as diverse applications. This objective requires a careful design to simultaneously satisfy seemingly conflicting objectives such as being simple, easy-to-use for the users, and easy-to-adopt for the developers. Another challenge in ontology management is to provide ontology-enhanced industrial applications with a system that is scalable (supporting thousands of simultaneous distributed users), available (running 365x24x7), fast, and reliable. These nonfunctional features are essential not only for the initial development and maintenance of ontologies, but also during their deployment.

To provide a holistic management support for the entire lifecycle of ontological information, including ontology creation, storage, search, query, reuse, maintenance, and integration, an ontology management system needs to address a wide range of problems: ontology models, ontology base design, query languages, programming interfaces, query processing and optimization, federation of knowledge sources, caching and indexing, transaction support, distributed system support, and security support, to name a few. While some of these areas are new challenges for ontology management systems, some are familiar and there have been active studies, particularly in relation to traditional studies on knowledge representation, or recent studies on Semantic Web standards. Our approach to the ontology management support is a pragmatic one, that is, we identify missing pieces in this picture, and engineer and synthesize them with prior work for providing a holistic management system for ontological information.

Figure 4 shows a schematic overview of the SnoBase ontology management system. Conceptually, the application programs interact with the JOBC API that provides high-level access to ontology resources and the ontology engine. The application program interacts with the JOBC API that provides an access to an implementation of the API via an ontology base driver. In this case, our driver is the SnoBase driver. In this section, we will describe the each component of the SnoBase ontology management system.

Copyright © 2006, Idea Group Inc. Copying or distributing in print or electronic forms without written permission of Idea Group Inc. is prohibited.

Figure 4. SnoBase ontology management system architecture

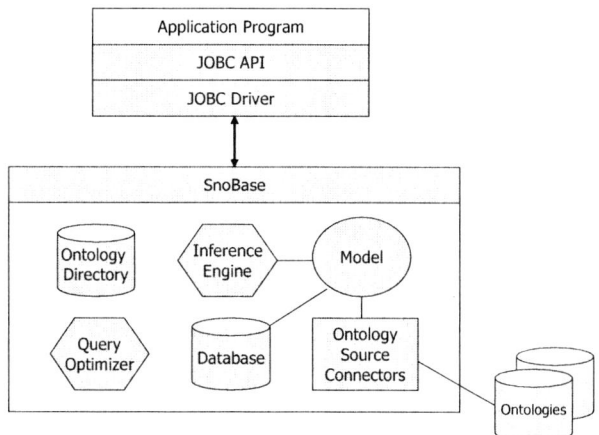

JOBC API

The SnoBase system provides a Java API referred to as Java Ontology Base Connector (JOBC), which is the ontological equivalent of the Java Data Base Connector (JDBC). The JOBC API follows the design patterns of JDBC, with several alterations. Just like JDBC, JOBC provides a connection-based interaction between applications and ontology sources. Also, JOBC provides JDBC-style, cursor-based result sets for representing query results. The similarity of JOBC to JDBC was a design decision to help application developers of SnoBase can quickly learn the programming style of JOBC from their previous experience of the popular JDBC protocol. One difference between JOBC and JDBC is that JOBC allows connections to be made without reference to a particular base ontology. Such connections provide an access to default ontologies of the top-level definitions of XML-based ontology languages such as OWL, RDF, RDF schema and XML schema. These definitions are required in order to process any ontological information.

SnoBase Driver

This component is an IBM driver for the JOBC interface that is equivalent to the IBM DB2 driver for JDBC. The SnoBase driver consists of Java classes that will provide an implementation of the JOBC API, and contains of a number of

Copyright © 2006, Idea Group Inc. Copying or distributing in print or electronic forms without written permission of Idea Group Inc. is prohibited.

components: a local ontology directory, an inference engine, a working memory, a query optimizer and a set of connectors, and other infrastructure needed to support ontology management.

Ontology Directory

This component provides the metalevel information about ontologies that are available to the SnoBase driver. By default, the ontology directory contains the references to the top-level definitions of OWL, RDF, RDF Schema, XML Schema, and similar definitions for the set of XML-based ontology languages supported. In addition, the ontology directory provides metadata such as deployment information and additional sources of ontology information. For each ontology source, the directory will need to store the URI, but may additionally store information about the contents of the ontology source to aid in query optimization.

Inference Engine

This component provides a mechanism for interpreting the semantics of an ontology language, represented as a set of language specific rules. The rules are used to answer queries, when the requested fact is not immediately available, but must be inferred from available facts. For example, if the application requests the childrenOf an individual, but the working memory only contains parentOf relations, the inference engine can use the inverse property statements about childrenOf and parentOf to identify the correct response. The details of this component, different approaches to implementing this component and issues of the scalability and performance will be discussed in section 6.

Query Optimizer

For applications that connect to large databases and/or ontologies, it will not be feasible to load the entire set of available information into working memory. Instead, the driver will query the ontology source for appropriate information as it is needed. In addition, the task of the query optimizer is to not only optimize the retrieval of information from ontology sources, but also coordinate queries that span multiple sources.

Copyright © 2006, Idea Group Inc. Copying or distributing in print or electronic forms without written permission of Idea Group Inc. is prohibited.

Ontology Source Connectors

These connectors provide a mechanism for reading, querying, and writing ontology information to persistent storage. The simplest connector is the file connector that is used to store information to the local file system. In addition, there will be connectors for storing ontological information in remote servers. Also, the connectors are used to implement caching of remote information to cache the definitions of the top-level ontology definitions OWL, RDF, RDF schema, and XML schema to allow the system to work if the W3C Web site were inaccessible.

Query Languages

Currently, the SnoBase system supports a variant of OWL Query Language (OWL-QL) as an ontological equivalent of SQL (Structured Query Language). OWL-QL is a language and protocol supporting agent-to-agent query-answering dialogues using knowledge represented in OWL. It precisely specifies the semantic relations among a query, a query answer, and the ontology base(s) used to produce the answer. It also supports query-answering dialogues in which the answering agent may use automated reasoning methods to derive answers to queries. An OWL-QL query contains a query pattern that is a collection of OWL sentences in which some literals and/or URIs have been replaced by variables. A query answer provides bindings of terms to some of these variables such that the conjunction of the answer sentences — produced by applying the bindings to the query pattern and considering the remaining variables in the query pattern to be existentially quantified — is entailed by a knowledge base (KB) called the answer KB.

This design provides a simple but expressive query model. To make a query, a program simply describes the concept it is searching for, indicating with variables which aspects of matching concepts it is interested in receiving as part of a reply. This query model is similar to the concept of query-by-example, but with the advantage that the ontology language allows a richer method for describing the examples. Another advantage of the OWL-QL query approach is that the mechanism is easily adaptable to a variety of ontology representation languages. In addition to OWL-QL, SnoBase also supports another ontology query language, RDQL, whose specification was submitted to W3C for a possible recommendation. RDQL is similar to OWL-QL in its underlying query mechanism.

Copyright © 2006, Idea Group Inc. Copying or distributing in print or electronic forms without written permission of Idea Group Inc. is prohibited.

JAVA Ontology Base Connector

As described earlier, we designed the JOBC API for SnoBase as an ontological equivalent of the Java Data Base Connector (JDBC). The API is implemented using the abstract factory pattern (Grand, 1989). An abstract factory class defines methods to create an instance of each abstract class that represents a user interface widget. Concrete factories are concrete subclasses of an abstract factory that implements its methods to create instances of concrete widget classes for the same platform. The DataManager class provides a method that is used to construct a connection, based on the URI used to initiate the connection. There is a mechanism in the DataManager that uses the database type specified in the URI to identify and load the correct driver. This driver is then used to create a connection of the appropriate type. The connection then acts as a factory to produce objects, such as statements. The objects created implement interfaces defined in the JDBC package, but have implementations that are provided by the driver that is loaded. We follow a similar design pattern in the implementation of JOBC, with several alterations, as described above. The following code sample illustrates the use of JOBC.

In this example, the code first gets a connection. This connection is then used to create resources and a statement (john isA researcher). This statement is then asserted into the inference engine. The code then creates a simple query for the asserted fact and retrieves it from the working memory of the inference engine. More complex queries can be implemented using variables. For example, the following query requests all who are researchers.

The results of such a query are a set of triples that include the binding(s) of the variable(s) in the query. In this case, the variable X is bound to john. Using these basic APIs, SnoBase programmers can build more complicated queries. For

Figure 5. A JOBC statement

```
/* We connect to an ontology resource. */
Connection connection = DriverManager.getConnection();
RDFResource john = connection.createRDFResource("John");
RDFProperty isA = connection.createRDFProperty("isA");
RDFClass researcher = connection.createRDFClass("researcher");

/* We assert a statement in inference engine: John isA researcher. */
Statement statement = connection.createStatement(John, isA, researcher);
connection.assert(statement);

/* We create a simple query. */
StatementCollection query = connection.createStatementCollection();
query.addStatement(statement);
ResultSet resultSet = connection.select(query);
```

Copyright © 2006, Idea Group Inc. Copying or distributing in print or electronic forms without written permission of Idea Group Inc. is prohibited.

Figure 6. A JOBC query

```
/* We form a query: show me all who isA researcher. */
Variable X = connection.createVariable("?X");
Statement queryStatement = connection.createStatement(X, isA, researcher);

/* We create a simple query. */
query.addStatement(queryStatement);
resultSet = connection.select(query);
```

example, a query may contain multiple variables and multiple (query) statements. Also, note that SnoBase does not simply retrieve information previously stored for queries. Instead, by using an inference engine, it infers for answering facts that are not immediately available.

Direct Representation of Ontologies in a Database

State of the art inference engines for ontology management systems are either memory-based or use ad-hoc solutions for persisting data (for a survey of ontology storage systems, see Magkanaraki et al., 2002). While this is adequate for dealing with the class hierarchies in small to medium-size ontologies, it does not scale for applications that involve large amounts of instance data. This is due to the emphasis that is placed on the metadata (hierarchy of classes or concepts) as first-class citizen as opposed to the data (instances of classes).

However, many new application domains of enterprise computing such as e-commerce and social networks, and bioinformatics deal with large amounts of preexisting data that needs to be linked to an ontology. Current solutions recommend migration of the existing data into the ontology data structures. However, if other applications still use that data, this approach requires constant replication to keep the two versions in sync. Moreover, typical ad-hoc storage solutions do not provide the same level of support for data integrity, concurrent access, and recovery as a mature database management system. We believe that approaches that create custom storage systems may suffer from severe limitations in the long run, such as problems arising from the need to integrate different ontologies, or ontologies with existing data in terms of scalability and flexibility.

Copyright © 2006, Idea Group Inc. Copying or distributing in print or electronic forms without written permission of Idea Group Inc. is prohibited.

To overcome these limitations, we propose exploring database-centric architectures for storing and manipulating ontological data. Such solutions will take advantage of existing standards for data management and the DBMS features that have been optimized over the years (robustness, concurrency control, recovery, scalability). This is, however, just the first step towards leveraging relational database systems for large-scale ontology data management (ODM). Due to the inherent complexity of ontological queries, a straightforward database implementation of an ontology system may not perform optimally. The reason is that database systems are not optimized for this type of application. We identify several promising directions of future research with the hope of stimulating the interest of the database community in supporting efficient management of ontological data.

In this section, we introduce a solution to the lack of scalability and performance problem of the traditional approach to reasoning in ontology management systems. This solution improves reasoning for ontologies by directly representing facts in relational database tables and representing action rules in SQL triggers. This solution provides a single table called the Fact Table, which stores facts asserted by ontology. The table is designed to store facts expressed in N3 notation. Also, this solution provides a set of triggers which are fired when a statement is inserted, updated, or deleted from the Fact Table. Facts added by the triggers are called derived facts, as opposed to asserted facts which are originated from the ontology. The Fact Table stores these derived facts along with the keys (IDs) of statements that caused their existence. Those keys are called justifications. Note that the justification fields of asserted facts are null. The justification fields are important because they are used to delete derived statements when their justification statements are deleted from the Fact Table. The deletion of derived facts is also automatically managed by using triggers.

Figure 7 displays the data schema of the Fact Table, which shows the fields for three parts of statements: subject, verb, and object; justification fields; along with the key field. Note that the fields for subject, verb, and object are foreign keys to an ancillary table which stores actual strings. The model field represents the scope of the given statement, but it is not directly related to this discussion.

Figure 8 displays a couple of triggers that enforce the semantic of subClassOf relations, namely, the subClassOf relation is transitive. If a statement inserted into the Fact Table completes the rule's antecedent, then the trigger adds a derived fact as a consequent. There can be many rules represented in plain SQL triggers in the system to implement the meaning of constructs of RDF and OWL.

Figure 9 displays a few facts derived by the triggers shown in Figure 8. In this simple example, it is relatively straightforward to see the justifications of each derived statement, as shown in Figure 10. Note that firing of a rule sometimes can cause chains of trigger firing. Also, note that a statement can be derived by

Copyright © 2006, Idea Group Inc. Copying or distributing in print or electronic forms without written permission of Idea Group Inc. is prohibited.

Figure 7. Data schema for Fact Table

```
// Create fact_table

CREATE TABLE FACT_TABLE
(
        FACT_ID BIGINT DEFAULT AUTOINCREMENT INITIAL 1 INCREMENT 1 NOT NULL,
        SUBJECT BIGINT NOT NULL,
        VERB BIGINT NOT NULL,
        OBJECT BIGINT NOT NULL,
        MODEL VARCHAR(256) NOT NULL,
        JUSTIFICATION_1 BIGINT,
        JUSTIFICATION_2 BIGINT,
        JUSTIFICATION_3 BIGINT
)
```

Figure 8. Triggers for deriving implied facts from asserted facts (part)

```
// (A subClassOf B) and (B subClassOf C) => (A subClassOf C)

CREATE TRIGGER SUBCLASSCLASS1
        AFTER INSERT
        ON FACT_TABLE
        REFERENCING NEW ROW AS inserted_fact
        FOR EACH ROW
        INSERT INTO fact_table (subject, verb, object, model, just_1, just_2)
        SELECT DISTINCT inserted_fact.subject, verb, object, model, inserted_fact.fact_id, fact_id
        FROM fact_table
        WHERE verb = 2
                AND verb = inserted_fact.verb
                AND subject = inserted_fact.object
                AND model = inserted_fact.model

CREATE TRIGGER SUBCLASSCLASS2
        AFTER INSERT
        ON FACT_TABLE
        REFERENCING NEW ROW AS inserted_fact
        FOR EACH ROW
        INSERT INTO fact_table (subject, verb, object, model, just_1, just_2)
        SELECT DISTINCT subject, verb, inserted_fact.object, model, fact_id, inserted_fact.fact_id
        FROM fact_table
        WHERE verb = 2
                AND verb = inserted_fact.verb
                AND object = inserted_fact.subject
                AND model = inserted_fact.model
```

multiple combinations of justification statements. Therefore, the solution should provide a mechanism that prevents a statement from being added multiple times to the Fact Table.

With statements both asserted and derived in relational database tables, queries to the ontology management system become SQL queries to the relational database tables. Most ontology management systems available today uses certain query languages which express queries with a collection of N3 state-

Copyright © 2006, Idea Group Inc. Copying or distributing in print or electronic forms without written permission of Idea Group Inc. is prohibited.

Figure 9. Derived facts of the university ontology (part)

```
101    http://www.w3.org/2000/01/rdf-schema#subClassOf(
       http://rgoodwin.watson.ibm.com/snoscape/university.owl#Employee,
       http://rgoodwin.watson.ibm.com/snoscape/university.owl#Object)

102    http://www.w3.org/2000/01/rdf-schema#subClassOf(
       http://rgoodwin.watson.ibm.com/snoscape/university.owl#AcademicStaff,
       http://rgoodwin.watson.ibm.com/snoscape/university.owl#People)

103    http://www.w3.org/2000/01/rdf-schema#subClassOf(
       http://rgoodwin.watson.ibm.com/snoscape/university.owl#AcademicStaff,
       http://rgoodwin.watson.ibm.com/snoscape/university.owl#Object)

104    http://www.w3.org/2000/01/rdf-schema#subClassOf(
       http://rgoodwin.watson.ibm.com/snoscape/university.owl#Researcher,
       http://rgoodwin.watson.ibm.com/snoscape/university.owl#Employee)

105    http://www.w3.org/2000/01/rdf-schema#subClassOf(
       http://rgoodwin.watson.ibm.com/snoscape/university.owl#Researcher,
       http://rgoodwin.watson.ibm.com/snoscape/university.owl#People)

106    http://www.w3.org/2000/01/rdf-schema#subClassOf(
       http://rgoodwin.watson.ibm.com/snoscape/university.owl#Researcher,
       http://rgoodwin.watson.ibm.com/snoscape/university.owl#Object)

107    http://www.w3.org/2000/01/rdf-schema#subClassOf(
       http://rgoodwin.watson.ibm.com/snoscape/university.owl#PhDStudent,
       http://rgoodwin.watson.ibm.com/snoscape/university.owl#AcademicStaff)

108    http://www.w3.org/2000/01/rdf-schema#subClassOf(
       http://rgoodwin.watson.ibm.com/snoscape/university.owl#PhDStudent,
       http://rgoodwin.watson.ibm.com/snoscape/university.owl#Employee)

109    http://www.w3.org/2000/01/rdf-schema#subClassOf(
       http://rgoodwin.watson.ibm.com/snoscape/university.owl#PhDStudent,
       http://rgoodwin.watson.ibm.com/snoscape/university.owl#People)

110    http://www.w3.org/2000/01/rdf-schema#subClassOf(
       http://rgoodwin.watson.ibm.com/snoscape/university.owl#PhDStudent,
       http://rgoodwin.watson.ibm.com/snoscape/university.owl#Object)

...
```

Figure 10. Justifications of derived facts

```
101:    (3, 5)
102:    (5, 7)
103:    (7, 101), or (3, 102)
104:    (7, 8)
105:    (8, 102), or (5, 104)
106:    (8, 103), (101, 104) or (3, 105)
107:    (8, 10)
108:    (10, 104), or (7, 107)
109:    (10, 105), (102, 107), or (5, 108)
110:    (10, 106), (103, 107), (101, 108), or (3, 109)
```

Copyright © 2006, Idea Group Inc. Copying or distributing in print or electronic forms without written permission of Idea Group Inc. is prohibited.

ments in which some parts are replaced by variables. A query answer provides bindings of terms to some of these variables. As explained earlier, the query language of the IBM's SnoBase system which is loosely based on OWL-QL follows this query pattern. The solution presented in this section provides a query processor which transforms ontology queries into SQL queries against the Fact Table.

Provision of an inference mechanism which is purely based on mature relational database technology and SQL would resolve the problems with scalability and performance of the traditional approach based on rule-based expert systems. In addition, query optimization techniques known to relational databases such as indexing, schema optimization, and performance tuning can be applied to this proposed approach and further improve the performance of ontology query execution.

Related Work

One of the notable studies in ontology management is the Open Knowledge Base Connectivity (OKBC) (Chaudhri et al., 1989), which shares somewhat similar objectives with JOBC. It is a protocol for accessing knowledge bases stored in Knowledge Representation (KR) Systems. The API has been developed to address the issue of knowledge base tool reusability and was implemented in various languages including Java. It provides a set of operations with a generic interface to underlying KR systems, so that it provides applications with independence from the idiosyncrasies of specific KR system and enables the development of generic tools.

Other work on ontology management includes KAON (Oberle et al., 2004), SymOntoX (Missikoff, 2003), pOWL (2004), and Jena (2003). KAON is a tool suite for ontology management, providing an editor, a Web interface, an API, and an inference engine based on a deductive database approach. SymOntoX is a Web-based ontology editing environment which supports collaborative and distributed ontology authoring activities. Similarly, pOWL also provides a Web-based ontology editing environment. Jena is a collection of RDF tools written in Java that includes a Java model/graph API, an RDF parser, a query system, support classes for OWL ontologies, and persistent and in-memory storage. Jena provides statement-centric methods for manipulating an RDF model as a set of RDF triples and resource-centric methods for manipulating an RDF model as a set of resources with properties.

Additionally, there is active work for ontology integration, which enables to detect and resolve ontological differences and mismatches among existing

Copyright © 2006, Idea Group Inc. Copying or distributing in print or electronic forms without written permission of Idea Group Inc. is prohibited.

models and ontologies. Systems such as PROMPT (Noy & Musen, 2000) and Chimera (McGuiness, 2000) belong to this category. A more comprehensive survey on ontology tools can be found in Denny (2004). This survey focuses on the ontology editors and reports 94 existing tools. Additionally, the study reports on management functions of the existing systems such as reasoning and problem solving facilities, standard ontology language support, version control, visual navigation support, ontology alignment, natural language support, collaborative development support, and so forth.

There are many activities going on in the areas of ontology query. In addition to OWL-QL described in section 4, there are a few other ontology query languages of interest. RDQL (RDF Query Language) is a typed, declarative query language for querying RDF description bases. TRIPLE is an ontology query, inference, and transformation language for the Semantic Web (Sintek & Decker, 2001). It allows the semantics of languages on top of RDF to be defined with rules.

The database-centric architecture for storing and managing ontological data presented in section 6 is related to deductive databases (Gardarin, 1985; Coral, 2001; Datalog, 1995). The concept of the deductive database was implemented in several systems including SABRE (Gardarin, 1985), CORAL Deductive Database (Coral, 2001), and Datalog (Datalog, 1995). They often provided a logic-based, declarative query language for the relational model, and imperative constructs such as update, insert, and delete rules. In addition, CORAL provided C++ API to combine declarative and imperative programming.

One aspect that separates our approach from the deductive databases is that it teaches rules are written in the standard SQL language as SQL triggers, while the prior art (CORAL, SABRE, Datalog) teaches that rules are written in a logic-based, declarative query language, which is not standard SQL. In addition, unlike prior art, this approach takes advantage of the mature standard SQL technology for managing rules (for update, assert, and retract) instead of some other means, which is not standard SQL.

Another database-centric approach related to ontology support was discussed in (Das et al., 2004). This approach expanded SQL by adding new operations for supporting ontology-based semantic matching in SQL queries. It allows users to write a SQL query by using the new ontology operators that returns not only syntactically matching rows of data, but also rows of data semantically related as defined in the ontology also specified in the query. It stores ontologies in system-defined tables instead of user tables, and so, they are invisible to the users. Also, inference for queries is hidden in the implementation of the ontology operators. Therefore, it is invisible to the users. The approach additionally discussed a special indexing scheme for speeding up the ontology-based semantic matching queries. This work focused on extending SQL to support

Copyright © 2006, Idea Group Inc. Copying or distributing in print or electronic forms without written permission of Idea Group Inc. is prohibited.

ontology-based semantic matching. They placed the actual implementation of the inference and ontology storage mechanisms in a lower level of the system. In contrast, the database-centric approach presented in this chapter focuses on the implementation of the inference and ontology storage mechanisms by using plain SQL constructs (triggers) and user tables. The user queries ontologies through OWL-QL which SnoBase supports and the OWL-QL queries are automatically translated into SQL queries.

Concluding Remarks

An increasing range of applications require an ontology management system that helps externalize ontological information for a variety of purposes in a declarative way. The primary objective of ontology management systems is to provide holistic control over management activities for ontological information by externalizing them from application programs. Ontology management systems provide ontology independence to applications in a similar way that database management systems provide data independence. One of the pragmatic challenges for ontology management system research is how to create missing component technology pieces, and to engineer them with existing results from prior research work for providing a holistic management system.

In this chapter, we presented a project for developing an industry-strength ontology management system that will provide reliability, scalability, and performance for enterprise uses, and also provide functionality robust and sufficient for different levels of practitioners of ontological information. We described the design and implementation of the SnoBase ontology management system, which is under development at IBM T.J. Watson Research Center. The programming interface of the SnoBase system provides a Java API, Java Ontology Base Connector, which is the ontological equivalent of the Java Data Base Connector. Similarly, this system supports a variant of OWL Query Language (OWL-QL) as our ontological equivalent of SQL (Structured Query Language), and will support a few other ontology query languages in the future.

As ontologies are becoming central to an increasing number of enterprise applications such as business process and information integration, and information search and navigation which require scalability and performance, efficient storage and manipulation of large scale ontological data is going to be essential. In this chapter, we introduced an architecture that leverages relational database systems for storing ontologies. We described a direct representation of ontologies in a relational database using a vertical table for storing ontology triples and triggers for automatic maintenance of inferred facts. This solution is primarily

Copyright © 2006, Idea Group Inc. Copying or distributing in print or electronic forms without written permission of Idea Group Inc. is prohibited.

concerned with the storage of classes and their relations and less with instance data. In order to accommodate large volumes of data, we are working on an architecture that leaves the data in place, and provides a virtualization layer that effectively links the data triples to ontology classes. The database-centric approach to ontology management support is still in its infancy. For this approach to scale, a number of technical challenges need to be addressed. Some directions for further investigation include:

- Improving trigger evaluation by exploiting the fact that many inference rule-specific triggers share considerable portions of their conditions.

- Exploring query rewriting possibilities by analyzing the characteristics of vertical virtual views used in the database-centric inference scheme.

- A specialized metadata, ontology indexes to guarantee short response times when supporting an increasing number of enterprise applications with large volumes of instance data.

- Extensions to SQL for ontological queries over classes, relations and instances. For example, it might be useful to allow path patterns over the labeled graph of classes and relations.

- Exploring hybrid approaches to storage ranging from fully materialized vertical tables to fully virtual views over existing tables. Cost measures can take into account query work-load and physical distribution of data.

- Reasoning about the cardinalities of the relations between classes and tables. For example, one table corresponds to one and only class, one table stores instances of many classes, a single class can span multiple tables, and finally many-to-many class-table relations.

- Lazy vs. Eager computation of derived facts: some inference rules involve transitive closure and are harder to evaluate at query time; for others, simpler rules (non-recursive) can be added to a user query automatically and evaluated at query time.

References

Akkiraju, R., Verma, K., Goodwin, R., Doshi, P., & Lee, J. (2004). *Executing abstract Web process flows*. The 14th International Conference on Automated Planning and Scheduling (ICAPS), Whistler, British Columbia, Canada.

Copyright © 2006, Idea Group Inc. Copying or distributing in print or electronic forms without written permission of Idea Group Inc. is prohibited.

Brownston, L., Farrell, R., & Kant, E. (1985). *Programming expert systems in OPS5: An introduction to rule-based programming.* The Addison-Wesley series in Artificial Intelligence.

Chaudhri, V., Farquhar, A., Fikes, R., Karp, P., & Rice, J. (1989). *OKBC: A programmatic foundation for knowledge vase interoperability.* AAAI-98.

Coral Database Project. (2001). Retrieved from http://www.cs.wisc.edu/coral

Das, S., Chong, E. I., Eadon, G., & Srinivasan, J. (2004). *Supporting ontology-based semantic matching in RDBMS.* The 30th International Conference on Very Large Data Bases, Toronto, Canada.

Deductive databases: Datalog approach. (1995). Retrieved from http://goanna.cs.rmit.edu.au/~zahirt/Teaching/subj-datalog.html

Denny, M. (2004). *Ontology tools survey.* Retrieved from http://www.xml.com/pub/a/2004/07/14/onto.html

Fikes, R., Hayes, P., & Horrocks, I., (2003*). OWL-QL — A language for deductive query answering on the Semantic Web.* CA: Stanford University, Knowledge Systems Laboratory.

Forgy, C. (1982). Rete: A fast algorithm for the many pattern/many object pattern match problem. *Artificial Intelligence, 19,* 17-37.

Gardarin, G., & Pasquer, F. (1985). *Design and implementation of SABRE: A deductive and parallel database machine.* Database Machines, NATO: Pergamon Press.

Grand, M. (1989). *Patterns in Java, Volume 1: A catalog of reusable design patterns illustrated with UML.* John Wiley & Sons.

Horrocks, I. (2000). *University ontology.* Retrieved from http://protege.stanford.edu/plugins/owl/owl-library/ka.owl

J2EE JDBC Technology (2002). Retrieved from http://java.sun.com/products/jdbc/

Jena 2.0: A Semantic Web Framework (2003). Retrieved from http://www.hpl.hp.com/semweb/jena.htm

Krebs, V. (2002). *An introduction to social network analysis.* Retrieved from http://www.orgnet.com/sna.html

Lee, J., & Goodwin, R. (2005). *The Semantic Webscape: A view of the Semantic Web.* The International World Wide Web Conference, Chiba, Japan.

Magkanaraki, A., Karvounarakis, G., Anh, T. T., Christophides, V., & Plexousakis, D. (2002). *Ontology storage and querying.* ICS-FORTH Technical Report, No 308.

Copyright © 2006, Idea Group Inc. Copying or distributing in print or electronic forms without written permission of Idea Group Inc. is prohibited.

McGuiness, D. L. (2000). Conceptual modeling for distributed ontology environments. *Proceedings of the 8th International Conference on Conceptual Structures Logical, Linguistic, and Computational Issues (ICCS 2000)*, Darmstadt, Germany.

Missikoff, M., & Taglino, F. (2003). *SymOntoX: A Web-ontology tool for eBusiness domains*. The 4th International Conference on Web Information Systems Engineering (WISE'03), Rome, Italy.

Noy, N. F., & Musen, M. A. (2000). *PROMPT: Algorithm and tool for automated ontology merging and alignment*. The 17th National Conference on Artificial Intelligence (AAAI-2000).

Oberle, D., Volz, R., Motik, B., & Staab, S. (2004). An extensible ontology software environment. In S. Staab, & R. Studer (Eds.), *Handbook on ontologies* (pp. 311-333). Springer.

pOWL: Semantic Web Development Platform (2004). Retrieved from http://powl.sourceforge.net/

Primer: Getting into RDF & Semantic Web using N3. (2001). Retrieved from http://www.w3.org/2000/10/swap/Primer.html

Protégé Ontology Editor. (2005). Retrieved from http://protege.stanford.edu/

RDF Data Query Language (RDQL). (2004). *RDQL — A query language for RDF*. Retrieved from http://www.w3.org/Submission/2004/SUBM-RDQL-20040109/

Resource Description Framework (RDF). (1999). Retrieved from http://www.w3.org/RDF/

Resource Description Framework (RDF): Projects and applications. (2004). Retrieved from http://w3c.org/RDF/#projects

Semantic Web (2001). Retrieved from http://www.w3.org/2001/sw/

Sintek, M., & Decker, S. (2001). *TRIPLE — An RDF query, inference, and transformation language for the Semantic Web*. International Semantic Web Conference.

SnoBase: IBM Ontology Management System (2003). Retrieved from http://alphaworks.ibm.com/tech/snobase

Web Ontology Language (OWL). (2004). Retrieved from http://www.w3.org/2004/OWL/

Web Ontology Language (OWL): Tools, projects and applications. (2004). Retrieved from http://w3.org/2004/OWL/#projects

Zhu, J., Tian, Z., Li, T., Sun, W., Ye, S., Ding, W., et al. (2004). Model-driven business process integration and management: A case study with the Bank SinoPac Regional Service Platform. *IBM Journal of Research and Development, 48*(5/6).

Copyright © 2006, Idea Group Inc. Copying or distributing in print or electronic forms without written permission of Idea Group Inc. is prohibited.

Chapter V

From Ontology Phobia to Contextual Ontology Use in Enterprise Information Systems

Rami Rifaieh,
University of California San Diego, USA

Aïcha-Nabila Benharkat,
National Institute of Applied Science of Lyon, France

Abstract

Shared understanding in an enterprise is necessary to permit a unifying framework serving as the basis of communication between people, interoperability between systems, and other system engineering benefits such as reusability, reliability, and specification. Bringing systems to work together is increasingly becoming essential for leveraging the Enterprise Information Systems (EIS) and reaching common goals. Currently, enterprises develop their systems independently with low consideration for the collaboration that systems can play with other systems. Certainly, semantic sharing represents the daunting barrier for making these systems work together through common shared understanding. In the last decade,

Copyright © 2006, Idea Group Inc. Copying or distributing in print or electronic forms without written permission of Idea Group Inc. is prohibited.

theoretical research such as ontologies and context were suggested separately as formal support for treating the semantics-sharing problem. In order to resolve this main problem, we intend to pair up the two notions of Context and Ontologies. Typically, contextualization can be seen at the ontology level in order to enable the multiple views and multi-representation requirements. Hence, the formal representation of contextual ontologies should preserve adequate reasoning mechanisms. A machine understandable semantics and interpretation should be also given for information in a context, according to a specific system's point of view. However, we perceived a growing ontology phobia in many enterprises. This fear is based on misunderstanding of ontologies' advantages and lack of practical applications for theoretical proposals. The aim of this chapter is twofold. On one hand, it concentrates on studying the application of tightening together context and ontologies which can serve as formal background for reaching a suitable EIS environment. It invests in resolving the semantic-sharing problem between these systems. It focuses on suggesting a formalism for contextual ontologies based on a combination of Description Logics and Modal Logics. On the other hand, it investigates issues and arguments helping to overcome the ontology phobia. It shows with examples the usefulness of these contextual ontologies for resolving the semantic-sharing problem between some EIS.

Introduction

The introduction of computers into business in the 1950s has had a strong impact on corporations. In today's conquest of market places, enterprises call on their systems to meet with the strategic requirements and competitive challenges. These systems become responsible for supplying employees with the needed tools to communicate, to cooperate, to answer their queries, and to support their decisions. In this perspective, Enterprise Information Systems cover a set of information systems used by organizations in order to fulfill their operational needs and to support their strategic goals. This includes the integration of systems within the organization and the linking of the internal processes electronically to those of other organizations. In essence, this raised area of research is concerned with the study of EIS as tools supporting the business processes and management of organizations.

For these EIS the emphasis is put on the properties of openness, scalability, and autonomy. Architecture of distributed components and their shared understanding are necessary to permit the *reuse* of these components in different contexts,

Copyright © 2006, Idea Group Inc. Copying or distributing in print or electronic forms without written permission of Idea Group Inc. is prohibited.

and the *interoperability* between heterogeneous systems developed by different organizations (Benaroach, 2002). Lack of interoperability, caused by heterogeneity, compromises enterprises collaboration and knowledge sharing. Regardless of these facts, many EIS are developed independently, based on specification with an intention of monorepresentation of the domain of interest (Arara, 2004).

By analyzing the preceding issues, we can distinguish that the problem originates in semantics heterogeneity which degrades interoperability and reusability for EIS. For instance, semantic interoperability is fundamentally driven by the purpose of communication (Obrst, 2003). To make systems interoperate, semantics sharing is essential for the process and should be guided by the interpretation within a particular context and from a particular point of view. Therefore, the way to address the problem of semantics heterogeneity is to reduce or eliminate conceptual and terminological mismatches.

Ontology is the term used to refer to the shared understanding of some domain of interest which may be used as a unifying framework to solve the problem of semantic integration and semantic sharing. An ontology necessarily entails or embodies some sort of world view with respect to a given domain. A domain is a specific subject area or areas of knowledge, like medicine, tool manufacturing, real estate, automobile repair, financial management, and so forth. Furthermore, ontologies started to be more aware that having a single representation is sometimes inadequate, matter of levels of granularity and context dependent characteristics. Therefore, they should also provide the ability for integrating different user perspectives and representations (i.e., contexts).

Ontologies and context each separately played a major role in many AI applications, integration for heterogeneous and distributed systems, and system engineering (Gruber, 1993; Nuseibeh, 1994). On one hand, local information systems ontologies are foreseen to play a key role in partially resolving the semantic conflicts and differences that exist among systems. On the other hand, the notion of context carried through views, aspects, and roles can support development processes of complex systems. It is defined as a locally managed object, which encapsulates partial knowledge about a particular system.

This chapter examines contemporary themes in enterprise information systems. It investigates research and practice in ontologies and context paradigms for improving semantics sharing among these systems, and attempting to surpass the ontology phobia.

The remaining parts of this chapter are organized as follows:

First, it focuses on unveiling how ontology and context can help for resolving semantic-sharing problems. It studies these notions separately and investigates their usefulness with EIS. It identifies, as well, the limits of simple ontology to

Copyright © 2006, Idea Group Inc. Copying or distributing in print or electronic forms without written permission of Idea Group Inc. is prohibited.

answer EIS needs, especially facing the multiple views and multirepresentation problems. Therefore, it suggests pairing up the two notions of ontologies and context to overcome these problems.

Next, it seeks formalism for defining the paradigm of contextual ontology. This formalism allows each ontological concept to be identified by a different manner and to have a different role depending of the context of use. It presents the contextual ontology approach with the stamping mechanism and contextual bridge rules mechanism. Some examples are suggested to show the multirepresentation problem. Later on, these examples are resolved using the suggested formalism.

The discussion concentrates in identifying the factors that contributed to the ontology phobia in Enterprises. It shows at first the state of the using ontologies in a productive environment with enterprise ontology and local information systems ontologies. It concludes a discussion about overcoming IT decision makers' fears of using ontologies (i.e., our contextual ontologies).

Ontologies in the Service of Enterprise Information Systems

Ontology is a word that has been the subject for many studies and a puzzle for many philosophers and scientists as well. In the last decade, this debated subject has contaminated the computer science community. Indeed, the research about ontological issues has been widely active in various areas such as knowledge engineering, intelligent information integration, knowledge management, cooperative information systems, and more.

This part presents the origin of the ontology research and shows different definitions for this topic in computer science field. It covers an overview of ontology research applications, languages, and formalism. It suggests, finally, pairing up the notion of context and ontologies in order to overcome the semantic-sharing difficulties within Enterprise Information Systems.

Background: What is an Ontology About?

To put it in a nutshell, the computer's research in ontology has been carried out in the AI domain (Borgida, 1989) and with the project ARPA Knowledge Sharing Effort (Gruber, 1991). Since then, many other computer communities have begun using ontologies to establish a joint terminology between human and machines as members of interest. The most referred-to definition for ontology is given by Gruber (1993), baptizing ontology as specification of conceptualization. It is

Copyright © 2006, Idea Group Inc. Copying or distributing in print or electronic forms without written permission of Idea Group Inc. is prohibited.

explicit, which means that it can not be implicitly assumed and should be processable by machine (Welty, 2003). This definition has been criticized by Guarino (1995). After examining many possible interpretations of ontology Guarino concluded that an ontology describes a hierarchy of concepts related by subsumption relationships; in more sophisticated cases, suitable axioms are added in order to express other relationships between concepts and to constrain their intended interpretation. Sowa (2000) gave the definition for ontology as the study of the categories of things that exist or may exist in some domain.

To resume, ontology provides a shared vocabulary for common understanding of a domain. It includes computer-usable definitions of basic concepts in the domain and the relationships among them. Ontology can be used for human and machine communication, interoperability between systems, software engineering, and so forth. On one side, ontologies can be used in a very shallow or general sense, namely as taxonomical structure. On the other side, ontologies can be very formal with a clear semantics and machine-processable format along with an efficient reasoning support.

Hereafter, we adopt Gruber's definition referring to ontology as an explicit (not implicit) specification of conceptualization, with Guarino's interpretation of conceptualization.

Formal Ontology and Information Systems

An explicit ontology may take a variety of forms, but imperatively it will include a vocabulary of terms and specification of their meaning (i.e., definitions). Ontologies are distinguished along a spectrum of formality and referring to a different degree of formality by which a vocabulary is created. In the literature, ontologies fall in many types based on different spectrum (Guarino, 1998; Sowa, 2000; Obrst, 2003). Therefore, the same term *ontology* can be used to describe models with different degrees of structure. Figure 1 shows a snapshot of this spectrum, with respect to Obrst (2003), and can be understood as follows:

- **Highly informal:** expressed loosely in natural language. An informal ontology contains a list of types that are either undefined or defined only by statements in a natural language.

- **Semi-informal:** expressed in a restricted and structured form of natural language. In this zone, we identify weak structure such as taxonomies, Yahoo hierarchy, biological taxonomy, and so forth.

- **Semi-formal:** expressed in an artificial formally defined language. This zone includes elements with more structure: in the bottom database schemas and metadata schemes (ICML, ebXML, WSDL), on the top

Copyright © 2006, Idea Group Inc. Copying or distributing in print or electronic forms without written permission of Idea Group Inc. is prohibited.

Figure 1. Ontology spectrum

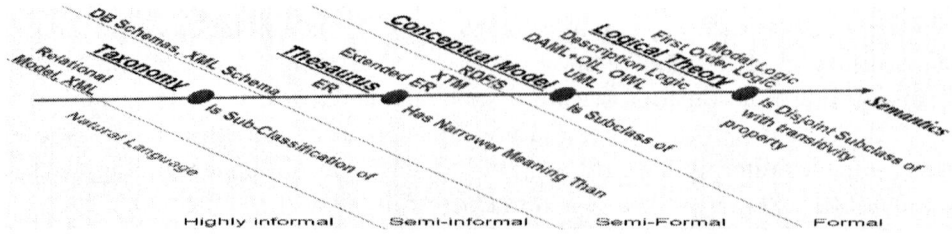

conceptual models (OO models, UML, XML topic maps, etc.), and in-between thesaurus (WordNet).

- **Rigorously formal:** usually expressed in a logic-based language (first order logics) with theorems and proofs mechanism. Fundamentals of formal ontology include axiomatic theories organized into partial order (lattice). Axiomatic theories or simply axioms use first order logic for a rich expression of the constraints between the entity and relation types. We distinguish between theorems and theories. Theorems are licensed by a valid proof using inference rules. Theories are possible other theorems, as yet unproven. Using axiomatic-deductive method provides deduction and derivation (nonmonotonic reasoning) which are based on completeness and decidability.

According to the level of granularity in Guarino (1998) and Gómez-Pérez (1999), ontologies levels of abstraction can be classified as:

- Top-level ontologies describe very general concepts such as space, time, matter, object, event, action, etc., which are independent of a particular problem or domain.

- General ontologies define a large number of concepts relating to fundamental human knowledge.

- Domain ontologies, which describe the vocabulary related to a specific domain (like medicine or automobiles).

- Task ontologies, which define concepts, related to the execution of a particular task or activity, such as diagnosing or selling.

- Application ontologies define concepts essential for planning a particular application.

Copyright © 2006, Idea Group Inc. Copying or distributing in print or electronic forms without written permission of Idea Group Inc. is prohibited.

- Meta-ontologies, generic, or core ontology define concepts which are common across various domain.

What Ontologies Can Do for Enterprise Information Systems

Formal ontologies can be used at development time or at runtime, therefore, we should distinguish between ontology-aware IS and ontology-driven IS. Ontologies used in runtime can be considered essentially with communication and interoperability. Reusability, reliability, and specification can be considered as development-time use of ontologies.

Communication

In particular, ontology-based human communication aims at reducing and eliminating terminological and conceptual confusion by defining a shared understanding, that is, a unifying framework enabling communication and cooperation amongst people for reaching better enterprise organization. Presently, one of the most important roles an ontology plays in communication is that it provides unambiguous definitions for terms used in a software system.

More examples of case uses have been developed in many areas such as:

- Disease Ontology project intends to create a comprehensive hierarchical and controlled vocabulary for human disease representation (http://diseaseontology.sourceforge.net);
- FAO (Food and Agriculture Organization of the United Nations), is committed to help information dissemination by providing consistent access to information for the community of people and organizations through the Agricultural Ontology Service (AOS) project (http://www.fao.org/agris/aos/);
- Open EDI ontology defines an ontology for data management and interchange between enterprises known as ISO/IEC JTC 1/SC 32 (http://www.jtc1sc32.org/).

Interoperability

Interoperability is considered the ability of many systems to cooperate for a common purpose. It includes the opportunity to exchange data between systems and services call across many platforms and systems. The main problem to achieve interoperability is the heterogeneity that includes structural heterogeneity (schema) and semantics heterogeneity (Kashyap, 1997).

Copyright © 2006, Idea Group Inc. Copying or distributing in print or electronic forms without written permission of Idea Group Inc. is prohibited.

Ontologies can effectively help in resolving the problem of semantic heterogeneity and reaching a well-established interoperability process. They can act as a conceptual model representing enterprise consensus semantics through an integrating environment (e.g., global architecture). This environment, covering enterprise-wide systems, enables different software tools and information systems to collaborate.

System Engineering

This section considers applications of ontologies that support the design and development of the software systems themselves (i.e., at the development time).

- **Specification:** Ontology covers, by definition, the specification and the conceptualization. The basic idea is to use ontology to model the application domain and to provide a vocabulary for specifying requirements. This ontology will help considerably during the development of the software application. An approach consisting of specifying local ontologies in support of semantic interoperability was suggested by Benaroach (2002). This use of ontology as assistance for identifying specification is also considered in many research works such as Cranefield (1999), Osterwalder (2002), etc. In Jasper (1999), a set of scenarios for the use of ontologies are presented, along with a classification for ontologies applications. One of these scenarios consists of using ontology as specification by building a conceptual model of the domain in UML, which will then comprise the explicit ontology for the application.

- **Reliability:** Reliability defines the ability of a system or component to perform its required functions under stated conditions. We can differentiate between the roles that an ontology might play for reliability of software systems. Indeed, informal ontologies can improve the reliability by serving as a basis for manual checking of the design against the specification. Formal ontologies enable the use of semi-automated consistency checking of the software system with respect to the declarative specification.

- **Reusability:** The accomplishment of reusability largely depends on the sharing of a similar conceptualization. By characterizing classes of domains and tasks within these domains, ontologies provide a framework for determining which aspects of a system are reusable between different domains and tasks. Indeed, a clear semantic is needed of the concepts-related components being reused. Thus, ontology-driven IS can import and export modules and components between their systems (Mizoguchi, 1997).

Copyright © 2006, Idea Group Inc. Copying or distributing in print or electronic forms without written permission of Idea Group Inc. is prohibited.

Figure 2. Ontology architectures

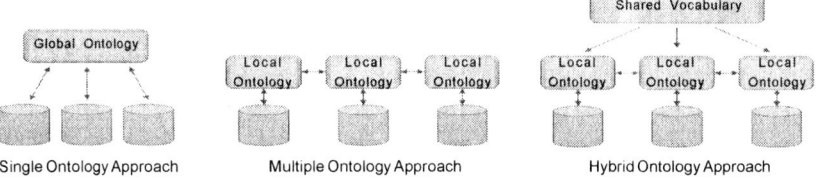

How Can We Use Ontologies for Semantics Sharing?

Ontologies can also be used for the identification and association of semantically corresponding information concepts. In order to establish this correspondence and sharing semantics, there are many ways that ontologies may be employed. In general, three different directions can be identified: *single ontology approaches, multiple ontologies approaches*, and *hybrid ontologies approaches*. Figure 2 gives an overview of the three main architectures as described in Wache (2001). We discuss, in the following, these three architectures.

* **Single Ontology Approach:** this approach consists of using a global ontology for providing a shared vocabulary and specification of the semantics. Thus, all information systems are mandatory related to this global ontology. Sometimes, it is necessary to combine (import) new ontology in order to take into consideration a new system plugged to the global architecture. Nevertheless, this process is possible when the imported ontology does not have a different view on the domain (e.g., different granularity, perspectives). In this approach, maintaining the global ontology and the minimal ontological commitments is a hard task.

* **Multiple Local Ontologies Approach:** this approach consists of defining a local ontology for each information system. These ontologies do not necessarily share the same semantics and no common minimal ontological commitment is needed. The local ontologies can be developed independently and can be simply added or removed from the architecture without affecting any change. On the other hand, the lack of a common vocabulary makes it difficult to make the systems work together. To overcome this problem, an additional representation formalism defining the interontology mapping is needed.

Copyright © 2006, Idea Group Inc. Copying or distributing in print or electronic forms without written permission of Idea Group Inc. is prohibited.

- **Hybrid Approach:** it brings together the single and the multiple approaches. Therefore, the semantics of each system is defined locally as in the multiple approach. In order to make the local ontologies comparable to each other, they are built from a global shared vocabulary (Goh, 1997). The shared vocabulary contains basic terms (the primitives) of a domain which are combined in the local ontologies in order to describe more complex semantics. Sometimes the shared vocabulary is also an ontology as described in Stuckenschmidt (2000).

To conclude, we referred in this section the word *mapping* to signify the relation or connection between ontologies or parts of an ontology to another ontology. This notion has been considered differently in many studies such as alignment between ontology (Sowa, 2000), defined mappings (Peerce, 1999), lexical relations (Mena, 1996), and top-level grounding (Calvanese, 2001).

Overview of Libraries, Languages, and Development Tools

During the last decade, an uncountable number of studies have been lead regarding ontologies research. After importing the word from philosophy and identifying its meaning for the computer science community; the researchers studied representation languages and implementation tools along with ontology engineering methods. The second interest of the community was to identify a set of examples of developed ontology, many projects such as CYC upper ontology (http://www.opencyc.org/), LinkBase (1 million medical concepts, 420 link types) (http://www.landcglobal.com/pages/linkbase.php), or Towntology project (Keita, 2004) have goals to provide useful and ready-to-use ontologies. On the other hand, a set of libraries becomes available to access, to reference, or to import during the development cycle. We enumerate DAML ontology library (http://www.daml.org/ontologies/), Ontolingua (http://www.ksl.stanford.edu/software/ontolingua/), OWL ontology library (http://protege.stanford.edu/plugins/owl/owl-library/index.html), and Kactus library (http://web.swi.psy.uva.nl/projects/NewKACTUS/library/library.html).

As about languages, there is an overwhelming dominance of systems using some variants of description logics as an ontology representation language. We find in this category Loom (MacGregor, 1991), CLASSIC (Borgida, 1989), CARIN (Goasdoué, 1999), OIL (Fensel, 2001), and so forth. Last but not least, description logics were chosen as underlying formalism for OWL (http://www.w3.org/TR/owl-features/) (Ontology Web Language) suggested by W3C for creating the Web semantics paradigm. However, these languages vary by their constructors (e.g., logical operators, slot constraints, axioms, etc.), the efficiency of their reasoning support, and the soundness and completeness of their algorithms. A

Copyright © 2006, Idea Group Inc. Copying or distributing in print or electronic forms without written permission of Idea Group Inc. is prohibited.

comparison between the features and the expressiveness of a set of these languages can be found in Wache (2001).

However, other languages such as Frame-Logic (Flogic) (Kifer, 1995), KIF (Knowledge Interchange Format)-based Ontolingua (Farquhar, 1997), and OKBC (Open Knowledge Based Connectivity) (Chaudhri, 1998) are used for describing ontologies. They offer, on one hand, common elements for the definition of concepts and relations. On the other hand, they are based on First Order Logic axioms and differ in expressiveness and computational properties. Only OKBC does not provide an axiom language that is sufficient for the description of terminological axioms (Wache, 2001). Yet, we should refer to Corcho (2000) as providing an evaluation of the ontology languages mentioned above. Although Cranefield (1999) suggested the use of UML and OCL as ontology language, this technique lacks the support of reasoning services. For indication, Table 1 shows a comparison between the expressivity of ontology languages using XML syntax.

Ontologies are usually developed using special tools that can model rich semantics and to assist team's members in building ontologies (e.g., Protégé http://protege.stanford.edu/). Currently, there are no available tools, which support the complete development cycle, but there are many existing tools for tasks such as ontology capturing, representation, visualization, and editing (Arara, 2002). In Duineveld (2000), the authors gave a comparative study of existing ontological engineering tools.

Ontology Applications

It is almost impossible to enumerate the applications where ontologies have been exploited, but we may be able to classify them in some manner according to some key aspects of technological and technical trends as follows:

Table 1. Comparison between XML-based ontology languages

	Data Types		Types of properties		Property Element		Classes	
	Primitive data-type	Numeric min, max	Transitive	Inverse	Import element	Individual element	Negation / Disjoint classes	Inheritance
XOL	Yes	Yes	No	Yes	No	Yes	No	No
OIL	No	No		Yes	Yes	No	Yes	Yes
RDFS	No	No	No	No	No	Yes	No	Yes
DAML+ OIL	Yes	Yes	Yes	Yes	No	Yes	Yes	Yes
OWL	Yes	Yes	Yes	Yes	Yes	Yes	Yes	Yes

Copyright © 2006, Idea Group Inc. Copying or distributing in print or electronic forms without written permission of Idea Group Inc. is prohibited.

- **Interoperable Ontology-Based Information Systems:** Many research works have been focusing on ontologies and semantic interoperability applications. The Fonseca (1999) paper introduces, for instance, a geographic information system architecture based on ontologies and studies interoperability for a proposed system. Ontobroker project (Decker, 1998) studies the interoperability for semistructured data in the context of the Web.

- **Information Searching and Semantic Web:** Ontologies allow a content-based search instead of key word-based search. Many search techniques in the Web (e.g., Yahoo, Google, etc.) are ontology-based searches. The use of ontologies for reaching the semantic Web is currently a very active research area (e.g., OWL) (Berners-Lee, 2004).

- **Intelligent Metadata and Specification of Information Technology (IT) Systems:** One important role played by an ontology is the explicit descriptions of specification for information systems in Cranefield (1999) and applied in the domain of management software tools in Osterwalder (2002). Ontology can easily be used to describe metadata for some applications systems. In Van Zyl (1999), an ontology based metadata representation is presented for data warehouse systems.

- **Query and Reasoning Support Services:** Ontology-based queries allow for rich semantics that will lead to precise query results. Formal ontologies based on description logics have support for reasoning services provision and provides users with inferred results for their queries. For instance, in *observer project* (Mena, 1996) queries processing in global information systems is studied, where separated ontologies are used for describing information sources. Some scenarios of using ontologies for query answering are presented in Ceusters (2003).

The above categories show the widely diverse spectrum of possible applications of ontologies. In addition, a framework for understanding and classifying ontology application was presented by Jasper (1999). A survey of existing approaches and applications for integration of information using ontologies is also presented in Wache (2001). It should also be noted that applications have similarities and common factors. For instance, they all need to share the meaning of terms in a given domain, and they all strive for working and communicating in cooperative environments. By use of ontology, barriers among applications in a specific domain of interest can be drastically reduced and semantics can be more easily shared.

Copyright © 2006, Idea Group Inc. Copying or distributing in print or electronic forms without written permission of Idea Group Inc. is prohibited.

Context and Ontology

Likewise ontology, the notion of context has emerged in many disciplines, such as philosophy, linguistics, and cognitive psychology, before arriving to computer science. Focusing on knowledge representation in AI, contexts appear as a mean of partitioning knowledge into manageable sets (Hendrix, 1979), or as logical constructs that facilitate reasoning activities (Guh, 1991). McCarthy (1993) has introduced contexts as abstract mathematical entities with properties, which constituted one of the first examples of formalization of context.

The Notion of Context in Information Systems

In information systems, the notion of context was carried through views, aspects and roles, and workspaces. In particular, context is defined in multiple databases as a key component that captures the semantics related to an object's definition and its relationships to other objects. In software engineering, this notion of viewpoints (i.e., context) has been used to support the actual development process of complex systems. In this respect, a viewpoint is defined as a locally managed object or an agent, which encapsulates partial knowledge about the system and its domain. Views are also used in requirements engineering (Nuseibeh, 1994) "as a vehicle for separation of concern".

The advantages of using the notion of context in several disciplines have been studied in Akman (1996). They can be summed up in economy of representation: efficiency of reasoning, allowing inconsistencies and contradicting information, resolving lexical ambiguity, and flexible entailment. In particular, the notion of context is becoming more interesting with the proliferation of loosely coupled systems and Peer-to-Peer (P2P) technology where context is used to represent the peer knowledge (Gold, 2001). Recently, in context-aware computing, formal approaches are taken to model context based logical reasoning mechanism. Context reasoning is used to achieve the consistency of context and the deduction of a high-level, implicit context from the low-level explicit ones (Wang, 2004).

However, we are interested in this chapter of using context for solving the problem of multirepresented elements and their manipulation. There is currently a crucial need for an abstraction mechanism that deals with an explicit description of multirepresented concepts (Balley, 2004) and semantics of context dependent objects (Serafini, 1997). For multirepresented concepts, the same data element might be used by different entities of different applications to mean different things. Different data element names could also be used to represent the same things, potentially creating hundreds of instances of the same data all

Copyright © 2006, Idea Group Inc. Copying or distributing in print or electronic forms without written permission of Idea Group Inc. is prohibited.

inconsistently named. In addition, context semantic sharing considers the notion of context as a way of restructuring and organizing information elements according to the context in which they occur.

Contextualizing Local Ontologies (Pairing Up Ontology and Context)

In multiple and hybrid approaches for ontologies presented previously, the interesting point is how the local ontologies are described. The local description of information is called context as used in COIN (Goh, 1997) as an attribute value vector. Hence, the role of the ontology is to describe the terms and the structure of the concepts. Context is used to label the belonging of information elements when mapping is performed between the two representations.

The paradigm illustrating the combination of ontology and context can achieve an explicit specification of conceptualization from different perspectives or contexts. Contextualizing local ontologies, as an abstraction mechanism, can be resumed by contextualizing ontological concepts and attributes. In such a way, that ontological information is provided according to specific interest, purpose, and level of details. Thus, the notion of context is used as a form of views to deal with concepts from different perspectives. Context can be used to handle inconsistent and contradictory concepts in the same ontology base, as long as they are treated in different contexts.

We consider that contextual ontology is defined as ontology in which its concepts can be seen from different points of view (multirepresented concepts). This suggested paradigm provides a wider support to access heterogeneous information sources by various applications in a domain. It essentially treats the problem of multirepresentation of concepts and multiviews requirements in the various contexts of use in order to ensure a clear and clean semantic sharing between many systems.

Therefore, this suggestion to pair up the two essential elements (context and ontology) is mainly proposed to deal with the problems in information modeling and semantic heterogeneity, in particular, the multiple views and multirepresentation problems. In other words, this approach is different from those that use context modeling in the sense that ontology with multirepresented concepts and context with multiple views are our prime concern. The term Contextual Ontologies will be used to indicate that the ontology we are dealing with is context-based referring to this illustrated paradigm.

Copyright © 2006, Idea Group Inc. Copying or distributing in print or electronic forms without written permission of Idea Group Inc. is prohibited.

Requirements for Contextual Ontologies in EIS

From the previous discussion, the use of ontologies as well as contextual ontologies can reduce the loss of semantics in information exchange among heterogeneous applications; improve reusability (e.g., objects, components, etc.); and provide a global view over enterprise architectures, infrastructures, and applications. Therefore, contextual ontologies are highly placed to help resolving semantic sharing among EIS. We argue that enabling the semantics sharing can be done through Contextual Ontologies paradigm considering:

- Establishing the context semantic representation via local ontologies and expressing their belonging using a special mechanism.

- Defining semantic mappings mechanism (i.e., bridges rules or coordination rules) among Contextual Ontologies' concepts, which are multirepresented, and treating these mappings with logical theories.

- Defining a global inference mechanism, this is based on local ontologies knowledge and the semantic mappings.

What Problems Do Contextual Ontologies Help Solve in EIS?

Likewise ontologies, contextual ontologies aim at first to help with heterogeneous systems problems (e.g., heterogeneous databases). Whereby, different organizational units, service providers have a radical difference between their systems including different syntactical (i.e., the format), structural (i.e., schema), and semantic (i.e., the meaning or the interpretation). They all speak different languages for access, description, schemas, and meaning. We strongly argue that using the contextual ontology as a common description for EIS can give systems the ability to be reusable and interoperable, along with a global query answering service. This contribution broadly appears in the enterprise-wide system interoperability, currently system-of-systems, and vertical stovepipes, where contextual ontologies act as conceptual model representing enterprise shared semantics. In addition, contextual ontology can be used to improve systems reusability by identifying semantically similar elements and revealing their reusable bit code. Finally, it can provide a global understanding to be used for query answering to obtain only meaningful relevant information.

In the next part, we will seek formalism for contextual ontology. This formalism should cover the mechanisms required by the Contextual Ontologies. It should also allow explicit semantics, machine understandable, and logic inference. We will summarize also the related works to contextual ontologies and their formalisms.

Copyright © 2006, Idea Group Inc. Copying or distributing in print or electronic forms without written permission of Idea Group Inc. is prohibited.

Contextual Ontologies Formalism

In essence, semantics sharing between users of large communities with diversified perspectives is a challenging direction of research that requires more attention. At present, concepts are expressed formally as a single representation, in the sense that the representation language is characterized by defining a unique concept and its properties as a fixed data set. In contrast to this assumption, a real world entity is unique but it can have several representations (Benslimane, 2003a) due to various interests, purposes, or perspectives. In fact, the multirepresentation phenomenon becomes normative rather than exceptional, if interoperations among systems are sought. The notion of context is used as a form of views to deal with concepts from different perspectives (Arara, 2004a). The contextual ontologies are defined at the abstraction level to take into account the diverse points of view and multirepresentation.

This part of the chapter introduces the multirepresentation through examples from Enterprise Information Systems. It studies, after that, the three mechanisms required for contextual ontologies (stamping mechanism, semiglobal interpretation mechanism, and semantic similarity mechanism). It suggests, then, using the modal description logics formalism for coping with contextual ontologies requirements. Similarly to the use of the description logics for expressing ontologies, it shows the use of modal description logics for expressing contextual ontologies. More specifically, the language based on the ALCN augmented by modal operators is going to be used for this purpose. Finally, it gives an overview of the relevant related works to contextual ontologies and multirepresentation issues.

Examples of Multirepresentation

Being inspired by e-commerce, a product can have different definitions, scales, and prices (e.g., with respect to currency). Furthermore, this same product needs to be marketed by many enterprises where each has its own representation, which varies in properties and possibly structures from other representations. In brief, the multiple representations are rooted to the abstraction mechanism where several conceptualizations are associated with the same object due to several other factors such as viewpoints, contexts, special interest, and so on.

Contextual ontologies permit us to overcome this problem by allowing the multiplicity of a concept representation. A set of context-dependent ontologies can be defined and put together without being integrated for this purpose. Hence, contextual ontology is an ontology that is kept locally, but inter-related with other ontologies through a mapping relationship between similar concepts. We argue

Copyright © 2006, Idea Group Inc. Copying or distributing in print or electronic forms without written permission of Idea Group Inc. is prohibited.

Figure 3. A side of UML model for HRIS

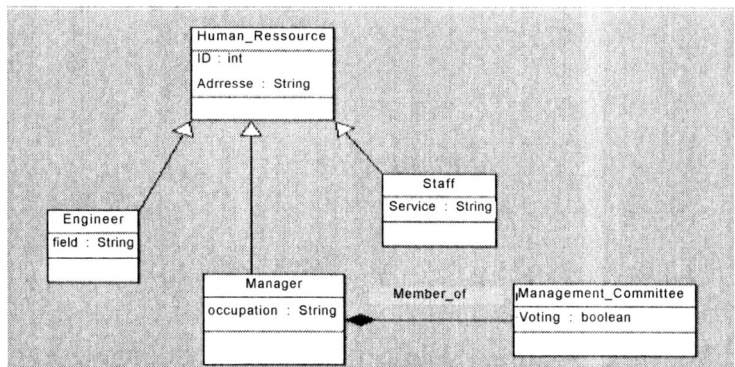

Figure 4. A side of UML model for PMIS

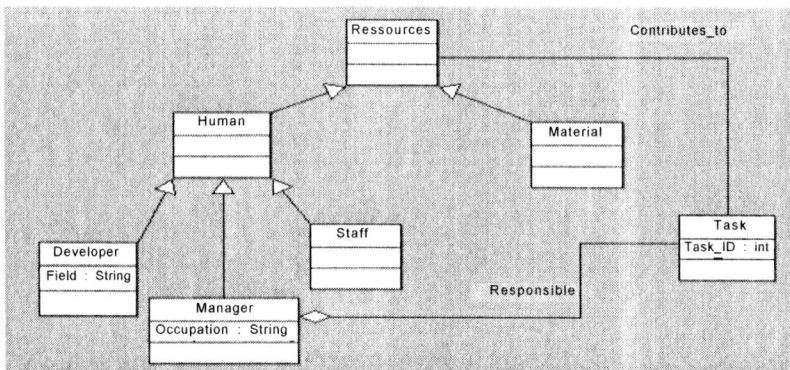

that using the contextual ontologies as a common description for EIS conceptualization (based on multiple view of system specification) can give systems the ability to share easily more semantics.

Example PMIS &HRIS (Example 1)

Let us consider two information systems used in an enterprise: PMIS (Project Management Information System) and HRIS (Human Resource Information System). The UML models, defined in Figure 3 and Figure 4, represent the monorepresentation of each system. These systems contain concepts using the same identifier and having the same meaning, such as *Manager* in PMIS and *Manager* in HRIS, or concepts that are different but having the same compo-

Copyright © 2006, Idea Group Inc. Copying or distributing in print or electronic forms without written permission of Idea Group Inc. is prohibited.

nents, structure and are semantically similar, such as *Engineer* in PMIS and *Developer* in HRIS. In this case, we can identify that the concepts *Manager*, *Engineer* are multirepresented in these systems.

Example EDI & DW (Example 2)

We have developed, in previous research projects (Rifaieh, 2003a, 2003b), two different systems using a model called the "mapping expression model" (Rifaieh, 2002b). The first implements the QELT (Query based Extraction Load and Transformation tool) metadata used to generate the transformation queries. The second implements the mapping guideline of the EDI-Translator, which is used to identify the mapping between EDI messages. Figure 5 and Figure 6 show a snapshot of the UML class model of the implemented systems. We can recognize that these two systems have many multirepresented concepts. For instance, the concepts *Mapping*, *Field*, and *Record* of the EDI Translator (Rifaieh, 2003a, 2003b) are respectively multirepresented, versus *Mapping*, *Attribute*, and *Entity* of QELT (Rifaieh, 2002).

Studying Contextual Ontologies Mechanisms

The approach of contextual ontologies aims at supporting several applications that are associated with several representations of the same real world entities. Our basic idea is to treat such systems as local views, where they preserve their local semantics, but we allow coordination (bridge rules) for expressing the communication among locally independent systems. Globally, the approach respects the representation for many points of view for the same concept. It provides, as well, the possibility to define global knowledge over local ontology.

In order to reach the advantages of contextual ontologies, we should consider three mechanisms allowing the previous features: stamping mechanism, semiglobal interpretation mechanism, and semantic similarity mechanism. We study these mechanisms briefly in the following.

Stamping Mechanism

First of all, we need to differentiate between concepts that belong to different contexts. For this reason, we recall the stamping or trade marking technique, which can be used to distinguish one representation of the same element from other representations. Thus, the multirepresentations stamps, or simply, stamps, are used to characterize several representations of a real-world phenomenon for

Copyright © 2006, Idea Group Inc. Copying or distributing in print or electronic forms without written permission of Idea Group Inc. is prohibited.

Figure 5. Snapshot of OWL representation for EDITranslator (Rifaieh, 2003b)

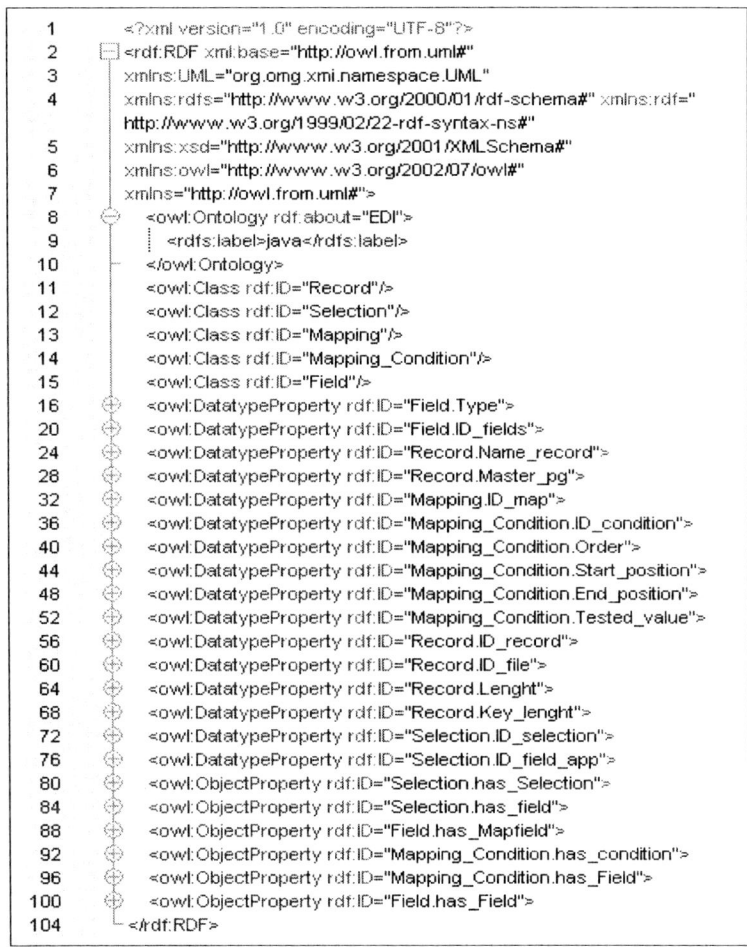

the multirepresentation paradigm. Hence, a concept can be used in one or more representation, but each representation of a concept is stamped or labeled differently. A stamping mechanism to characterize database elements in GIS applications was proposed in MADS system (Balley, 2004) to support several representations of data. To guarantee stamping consistency, stamps are obligatorily assigned to every piece of information in order to customize them.

Similarly, the stamping mechanism is used in our contextual ontology to stamp the components of each ontology and their operators and constructors. The usefulness of this technique consists of resolving the ambiguity for identifying a

Copyright © 2006, Idea Group Inc. Copying or distributing in print or electronic forms without written permission of Idea Group Inc. is prohibited.

Figure 6. A snapshot of OWL representation for QELT (Rifaieh, 2002a)

```
1      <?xml version="1.0" encoding="UTF-8"?>
2      <rdf:RDF xml:base="http://owl.from.uml#"
3        xmlns:UML="org.omg.xmi.namespace.UML"
4        xmlns:rdfs="http://www.w3.org/2000/01/rdf-schema#"
5        xmlns:rdf="http://www.w3.org/1999/02/22-rdf-syntax-ns#"
6        xmlns:xsd="http://www.w3.org/2001/XMLSchema#"
7        xmlns:owl="http://www.w3.org/2002/07/owl#"
8        xmlns="http://owl.from.uml#">
9        <owl:Ontology rdf:about="DW">
12       <owl:Class rdf:ID="Attribute"/>
13       <owl:Class rdf:ID="Entity"/>
14       <owl:Class rdf:ID="Function"/>
15       <owl:Class rdf:ID="Mapping"/>
16       <owl:Class rdf:ID="Transformation"/>
17       <owl:DatatypeProperty rdf:ID="Transformation.name">
21       <owl:DatatypeProperty rdf:ID="Transformation.description">
25       <owl:DatatypeProperty rdf:ID="Entity.name">
29       <owl:DatatypeProperty rdf:ID="Mapping.name">
33       <owl:DatatypeProperty rdf:ID="Mapping.description">
37       <owl:DatatypeProperty rdf:ID="Attribute.name">
41       <owl:DatatypeProperty rdf:ID="Attribute.type">
45       <owl:DatatypeProperty rdf:ID="Function.name">
49       <owl:DatatypeProperty rdf:ID="Function.function">
53       <owl:DatatypeProperty rdf:ID="connects.lower-boundary">
57       <owl:DatatypeProperty rdf:ID="connects.upper-boundary">
61       <owl:DatatypeProperty rdf:ID="participates.lower-boundary">
65       <owl:DatatypeProperty rdf:ID="participates.upper-boundary">
69       <owl:ObjectProperty rdf:ID="Function.has-arguments">
73       <owl:ObjectProperty rdf:ID="Function.has-function">
77       <owl:ObjectProperty rdf:ID="Function.has-targets">
81       <owl:ObjectProperty rdf:ID="Mapping.has-sources">
85       <owl:ObjectProperty rdf:ID="Mapping.has-targets">
89       <owl:ObjectProperty rdf:ID="Transformation.has-sources">
93       <owl:ObjectProperty rdf:ID="Transformation.has-targets">
97       <owl:ObjectProperty rdf:ID="Transformation.uses-transformations">
101    </rdf:RDF>
```

concept by its context. Hence, this labeling technique permits each concept C_i to be known (identified) by the context Ctx_i that it belongs to, e.g., $Ctx_1:C1$, $Ctx_1:C1 \cup Ctx_2:C2$, and so on. The stamped primitive concept is used to denote primitive concepts that are specifically available for some given contexts. If we consider the Example 1 as C_1, we can define the $C_1:Engineer$, $C_1:Manager$, $C_1:Staff$, and so forth.

The Description Logics stamping technique has been proposed in (Benslimane, 2003b), respecting the preceding characteristics, for stamping ontologies coded in this description logics formalism.

Copyright © 2006, Idea Group Inc. Copying or distributing in print or electronic forms without written permission of Idea Group Inc. is prohibited.

Semantic Similarity Mechanism

The second issue concerning the contextual ontologies is the potential to define the semantic relationships between multirepresented concepts. We need to be able to state that two concepts of two ontologies, though being contextually different, are related, because they both refer to the same object in the world. Therefore, a directional bridge rule asserting the type of semantic relationship is going to be expressed for these concepts (Rifaieh, 2004a).

The directionality of the bridge rules is important information that helps to understand the meaning of the rule. We can identify many types of bridge rules such as those defined in Borgida (2002), Huhn (2002), and Mitra (2000):

- The identity: a concept A of IS_1 is identical to B in IS_2 if $A \subseteq B$ and $B \subseteq A$.
- The subsumption: a concept A from IS_1 subsumes the concept B from IS_2 if every instance satisfying the description of B satisfies also the description of A, we note that $B \subseteq A$.
- The inclusion: all the instances of B in IS_2 have corresponding instances of A in IS_1 $(A \supseteq B)$.
- And so on.

In Example 1, we can define a bridge rule asserting that the concept *Employee* of PMIS subsume the concept *Manager* of HRIS, C_1:*Employee* $\supseteq C_2$ *Manager*.

In Example 2, we can define a bridge rule by asserting an identical relationship between the concepts *Mapping* of QELT and *Mapping* of EDI Translator, C_1:*Mapping* $\equiv C_2$ *Mapping*.

Semiglobal Interpretation Mechanism

The last issue concerns the definition of rules or axioms involving one or many context. These rules are very useful to build a semiglobal knowledge base for contextual ontology. We talk about semiglobal interpretation mechanism because these rules do not necessarily involve all the existing contexts. Therefore, the semiglobal interpretation considers the coherence in a subset of the available contexts. Whereas the local interpretation considers using concepts belonging to one context, semi-global interpretation defines rules with concepts belonging to many contexts (Rifaieh, 2004b).

Copyright © 2006, Idea Group Inc. Copying or distributing in print or electronic forms without written permission of Idea Group Inc. is prohibited.

In Example 1: using the existing multirepresented concepts can help us to create a new global concept. For instance the concept *Mangement_committee_member* can be defined with respect to a rule asserting that the members of the Management committee are the instances of *Manager* from HRIS and the instances of *Manager* from PMIS. Let us refer I_1="*John Smith*" as an instance of *Manager* in the system HRIS and J_1="*Thomas Green*" as an instance of *Manager* in the system PMIS. The semiglobal interpretation mechanism should offer the possibility to interpret that I_1 and J_1 are instances of *Mangement_committee_member.*

Overview of Description Logics and Modal Logics

In reality to use contextual ontologies, we encounter the following challenges: (1) finding a suitable representation model for these contextual ontologies that compromises expressivity and tractability, and (2) dealing with the required mechanisms studied before. In this section, we seek a formalism suitable to express these mechanisms. We have oriented our research toward the logics based approaches because these approaches have been successfully used in specifying explicitly the concepts and their relationships for a domain of interest (e.g., the use of Description Logics [DLs] for expressing ontologies).

Although DLs, are very expressive and machine processable, there are still several research problems encountering the use of DLs. For instance, the multiplicity of representations is not treated in DLs. Therefore, we should seek for an additional solution for expressing the multirepresentation issue. We suggest associating the DLs with the Modal Logics (MLs) (Lemmon, 1977) for expressing our contextual ontologies. In fact, the combination of DLs and MLs are already used with applications such as temporal databases, software specifications, and so forth.

Description Logics

In the last 30 years, terminological knowledge representation systems (TKRS) have been commonly used to represent facts about an application domain (Catarci, 1993). Ontologies were expressed using formalism such as CG (Conceptual Graph) (Sowa, 2000), frame logic (Farquhar, 1997), or KL-one based languages (DL family), etc. However, maturity of mathematical logics in general and in description logics in particular, has encouraged their use for expressing ontologies with applications in bioinformatics, e-commerce, environment, urban applications, and so on. Recently, they get more attention since it has been chosen as the basic formalism for Semantic Web ontology and the language OWL (Ontology Web Language).

Copyright © 2006, Idea Group Inc. Copying or distributing in print or electronic forms without written permission of Idea Group Inc. is prohibited.

Table 2. Syntax and semantic of ALCN

	Syntax		Semantic
$C, D \to T \mid \perp$		(Universal Concept,Bottom)	$T^I = \Delta^I$
	$A \mid$	(Atomic concept)	$\perp^I = \Phi$
	$C \cup D \mid$	(Conjunction)	$(C \cap D)^I = C^I \cap D^I$
	$C \cap D \mid$	(Disjunction)	$(C \cup D)^I = C^I \cup D^I$
	$\neg C \mid$	(Concept negation)	$(\neg C)^I = \Delta^I / C^I$
	$\forall R.C \mid$	(Universal Quantification)	$(\forall R.C)^I = \{ d_1 \in \Delta^I \mid \forall d_2 . (d_1, d_2) \in R^I \Rightarrow d_2 \in C^I \}$
	$\exists R.C \mid$	(Existential Quantification)	$(\exists R.C)^I = \{ d_1 \in \Delta^I \mid \exists d_2 . (d_1, d_2) \in R^I \wedge d_2 \in C^I \}$
	$\leq nR \mid$	(At most number restriction)	$(\leq nR)^I = \{ d_1 \in \Delta^I \mid \#\{ d_2 \mid (d_1, d_2) \in R^I \} \leq n \}$
	$\geq nR$	(At least number restriction)	$(\geq nR)^I = \{ d_1 \in \Delta^I \mid \#\{ d_2 \mid (d_1, d_2) \in R^I \} \geq n \}$

Description Logics view the world as being populated by individuals that can be grouped into classes, and that can be related to each other by binary relationships. A specific DL language (such as *ALC, ALCN, ALCNH*, etc.) provides a set of constructors and operators for building more complex concepts and roles. Complex description can be built from them inductively with concept constructors. Concept descriptions in *ALCN* are formed according to the syntax rules defined in the left side of Table 2. The semantic is given by the interpretation $I = (\Delta^I, \cdot^I)$, which consists of an interpretation domain Δ^I, and an interpretation function $\cdot^I$. The interpretation function assigns to every atomic concept A a set and to every atomic role R a binary relation. The interpretation function is extended to a concept description by the inductive definition in the right side of Table 2. For instance, say that two concepts C, D are equivalent, and write, if for all interpretation I. The number restriction semantics are interpreted using "#{.}" which denotes cardinality of the set.

A DL knowledge base K is a pair <T,A> where T is a terminology (TBox) and A is an ABox (Figure 7). It distinguishes between two types of collection:

Figure 7. DL-based knowledge representation system

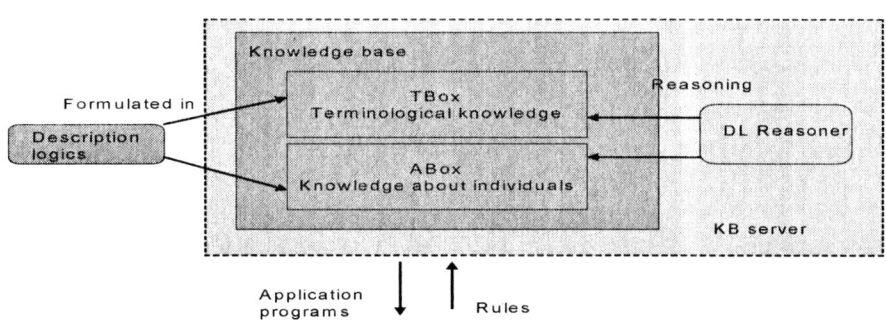

Copyright © 2006, Idea Group Inc. Copying or distributing in print or electronic forms without written permission of Idea Group Inc. is prohibited.

- The **TBox (Terminological Box):** Terminological axioms make statements about how concepts or roles are related to each other. It is a collection of subsumption assertions specifying the terminology used to describe some application domain, it is somehow similar to an IS schema (Baader, 2003).

- The **ABox (Assertional Box):** This second component of a knowledge base describes a specific state of affairs of an application domain in terms of concepts and roles. Some of the concept and role atoms in the ABox may be defined names of the TBox. In the ABox, one introduces individuals, by giving them names, and one assets properties of these individuals. Thus, it is a collection of assertions about individuals describing some state of worlds; it is somehow similar to a database of facts in an IS.

Modal Description Logics

Modal logic (Lemmon, 1977) is a form of logic which deals with expressions that are qualified by modalities such as possibly, necessarily, contingently, and others. Whereas traditional forms of order logic state only about assertions which are true or false, modal logic deals with logical relationships to express an assertion is true in a situation and false in other situation.

In modal logics, the semantics of expressions or formula is defined in terms of modalities as the truth of things in different worlds or contexts. Thus, Modal Logics distinguish modes of truth, which contrast with classical Description Logics, where things are just true or false, and not true in one situation (context) and false in another. For the generalized modal logic, modes of truth are explained by referring to (abstract) worlds, namely, truth in all (accessible) worlds, and truth in some (accessible) world. The Modal Logics can express also the temporal progression modality, belief modality, etc.

If a statement is true in all possible worlds, then it is a necessary truth. A statement that is true in some possible world (not necessarily our own) is called a possible truth. The possibility and necessity operators are represented respectively with $\Box C$ (is interpreted as the set of some possible worlds where C holds), $\Diamond C$ (is interpreted as the set of all the possible worlds where C holds). In this respect, Table 3 shows the syntax and semantics of modal operator in a multi-modal language.

In the work of Kripke (1959), these worlds can be related with so called accessibility relationships. The accessibility relations are set of binary relations involving two worlds. The set of accessibility relations define the Kripke's Structure of worlds. However, we can apply some restriction of these binary relations. For example, we can obtain the modal logic S4 by restricting the Kripke

Copyright © 2006, Idea Group Inc. Copying or distributing in print or electronic forms without written permission of Idea Group Inc. is prohibited.

Table 3. Syntax and semantic of modal operators

Syntax		Semantic
$\square_i C$	Necessity operator	$(\square_i C)^{I(w)} = \{x \in (\square_i C)^{I(w)} \text{ iif } \forall v \nabla_i w, x \in C^{I(w)}\}$
$\Diamond_i C$	Possibility operator	$(\Diamond_i C)^{I(w)} = \{x \in (\Diamond_i C)^{I(w)} \text{ iif } \exists v \nabla_i w, x \in C^{I(w)}\}$

structures to those where the accessibility relation is reflexive and transitive (Baader, 2003). Other modal logics restrict the accessibility relation to be symmetric, equivalence relation, etc. Moreover, the number of accessibility relations may be different from one. Then we are talking about multimodal logics, where each accessibility relation $_r$ can be thought to correspond to one agent, and is quantified using the modal operators $\Diamond_r$ and $\square_r$. We are interested in multimodal logics language because it permits to define many accessibility relationships between the identified worlds.

Syntax and Semantics of Modal Description Logics ($\mathcal{ALCN}_M$)

The combination between DLs and MLs can create Modal Description Logics languages (MDLs) such as $\mathcal{ALCN}_M$. The syntax of MDL $\mathcal{ALCN}_M$ consists of the classical ALCN constructs and the modal operators' $\square$ and $\Diamond$. If we consider the multimodal logics, then our modal operators become $\square_i$ and $\Diamond_i$. One must be careful in such a combination not to ruin the balance between expressiveness and effectiveness (Wolter, 1998).

The syntax permits to define expressions such as: $\Diamond C, \square C, C \wedge D, \neg C, \exists R.C$, etc., with the modal description language $\mathcal{ALCN}_M$. The intended semantic of ALC_M is a natural combination of standard Tarski-type semantic for the description part of ALC and the Kripke-type semantics for the modal part. In other words, the formal semantics of $\mathcal{ALCN}_M$ language are interpreted by:

- The conventional $\mathcal{ALC}$ interpretation (description logics) with Tarski's model where an interpretation is a pair $I = (\Delta^I, \cdot^I)$, such that the set Δ^I is the interpretation domain and $\cdot^I$ is the interpretation function that maps every concept to a subset of Δ^I and every role to a subset of $\Delta^I \times \Delta^I$.

- Kripke's structure stating what are the necessary relations between worlds and what are the formulas necessarily holding in some worlds. It defines where concepts are interpreted with respect to a set of worlds or contexts denoted by W. If $|W| = 1$ then our interpretation becomes the classical interpretation of Tarski-Model described above.

Copyright © 2006, Idea Group Inc. Copying or distributing in print or electronic forms without written permission of Idea Group Inc. is prohibited.

Having a set of contexts or worlds $W=\{w_1, ..., w_n\}$, the model of interpretation of concepts will require a Kripke's structure $< W, \nabla_r, I(w)>$, where ∇_r denotes a set of accessibility binary relations between contexts and $I(w)$ is the interpretation over W. A model of $\mathcal{ALCN}_\mathcal{M}$ based on a frame $F=<W, \nabla_0, \nabla_1, ...>$, is a pair $\mathcal{M}=<\mathcal{F},I>$ in which I is a function association with each $w \in W$ a structure

$$I(w) = < \Delta^{I(w)}, R_0^{I(w)}, ..., C_0^{I(w)},...,a_0^{I(w)},...> \text{ where}$$

$\Delta^{I(w)}$ is an interpretation domain in context $w \in W$, it's a non empty set, the domain of $\mathcal{M}$,

$R_i^{I(w)}$ is a set of roles that are interpreted in context w, in other words they are binary relations on $\Delta^{I(w)}$,

$C_i^{I(w)}$ is a set of concepts that are interpreted in context w (i.e., $C_i^{I(w)} \subseteq \Delta^{I(w)}$), and

$a_i^{I(w)}$ is a set of objects that are interpreted in context w.

The issue of satisfiability and algorithm's complexity in description logics with modal operators has been studied in Wolter (1998). The authors of Donini et al. (1996) proved that the reasoning tasks are reducible to the satisfaction problem for formulas, and they showed that satisfaction problem is decidable in the class of all $\mathcal{ALC}_\mathcal{M}$ models.

Expressing Contextual Ontologies with Modal Description Logics ($\mathcal{ALCN}_\mathcal{M}$)

We have studied separately the mechanisms of contextual ontologies and the modal description logics language $\mathcal{ALCN}_\mathcal{M}$. Likewise, DLs languages have been used with success for expressing ontologies, we suggest in this section the use of MDLs to express our contextual ontologies (Rifaieh, 2004b). Therefore, we present in this section the arguments that defend this proposal.

- Firstly, the formalism of MDLs offers the possibility to define a set of worlds where each world is an interpretation of classical description logics (Tarski interpretation) (Wolter, 1998). These worlds have coherent interpretations for their concepts, roles, and objects, which are identical to the interpretation of ontologies expressed with any DL language. Therefore, we can manage a local coherence for our contextual ontologies by assuming that each local ontology represents a world in the MDL formalism.

Copyright © 2006, Idea Group Inc. Copying or distributing in print or electronic forms without written permission of Idea Group Inc. is prohibited.

- Since each local ontology is represented with the simple DL formalism, we can use the stamping mechanism suggested in Benslimane (2003a) to differentiate between concepts belonging to different world (or local ontology). This extension of the DL helps us to put together without ambiguity the multirepresented elements of contextual ontologies.

- The mechanism of semantic similarity can be expressed using the accessibility relationships between worlds. In other words, the MDL formalism permits us to express the relationships between the local ontologies since each ontology is considered as world in this formalism. The accessibility relationship that exists between two worlds (i.e., local ontologies) is defined as a set of semantic bridge rules between the concepts of these ontologies Figure 8. Thus, if $\nabla_i = \{r_{i1}, r_{i2}, \ldots, r_{in}\}$ is the accessibility relationship between two ontologies, r_{ij} represents a bridge rule between two concepts of these ontologies.

- As we studied before, we need, in our contextual ontologies, a semiglobal interpretation mechanism. This mechanism should permit to express some rules or knowledge that are global (i.e., involving all the contexts) or semiglobal (i.e., involving a subset of the contexts). Since Kripke's structure provides an interpretation for a set of worlds with respect to their accessibility relationships and the modal operators. The semiglobal interpretation mechanism can be considered as Kripke's interpretation in MDL for these worlds with respect to the accessibility relationships defined previously between local ontologies (see Figure 9).

- In addition, we should respect, in the used MDL formalism, some parameters of design such as: rigid designators and finite aspect (Wolter, 1998). The first imposes referring to the same object in any world using the same designator. For instance, if we use *John Smith* as an object of the concept *Employee*, we consider that this designator represents the same object whatever the world we are in. This parameter is valid because our worlds represent the conceptualizations of a domain where the designators of object are unique. The second parameter imposes a finite number of worlds. This parameter is also valid in our case because the number of contextual ontologies representing the systems within the enterprise is not infinite.

- Finally, the language $ALCN_M$ offers the possibility of multi-modal logics for using the necessity and possibility operators: $\square_i$ and $\diamond_i$ where i refers to the accessibility relationship between two worlds. Therefore, we should understand that we have as much operators as we have accessibility relationships.

Copyright © 2006, Idea Group Inc. Copying or distributing in print or electronic forms without written permission of Idea Group Inc. is prohibited.

Figure 8. Accessibility relation between worlds or contextual ontologies

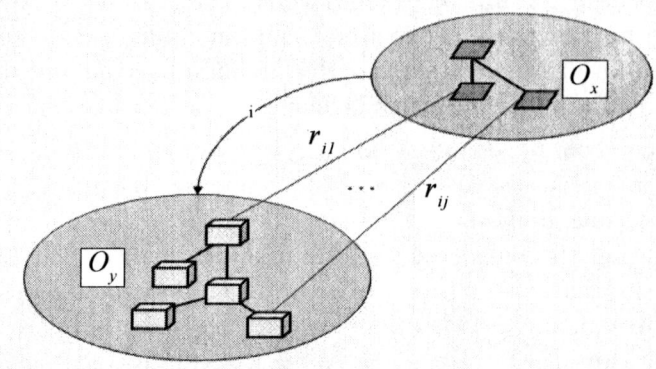

Figure 9. Labeled oriented graph representing Kripke's structure

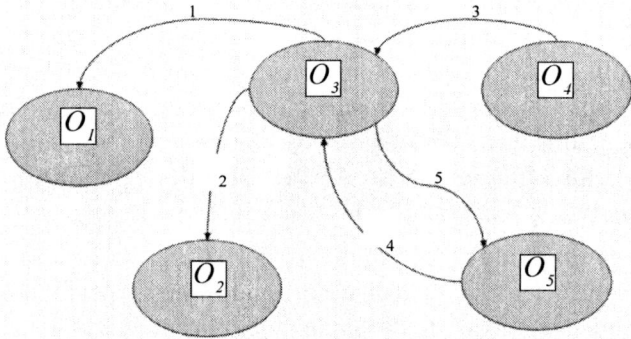

All these elements permit us to conclude that the MDLs languages (e.g., $\mathcal{ALCN}_{\mathcal{M}}$) are adequate for expressing our contextual ontologies. Indeed, using MDLs languages helped us to express all the mechanisms required by the contextual ontologies.

Revisited Examples

We consider in this section the reviewing of the preceding presented examples. In fact, we aim at showing the use of the MDL formalism with these examples.

Copyright © 2006, Idea Group Inc. Copying or distributing in print or electronic forms without written permission of Idea Group Inc. is prohibited.

Example 1

Let us consider the *Example PMIS & HRIS*, we assume that an ontological representation has been associated with each system and we need to define a new concept *Management-committee-member*. This concept is defined by using the new operators and with respect to the used mechanisms. Suppose that a *Management-committee-member* is every *manager* in HRIS and every *engineer* in HRIS having a management responsibility according to PMIS.

According to multirepresentation ontology paradigm, we can define for each contextual ontology the relevant concepts used by the system. Let us label the concepts of HRIS with *h* and concepts of PMIS with *p* to avoid any ambiguity.

We will consider the ontology of HRIS O_h with the definition of the following concepts:

h: Engineer $\subseteq$ h: Human_Ressource $\cap$ ($\forall$ field.String)

h:Manager $\subseteq$ h:Human_Ressource $\cap$ ($\forall$ occupation.String)

h:Staff $\subseteq$ ($\forall$ voting.String) $\cap$ ($\geq_h$ 1 *Member _ of^{-1}.Manager*)

The ontology of PMIS O_p includes the concepts:

p: Human $\subseteq$ p:Ressources

p:Developper $\subseteq$ p:Human $\cap$ ($\forall$ field.String)

p:Manager $\subseteq$ p:Human $\cap$ ($\forall$ occupation.String)

p:Task $\subseteq$ ($\forall$ Task_ID.Integer) $\cap$ ($\geq_p$ 1 *Responsible^{-1}.Manager*) $\cap$ ($\geq_p$ 1 *Contributes_to^{-1}.Re ssources*)

These ontologies have been contextualized by defining an accessibility relation ($\triangledown$i), including a set of bridges between their concepts. The relation $\triangledown$i = {r$_{i1}$, r$_{i2}$,... }, contains the bridge rule r$_{ij}$ attesting that *h:Employee* $\supseteq$ *p:Manager*.

The definition of a new global concept *Management*-committee-*member*, can be considered using the accessibility relation between the context ontologies O_p and O_h as following:

Manag _ committe _ member = h : Manager $\cup$ (*h : Engineer* $\cap$

The operator $\square_i$ applied to the concept p:Manager relies the objects of p:Manager in O_p, by necessity through the accessibility relation $\triangledown$i for being a subset ($\supseteq$) of

Copyright © 2006, Idea Group Inc. Copying or distributing in print or electronic forms without written permission of Idea Group Inc. is prohibited.

the concept h:Employee for O_h. In other words, the formula means: the members of Management Committee are the instances of h:Manager with the intersection between the instances of h:Engineer and p:Manager. These instances of p:Manager are by necessity obtained through the rule h:Employee $\supseteq$ p:Manager.

Example 2

Considering the example of EDI Translator and QELT, we can assume that two ontologies representation in DL have been associated to these systems. The multirepresented concepts *Mapping*, *Field*, and *Record* of EDI Translator correspond to the concept *Mapping*, *Attribute*, and *Entity* of QELT. If the ontology of EDI Translator is represented with O_h, using the stamping mechanism of contextual ontologies, we can express these concepts as follows:

$$h: Field \subseteq (\forall \ name.string) \cap (\forall \ IDField.String) \cap (\geq_h 1 \ hasMapField^{-1}.Manager)$$
$$\cap (\geq_h 1 \ hasField^{-1}.Selection) \cap (\geq_h 1 \ hasField.MappingCondition)$$
$$\cap (\geq_h 1 \ hasField^{-1}.Record)$$
$$h:Mapping \subseteq (\forall \ ID\_Mapping.String) \cap (\geq_h 1 \ hasMapField.Field)$$
$$\cap (\geq_h 1 \ hasSelection.Selection) \cap (\geq_h 1 \ hasCondition.Condition)$$
$$h:Record \subseteq (\forall \ name.String) \cap (\geq_h 1 \ hasield.Field)$$
...

Likewise, if the ontology of QELT is represents with Op, we can express the following concepts:

$$p:Mapping \subseteq (\forall \ name.String) \cap (\geq_p 1 \ hasSources.Entity) \cap (\leq_p 1 \ hasSources.Entity)$$
$$\cap (\geq_p 1 \ hasTarget.Entity) \cap (\leq_p 1 \ hasTarget.Entity)$$
$$\cap (\geq_p 1 \ useTransformation.Transformation)$$
$$p:Transformation \subseteq (\forall \ name.String) \cap (\geq_p 1 \ useTransformation^{-1}.Mapping)$$
$$\cap (\geq_p 1 \ hasTarget.Attribute) \cap (\leq_p 1 \ hasTarget.Attribute)$$
$$\cap (\geq_p 1 \ hasSources.Attribute) \cap (\leq_p 1 \ hasSources.Attribute)$$
...

These ontologies have been contextualized by defining an accessibility relation ($\triangledown i$), including a set of bridges between their concepts. The relation $\triangledown i = \{r_{i1}, r_{i2}, ... \}$, contains the bridge rule r_{ij} attesting that *h:Mapping $\equiv$ p:Mapping* (see Figure 10).

Copyright © 2006, Idea Group Inc. Copying or distributing in print or electronic forms without written permission of Idea Group Inc. is prohibited.

Figure 10. Graph-based representation for accessibility relationship ▽i with individual semantic similarities between some concepts of Example 2

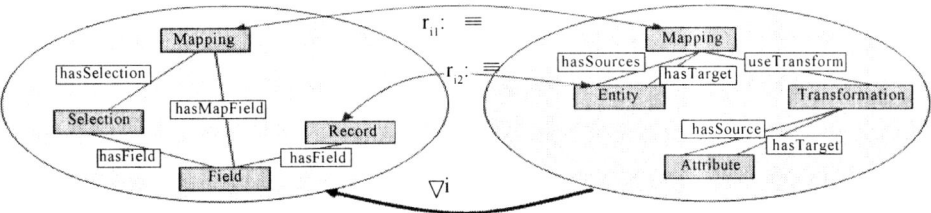

The definition of the bridge rules conforming to the mechanism of semantic similarity help us to use this information for crossing the inference between the two ontologies. For instance, if I_1 is an instance of the concept *h:Mapping* of QELT, we can deduce that I_1 can be also considered as an instance of *p:Mapping* with respect to the bridge rule r_{ij} of the accessibility relation ▽i.

Related Work

After studying the formalism tailored for contextual ontologies, we consider in this section the comparison of our approach to other approaches having similar goals. We distinguish between the alternative techniques and the alternative formalisms.

Comparison with Other Techniques

We consider a comparison with other ontology-based techniques that are relevant to information integration and semantic interoperability such as integrated ontology, Contextual Ontology, and P2P ontology. These techniques are resumed as follows:

- **Integrated Ontology:** is a representation of global semantics. It defines a global understanding based on an established consensus (Staab, 2004). The consensus is needed for the common integrated ontology, which needs to be renewed each time an update occurs. It suffers from the loss of original understanding or loss of information in the profit of the unified representation. Regrettably, to reach the consensus, a considerable effort, cost, methodology, and update time is needed. Therefore, this solution does

Copyright © 2006, Idea Group Inc. Copying or distributing in print or electronic forms without written permission of Idea Group Inc. is prohibited.

not conform to the dynamic aspect of EIS, which need to make accessible as fast as possible (even without any consensus) the business system with new functionalities.

- **Contextual Ontology:** provides local and global semantics. It allows a global view without losing original representation. Indeed, it adapts the models to coexist through the contexts relationships. The contexts are related to an interpretation with a predefined structure. Thus, contextual ontology provides a dynamic consensus, rather than a static consensus, which is offered by integrated ontology (Rifaieh, 2004c).

- **P2P Ontology:** considers that nodes (ontologies) are equipotential in terms of functionalities and capabilities. Each peer has different amounts of knowledge that depend on the interactions it has performed in the network of available ontologies. Each peer can acquire new knowledge and/or extend its knowledge only by querying peers, which have this information. (Castano, 2003). Therefore, P2P ontology offers autonomy and low cost updating, but no global view can be reached with this technique.

- **Locally Independent Ontologies:** offers local semantic and interpretation where each system is viewed as being separated from others. Note that no global view can be expected from independent ontologies. Table 4 shows a comparison between the preceding techniques. We identified a set of criteria to compare these approaches illustrated as follows:

> Global view identifies whether the approach provides a global understanding or global representation,

> Local view identifies whether the approach provides a local understanding or local representation,

> Consensus identifies whether a commitment should be established between community members,

> Updating complexity defines the computational complexity for updating the used elements of the approach,

> Autonomy defines whether a system depends on external information to answer local needs,

> Cost of adding new ontology defines the complexity of incorporating a new ontology to the approach,

> Query answering defines the ability to answer global or local queries,

> Reusability defines the ability to reuse existing components for creating a new system, and

> Interoperability defines the ability to cooperate for a common purpose.

Copyright © 2006, Idea Group Inc. Copying or distributing in print or electronic forms without written permission of Idea Group Inc. is prohibited.

Table 4. Comparative analysis

	Global view	Local view	Consensus	Updating complexity	Autonomy	Cost of adding new ontology	Query answering	Reusability	Interoperability
Integrated ontology	Yes	No	Yes	High	Very low	Very high	High	Low	Fair
Contextual ontology	Yes	Yes	No	Fair	High	Fair	High	High	Fair
Peer2Peer ontology	No	Yes	No	Fair	Very high	Fair	Fair	Low	Fair
Locally independent ontology	No	Yes	No	Low	Very high	Low	Locally high Globally none	Low	Low

The values of the comparison are deducted by analysis of the state of the art for each approach. However, to reinforce the table, we need to check these results using empirical and experimental tests.

Comparison with Other Formalism

In reviewing literature, we found some works that deal with context, ontologies, and logics. In fact, a survey of context modeling is given in Strang (2004) that stratifies context approaches in key-value models, markup scheme models, graphical models, object-oriented models, logic-based models, and ontology-based models. We are interested in showing logic-based models and ontology-based models.

Logic-Based Models

A logic defines the conditions on which a concluding expression or fact may be derived with the inference process. In a logic-based context model, the context is consequently defined as facts, expressions, and rules. In this respect, formalisms have been suggested in the logic to cope with the notion of context.

One of the first logic-based context modeling approaches has been researched and published as Formalizing Context in early 1993 by McCarthy. He introduced contexts as abstract mathematical entities with properties useful in artificial intelligence. He introduced, as well, the basic relation ist(c,p) and value(c,p): ist(c,p) is predicate which asserts that the proposition p is true in context c, and value(c,e) is a function that gives the value of the term e in the context c. In Guha

Copyright © 2006, Idea Group Inc. Copying or distributing in print or electronic forms without written permission of Idea Group Inc. is prohibited.

(1991), the author elaborates more on these basic relations by presenting the lifting rules. By lifting rules, we can relate propositions and terms in sub-contexts to possibly more general propositions and terms in the outer context.

We can identify also the work in Giunchiglia (1993) and Ghidini (2001), referred as Multi-Context Systems (MCS), which is less on context modeling than on context reasoning. It treats a context as a specific subset of the complete state of an individual entity that is used for reasoning about a given goal. An extension of this work is currently revived with MCS/LMS (MCS/Local Model Semantics) in Roelofsen (2004), which consider propositional modal logics for tackling the context reasoning and seeking satisfiability algorithms.

In Borgida (2002), formal semantics of distributed description logics and distributed first order logics (Ghidini, 1998) are presented to provide coordination among a set of distributed information systems. In this approach, the authors argue that there will be no single global view, but correspondence between objects in the local domains should be furnished through directed mapping using bridge rules. They also show that it is possible to translate distributed description logics (DDLs) reasoning to description logics (DLs) reasoning. They claim, as well, that DDLs extends the reasoning available on ordinary schemas to the cases of multiple schemas connected by arbitrary binary correspondence between individuals (objects).

In Goh (1997), the context is treated as a formal object expressed in first order logic (FOL) to serve as a frame reference for sentences that are relative to some context. The COIN (COntext INterchange) team in MIT has investigated the notion of context for representing context knowledge and a context mediation engine in Firat (2003) and Goh (1997). COIN is constructed on a deductive and object-oriented data models, and it combines the syntax and semantics of F-logics and predicate calculus.

To conclude, logic-based context models may be composed distributed, but partial validation is difficult to maintain. Their level of formality is extremely high, but without partial validation, the specification of contextual knowledge within a logic based context model is very error-prone. Incompleteness and ambiguity seem to be not addressed neither. Applicability to these approaches seems to be not carried out to very high implementation results.

Ontology Based Models

One of the first approaches of modeling the context with ontologies has been proposed in Ötztürk (1997). Recently, the subject of combining ontologies and context has gotten more attention, for different reasons and applications. Thus, some formalisms have been suggested to cope the issue of defining a framework for this combination.

Copyright © 2006, Idea Group Inc. Copying or distributing in print or electronic forms without written permission of Idea Group Inc. is prohibited.

In Bouquet (2003), a framework and a concrete language C-OWL (Contextual OWL) for contextualized ontology was proposed. The definition of contextualized ontology is *"to keep its contents local but they put in relation with contents of other ontologies via context mappings"*. The new C-OWL language is augmented with relating (syntactically and semantically) concepts, roles, and individuals of different ontologies. The global interpretation is divided to the local interpretation and to the interpretation of the mapping rules. The local interpretation has two parts: Tarski's well-known interpretation and a hole interpretation of axiom that can be defined thought other context and which can be unsatisfiable in the first.

Another approach, the Aspect-Scale-Context information (ASC) model, has proposed ranges for the contextual information called scales (Strang, 2003). The ASC model using ontologies provides an uniform way for specifying the model's core concepts as well as an arbitrary amount of subconcepts and facts, altogether enabling contextual knowledge evaluated by ontology reasoner. This work built up the core of a nonmonolithic Context Ontology Language (CoOL), which is supplemented by integration elements such as scheme extensions for Web Services. All ontology-based context models inherit the strengths in the field of normalization and formality from ontologies.

The CONON context modeling approach in Wang (2004) is based on the same idea of ACS/CoOL approach namely to develop a context model based on ontologies because of its knowledge sharing, logic inferencing and knowledge reuse capabilities. In CONON, an upper ontology captures general features of basic contextual entities and a collection of domain specific ontologies and their features in each subdomain. The first order logic and description logics are used as reasoning mechanism to implement the CONON model, allowing consistency checking, and contextual reasoning.

However, the ASC model and the CONON model inherently support quality metainformation and ambiguity (Strang, 2003). Applicability to different existing models is limited to the type of formalism that adapts.

Evaluation

The definition of contextual ontologies in C-OWL (Bouquet, 2003) is identical to our perception of contextual ontologies, but the formalism suggested in this chapter differs with the one suggested for C-OWL. Our approach, however, is different from those that use context modeling in the sense that ontology with multirepresented concepts is our prime concern. Context, however, is exploited to allow for contextual information capturing in order to enable the implementation of concepts with multiple perspectives (Arara, 2004b). Hence, our formal representation of contextual ontologies should also preserve adequate reasoning

Copyright © 2006, Idea Group Inc. Copying or distributing in print or electronic forms without written permission of Idea Group Inc. is prohibited.

mechanisms, namely: concept satisfiability, concept subsumption, concept consistency, and instance checking.

We compare our formalism for contextual ontology versus the other formalisms surveyed in this section. Table 5 examines the properties based on the following criteria:

- **Expressiveness:** defines the power of the language to describe a phenomena or situation, example: ALC has limited expressiveness to describe "an elephant has precisely four legs".

- **Extensibility:** behavior to add operators and constructors, a language can be extended to represent needed phenomena, e.g., extension of ALC for number restriction ALCN.

- **Tractability:** is about reasoning with rules, it evaluates the difficulty to reason correctly with a representational language. Typically, we say that a problem is tractable if (we know) there exists an algorithm, solving the problem, whose run-time is (at worst) polynomial. Otherwise, we call the problem intractable.

Table 5. Comparison between related works formalisms

	Expressiveness		Extensibility	Tractability of reasoning	Computational complexity	Purpose
	Syntax	Semantic				
DDL	DL syntax Bridge rules	Tarskian interpretation	Limited to DLs family	Tractable	Same as DLs (simulation to DL proved)	Distributed information systems
COIN	F-Logic (extended DataLog)	Horn-logics	Extension based on built-in predicate and function symbols	ALP (Abductive Logic Programming)	?	Loosely information integration in databases
MCS/LMS	Propositional logics	Deductive facts reasoning	?	CSAT: Massacci's tableaux based procedure for PLC and equivalence results with modal logics	Polynomial (CSAT procedure)	Information Integration
COOL (ASC)	DAML +OIL OWL	Horn-logic Tarski interpretation	Limited by DL and Flogic	Tractability of *Ontobroker* with FLogic	?	Contextual interoperability and web services
CONON	DL + First order logics	Tarskian interpretation	Limited by RDF and OWL	?	?	Pervasive computing
C-OWL	DL syntax (OWL)+ bridge rules	Local Tarskian interpretation with Hole	Limited with OWL (SHOIQ)	Tractable but with reservation of local holes inconsistency	Not studied yet	Semantic web
Contextual Ontology (our approach)	Modal DL	Local Tarskian + Kripke's structure	Extensible with accessibility relation	ALCN (tractable) with respect to tractability operator	Not studied yet Specific algorithm is needed	Semantics sharing

Copyright © 2006, Idea Group Inc. Copying or distributing in print or electronic forms without written permission of Idea Group Inc. is prohibited.

- **Computational complexity:** is related to language expressiveness and leads tractability. Algorithms computational complexity increases with increasing expressivity. Therefore, a tradeoff between expressivity and computational complexity is important.

- **Purpose:** it shows the foreseen applications for the formalism.

Our Vision for Ontologies within the Enterprise

Improving EIS is relative to two factors: the push from technology assisted by academic research, and the pull from industry. In the effective use of EIS to support business goals, a number of key challenges face all large enterprises, whether they are governmental, in the private sector, or in universities. First, enterprise information systems are growing due to high demand for efficiency and data quality. In this context of an evolving economy, one of the major problems for enterprises is to ensure coherent and fast partnership and to share a common understanding of a given application domain. Local enterprise ontologies are considered, versus enterprise ontologies, to specify systems and to support semantic sharing for distributed and interorganizational applications.

From the previous section's discussions, contextual ontologies can be considered essential for the growth and competition among enterprises. The use of contextual ontologies requires promoting them as efficient framework for solving semantic-sharing problem within EIS. In this respect, this section tries to identify the use of ontologies (i.e., contextual ontologies) in the enterprise. It aims at defeating the ontology phobia and presents some arguments in order to convince IT decision makers to consider ontologies in their future systems developments.

Difference Between Enterprise Ontology and Local Information Systems Ontology

We differentiate between two uses of ontologies in the context of enterprises. Firstly, ontologies can be used for modeling enterprise activities and its business organization. Secondly, local enterprise ontologies can be used for modeling enterprise information systems and capturing specification (ontology-driven IS). The aim of the latter is to support semantic sharing for distributed and interorganizational applications. As a matter of fact, a few of effective enterprise ontology exists today in a productive setting, due to novelty and complexity of the tools, methods and techniques (i.e., artificial intelligence and knowledge

Copyright © 2006, Idea Group Inc. Copying or distributing in print or electronic forms without written permission of Idea Group Inc. is prohibited.

management). The development of enterprise ontology is a challenging work confronted by domain, time, and cost constraints. All these added factors have contributed to the depreciation of enterprise to ontologies technology contributions.

Enterprise Ontologies

The objective of an enterprise ontology is the conceptualization of the common economic phenomena of a business enterprise unaffected by application-specific demands. We find in the research literature many projects that aim at modeling enterprise activities. We can enumerate Core Enterprise Ontology (Bertolazzi, 2003), Edinburgh Enterprise Ontology (Uschold, 1998), and TOVE ontology (http://www.eil.utoronto.ca/enterprise-modelling/tove/index.html). Moreover, some methodologies and languages were suggested for capturing enterprise ontologies such as IDEF5 (http://www.idef.com/), PIF (http://ccs.mit.edu/pif/), and BEM (Business Engineering Model). IDEF5 is presented as a methodology for capturing an ontology. PIF is a frame language based on KIF and presented as a language for describing processes. BEM is part of the Open Information Model (OIM), which concepts are described using UML class diagram. Other suggestions can not be considered as enterprise ontology, such as the MIT Process Handbook (http://ccs.mit.edu/ph/), which represents about 4,000 classified processes. This handbook can be considered very useful for identifying enterprise activities and business processes. Otherwise, REA ontology (Geerts, 2000) focuses on enterprise internal business process in detail, but not for interenterprise collaboration.

Specifying Local Enterprise Information System Ontology

In the traditional software engineering domain, for each new application to be built, a new conceptualization is developed. According to the contribution of ontologies in this domain, they can be used as specification for high-level reusable software, like domain models and frameworks (Girardi, 2003). The main idea resumes by providing a vocabulary for specifying requirements for one, or more, target applications (Jasper, 1999).

Unfortunately, applications developed using traditional IS specification methods do not make their local ontology explicit because their underlying semantic information is usually implicit in the software (Benaroch, 2002). Therefore, if we do not start with modeling and using ontology, it will be hard to extract the semantics from the implemented systems. In this case, we can proceed

Copyright © 2006, Idea Group Inc. Copying or distributing in print or electronic forms without written permission of Idea Group Inc. is prohibited.

according to three ways to find these semantics: reading source code, analyzing run time, and interviewing users. An ideal alternative would be to create applications based on an IS specification method that make their local ontologies explicit in the first place (ontology-driven IS).

For instance, UML defines several types of diagrams that can be used to model the static and dynamic behavior of a system. It is widely adopted in industry and has a very large and rapidly expanding user community. Designers of enterprise information systems will be more likely to be familiar with this notation than KIF or DL, which are not widely known outside the AI research community (Cranefield, 1999). In UML, the ontology information is usually modeled in class diagrams to depict the concepts in the domain: an object diagram to show particular named instances of those classes, and Object Constraint Language (OCL) to offer powerful expressiveness for constraints.

Local Enterprise Information System Ontology versus Global Enterprise Ontology

In general, a system implements a certain number of business processes, but with specific use and intention. Albeit, enterprise ontologies can be very useful when it comes to IS requirements engineering, but they are not easy to understand and manipulate. We resume in the following some drawbacks of enterprise ontology approaches:

- It is difficult to design a precise and comprehensive enterprise ontology. Modeling a specific industry sector or even a specific enterprise is a more difficult challenge than modeling a set of systems.

- Some enterprise ontologies are very generic and difficult to personalize for a specific activity.

- If we assume that these ontologies exist and they are efficient, no guideline is provided to use the predefined concepts for reaching the specification of the enterprise information systems.

- Systems designers prefer to work on their systems independently, with a basic vocabulary, rather than a very tight enterprise ontology.

- They respect, in majority, top-down approach, where bottom-up seems to be a more realistic method for reaching a system's requirements.

- These ontologies adhere to a single ontology approach for sharing semantics, whereas multiple or hybrid approaches are more promising.

Copyright © 2006, Idea Group Inc. Copying or distributing in print or electronic forms without written permission of Idea Group Inc. is prohibited.

In order to avoid the gap between top-enterprise design and specific-system design, we believe that using ontologies in the enterprise should consider the specification of the system and a common vocabulary (which can be an ontology), in other word, the hybrid approach of sharing semantics. Therefore, in order to reach ontology-driven EIS, the elicitation and modeling of software requirements can be accomplished by binding local ontology and global ontology. Thereby, local EIS system ontology can be identified based on top-level enterprise ontology and the particularities of specific requirements of the implemented application.

Promoting Ontologies Use within the Enterprise

Although, we showed previously the adequacy of ontologies for enterprise evolution and set of essential productivity goals, a small number of enterprises have yet confronted this adventure. Studies about ontology use in practical enterprise domains show that more attention should be given to this promising technique. In April of the year 2004, the market research firm Gartner (http://www4.gartner.com/Init) identified taxonomies/ontologies as one of the leading IT technologies, ranking it third in its list of the top 10 technologies forecast for 2005 (Denny, 2004).

However, the market interest is not up to these predictions, through 2006, more than 70% of firms that invest in unstructured information-management initiatives will not achieve their targeted return on investment (ROI), due to underinvestment in taxonomy/ontologies building (Knox, 2003).

Therefore, when putting ontology to work in an enterprise environment, we should take into account to change the way that we are thinking and acting. This mind shift seems not yet ready for IT decision makers. In the next section, we discuss how to defeat this fear of adopting ontologies techniques in the enterprises, and allowing them to take place in the production chain.

Is it Ontology Phobia or Ignorance?

Based on our experience in consulting and our discussions with IT executives and others consultants, we perceived a persistent ignorance about ontologies applications and their advantages. Currently, few IT decision makers even know that ontologies exist. And as for those that do know, they have developed a fear of ontologies as a promising technology. This exaggerated and inexplicable fear is translated to an ontology phobia, excluding ontologies' chances to take their place in the enterprise architecture.

Copyright © 2006, Idea Group Inc. Copying or distributing in print or electronic forms without written permission of Idea Group Inc. is prohibited.

Let us put in a nutshell the elements that contribute to this situation:

- Enterprises are goal-oriented with high aims of profit. They consider generally the ROI as the balance of every activity. During an uncertain economical environment, they tend to have short or medium term investments. Surprisingly, sometimes it is preferred to keep repetitively fixing existing systems, rather than reconceiving the whole system applications.

- Ontologies development is a complicated process where ontologists and computer scientists have to work together. They require mutual work to classify the concepts and conceive their structure and knowledge base rules. All these elements have cooperated to make ontology development a very expensive investment, and risky. Therefore, ontologies developments are still limited. In opposite recall for adapting ontologies, from existing ones, seems to be a more attractive approach.

- Enterprises that will use ontologies should adopt new architectures and conceive their systems in a different way than how they used to do. Sometimes, they may have to build their own solutions using tools from several vendors. At the largest scale, when the amount of information requires a huge amount of concepts and categories that change frequently, this exercise will be a laborious trial and often an erroneous process.

- Enterprises doubt the maturity of ontological technologies and their applications arguing that announced goals are not yet reachable and have low cost-effectiveness.

Defeating Ontology Phobia

The following two elements are proposed to defeat ontology phobia. Firstly, we assure the interested party that the maturity of the ontological approach is already attained, due to the existing components of ontology techniques. These researches and applications were developed in order to answer the basic requests. Other arguments can be cited such as:

- Well-organized ontologies and knowledge bases can save an enterprise money by decreasing the amount of employee time spent trying to find, integrate, share, and reuse information in the myriad of EIS.

- Many satisfying results have been attained in the last decade including formal ontologies, languages, engineering methodology, tools, and so forth. In particular, we have almost all the required elements to design, develop, and construct ontologies. Other research areas such as versioning, evolu-

Copyright © 2006, Idea Group Inc. Copying or distributing in print or electronic forms without written permission of Idea Group Inc. is prohibited.

tion, and maintenance of ontologies are considered as current and interesting issues for investigation.

- Efficient tools are available for the development of enterprise ontologies. A recent census of ontologies development tools, for many area of applications, has counted up to 105 tools (Denny, 2004).

Second, we should take the responsibility to make ontologies more attractive; therefore, a certain number of works should be realized:

- Academic research should promote the originality of these techniques and prove their efficiency. Recently many sessions and conferences consider enterprise application of ontologies becoming more and more in vogue.

- Developing ontologies and making them widely accessible, indeed, enterprises are more interested to adapt existing domain ontologies for their business with a "minor" process of adaptation rather than constructing ontology from scratch.

- Defining methodology for adapting ontologies and showing success stories in spread domains are not negligible arguing elements.

- Presenting enterprise architecture and case studies where ontologies techniques play the main role. Our research adheres to this goal: it takes the responsibility to argue the usefulness of using ontology-driven IS and show an architecture and case study as elements of arguments (this part is not described in this chapter).

Conclusion

The main idea we are arguing in this chapter consists of encouraging plans of evolving EIS, and considering theoretical research such as ontologies and contexts. Therefore, the discussions involved in this chapter combine together academic research and enterprise practices. Hence, one of the credits of this work is to create a space of expertise and ideas exchange between these two worlds, with respect to the push and pull factors.

After studying in the first part ontologies *per se* as a formal solution for EIS communication, interoperability, and reusability, we showed the position of these ontologies in enterprise architecture. Next, we argued the relevance of defining ontologies for improving EIS. A major theme for the use of ontologies in this domain includes the creation of a shared modeling and a common environment

Copyright © 2006, Idea Group Inc. Copying or distributing in print or electronic forms without written permission of Idea Group Inc. is prohibited.

for different information systems and software tools. Any information technology application should use these integrated enterprise models, spanning activities, resources, and services. We identified, as well, the limits of simple ontology to answer EIS needs, especially facing the multiple views and multirepresentation problems. Therefore, we suggested pairing up the two notions of ontologies and context to overcome these problems.

The second part of this chapter consisted of studying theoretical and logical techniques in order to pair up ontology and context. The approach is based on the well-founded theory of *modal description logics* to build a semantically rich conceptual model that supports scalability and extensionality. Moreover, the proposed formalism of contextual ontology supports *information accessibility*, *filtering*, and *reasoning services* based on their contextual information. We associated Description Logics and Modal Logics formalisms (i.e., ALCNM language) for expressing this union. This formalism offers a strong machine processable formalism that meets with EIS semantics sharing needs. It brings an adequate ontology-based solution to resolve the real state of multiple views and multirepresentation phenomena.

Finally, in order to promote the use of contextual ontologies, we discussed the ontology phobia and showed how and where ontologies can contribute to EIS.

Technically, we exploited the formalism through architecture and a prototype. A project called EISCO (Enterprise Information System Contextual Ontologies) has been partially developed to this purpose. This step is very useful from a practical point of view, because reaching a theoretical result is not the goal per se, but facilitating EIS users' tasks is the foreseen goal. We illustrated many scenarios as well for using the suggested architecture, and implemented a prototype with a case study to show the feasibility of these suggestions (this work is out of the scope of this chapter) (Rifaieh, 2005).

References

Akman, V., & Surav, M. (1996). Steps toward formalizing context. *AI Magazine, 17*(3), 55-72. Retrieved September 10, 2004, from http://cogprints.ecs.soton.ac.uk/archive/00000464/

Arara, A., & Benslimane, D. (2002, December). Towards ontologies building: A terminology based approach. *Proceedings of the 2nd IEEE ISSPIT*, Marrakesh, Maroco (pp. 11-16). IEEE.

Arara, A., & Benslimane, D. (2004, June). Multi-perspectives description of large domain ontologies. *Proceedings of FQAS 2004, Lecture Notes in Artificial Intelligence 3055*, Lyon, France (pp. 150-16). Springer.

Copyright © 2006, Idea Group Inc. Copying or distributing in print or electronic forms without written permission of Idea Group Inc. is prohibited.

Baader, F., et al. (2003). *The description logic handbook: Theory, implementation, and applications.* Cambridge, UK: Cambridge University Press.

Balley, C., Parent, C., & Spaccapietra, S. (2004, June). Modeling geographic data with multiple representations. *International Journal of Geographical Information Science, 18*(4), 327-352.

Benaroach, M. (2002, June). Specifying local ontologies in support of semantic interoperability of distributed inter-organizational applications. *Proceedings of the 5th International Workshop Next Generation Information Technologies and Systems, NGITS 2002, Lecture Notes in Computer Science 2382,* Caesarea, Israel (pp. 90-106). Springer.

Benslimane, D., & Arara, A. (2003). Towards a contextual content of ontologies. In *Foundations of Intelligent Systems, Lecture Notes in Computer Science 2871* (pp. 339-343). Heidelberg: Springer-Verlag.

Benslimane, D., Vangenot, C., Roussey, C., & Arara, A. (2003, June 16-20). The multi-representation ontologies: A contextual description logics approach. *Proceeding of the 15th Conference on Advanced Information Systems Engineering, Lecture Notes in Computer Science 2798,* Austria (pp. 4-15). Heidelberg: Springer-Verlag.

Berners-Lee, T. (2004). Keynote Speech in 13th International World Wide Web Conference 2004. Retrieved September 9, 2004, from http://www.w3.org/2004/Talks/0519-tbl-keynote

Bertolazzi, P., Krusich, C., & Missikoff, M. (2003). An approach to the definition of a core enterprise ontology: CEO. In *Managing globally with information technology* (pp. 104-115). Hershey, PA: Idea Group Publishing.

Borgida, A., Brachman, R. J., McGuinness, D. L., & Resnick, L. A. (1989). CLASSIC: A structural data model for objects. *Proceedings of the 1989 ACM SIGMOD International Conference on Management of Data,* Portland, OR (pp. 58-67). New York: ACM Press.

Borgida, A., & Serafini, L. (2002). Distributed description logics: Directed domain correspondences in federated information sources. *Proceedings of On the Move to Meaningful Internet Systems, 2002 DOA/CoopIS/ODBASE* (Vol. 2519, pp. 36-53). London: Springer Verlag.

Bouquet, P., Giunchiglia, F., van Harmelen, F., et al. (2003, Octobter). C-OWL: Contextualizing ontologies. *Proceedings of the 13th International World Wide Web Conference on Alternate Track Papers & Posters,* Sanibel Island, FL (pp. 270-271). New York: ACM Press.

Calvanese, D., De Giacomo, G., & Lenzerini, M. (2001, August). Ontology of integration and integration of ontologies. *Proceedings of the 2001 Description Logic Workshop (DL 2001),* Stanford University, CA.

Copyright © 2006, Idea Group Inc. Copying or distributing in print or electronic forms without written permission of Idea Group Inc. is prohibited.

Castano, S., Ferrara, A., & Montanelli, S. (2003, September 7-8). H-match: An algorithm for dynamically matching ontologies in peer-based systems. *Proceedings of SWDB'03, The First International Workshop on Semantic Web and Databases*, co-located with VLDB 2003 (pp. 231-250). Berlin: Humboldt-Universität.

Catarci, T., & Lenzerini, M. (1993). Representing and using inter-schema knowledge in a cooperative information systems. *International Journal of Intelligent and Cooperative Information Systems, 2*(4), 375-398.

Chaudhri, V. K., Farquhar, A., Fikes, R., Karp, P. D., & Rice, J. P. (1998). *Open knowledge base connectivity.* Technical Report. OKBC Working Group. Stanford University, Knowledge Systems Laboratory. Retrieved October 10, 2004, from http://www.ai.sri.com/~okbc/spec.html

Corcho, O., Fernández-López, M., & Gómez-Pérez, A. (2003, July). Methodologies, tools and languages for building ontologies: Where is their meeting point? *IEEE Transactions on Data & Knowledge Engineering, 46*(1), 41-64.

Cranefield, S., & Purvis, M. (1999, July 31-August 6). UML as an ontology modelling language. *Proceedings of the 16th International Joint Conference on Artificial Intelligence (IJCAI-99),* Stockholm, Sweden. San Francisco: Morgan Kaufmann.

Decker, S., Erdmann, M., Fensel, D., & Studer, R. (1998). Ontobroker in a nutshell. *Proceedings of the 2nd European Conference on Research and Advanced Technology for Digital Libraries, LNCS 1513,* Crete, Greece, (pp. 663-664). London: Springer-Verlag.

Denny, M. (2004). *Ontology tools survey-revisited.* XML.COM. Retrieved July 14, 2004, from http://www.xml.com/pub/a/2004/07/14/onto.html

Donini, F., Nardi, D., & Rosati, R. (1996). Ground nonmonotonic modal logics. *Journal of Logic and Computation, 7*(4), 523-548.

Duineveld, A. J., Stoter, et al. (2000). Wondertools? A comparative study of ontological engineering tools. *International Journal of Human-Computer Studies, 52*(6), 1111-1133.

Farquhar, A., Fikes, R., & Rice, J. (1997). The ontolingua server: A tool for collaborative ontology construction. *International Journal of Human-Computer Studies, 46*(6), 707-727.

Fensel, D., van Harmelen, F., Horrocks, I., McGuinness, D., & Patel-Schneider, P. F. (2001). OIL: An ontology infrastructure for the Semantic Web. *IEEE Intelligent Systems, 16*(2), 38-45.

Firat, A. (2003, August). *Information integration using contextual knowledge and ontology merging.* PhD Thesis (supervised by S. Madnick, B.

Copyright © 2006, Idea Group Inc. Copying or distributing in print or electronic forms without written permission of Idea Group Inc. is prohibited.

Grosof, M. Siegel), MIT. Retrieved August 10, 2004, from http://ebusiness.mit.edu/bgrosof/paps/phd-thesis-aykut-firat.pdf

Fonseca, F., & Egenhofer, M. (1999). Ontology-driven geographic information systems. *Proceedings of the 7th ACM International Symposium on Advances in Geographic Information Systems*, Kansas City, MO (pp. 14-19). New York: ACM Press.

Geerts, G., & McCarthy, W. E. (2000, August). *The ontological foundation of REA enterprise information systems.* Working Paper. Michigan State University. Retrieved August 10, 2004, from http://www.msu.edu/user/mccarth4/Alabama.doc

Ghidini, C., & Giunchiglia, F. (2001, April). Local models semantics, or contextual reasoning = locality + compatibility. *Artificial Intelligence Journal, 127*(2), 221-259.

Ghidini, C., & Serani, L. (1998). Distributed first order logics. In *Studies in logic and computation, Frontiers of combining systems 2* (pp.121-140). Berlin, Research Studies Press.

Girardi, R., & de Faria, C. G. (2003, September 24). A generic ontology for the specification of domain models. *Proceedings of the 1st International Workshop Component Engineering Methodology* (pp. 41-51). Erfurt, Germany (online).

Giunchiglia, F. (1993). Contextual reasoning. *Epistemologia Italian Journal for the Philosophy of Science, Special Issue on I Linguagii e le Macchine, 16* (1993), 345-364. (Also IRST-Technical Report 9211-20, IRST, Trento, Italy)

Goh, C. H. (1997). *Representing and reasoning about semantic conflicts in heterogeneous information sources.* PhD Thesis, MIT. Retrieved September 26, 2004, from http://context2.mit.edu/coin/publications/goh-thesis/goh-thesis.pdf

Gold, R., & Mascolo, C. (2001, October). Use of context-awareness in mobile peer-to-peer networks. *Proceedings of the 8th IEEE Workshop on Future Trends of Distributed Computing Systems (FTDCS 2001),* Bologna, Italy. Washington, DC: IEEE Computer Society.

Gómez-Pérez, A., & Benjamins, V. R. (1999). Applications of ontologies and problem-solving methods. *AI-Magazine, 20*(1), 119-122.

Gruber, T. (1991). The role of common ontology in achieving sharable, reusable knowledge bases. *Proceedings of the 2nd International Conference Principles of Knowledge Representation and Reasoning,* Cambridge, MA (pp. 601-602). Morgan Kaufmann.

Copyright © 2006, Idea Group Inc. Copying or distributing in print or electronic forms without written permission of Idea Group Inc. is prohibited.

Gruber, T. (1995, November-December). Toward principles for the design of ontologies used for knowledge sharing. *International Journal of Human-Computer Studies, 43*(5/6), 907-928.

Guarino, N., & Giaretta, P. (1995). Ontologies and knowledge bases: Towards a terminological clarification. In *Towards very large knowledge bases: Knowledge building and knowledge sharing* (pp. 25-32). Amsterdam, The Netherlands: ISO Press.

Guarino, N. (1998, June). Formal ontology and information systems. *Proceedings of the 1st International Conference*, Trento, Italy (p. 337).

Guha, R. V. (1991). *Contexts: A formalization and some applications.* PhD Thesis, Stanford University.

Hendrix, G. (1979). Encoding knowledge in partitioned networks. In N. Findler (Ed.), *Associative networks* (pp. 51-92). New York: Academic Press.

Jasper, R., & Uschold, M. (2004, October). A framework for understanding and classifying ontology applications. *The 12th Workshop on Knowledge Acquisition, Modeling and Management,* Banff, Canada. Retrieved September 24, 2004, from http://sern.ucalgary.ca/KSI/KAW/KAW99/papers.html

Kashyap, V., & Sheth, A. (1997). Semantic heterogeneity in global information systems: The role of metadata, context and ontologies. In M. P. Papazoglou, & G. Schlageter (Eds.), *Cooperative information systems: Trends and directions* (pp. 139-178). London: Academic Press.

Keita, A., Laurini, R., Roussey, C., & Zimmerman, M. (2004, October). *Towards an ontology for urban planning: The Towntology Project.* Accepted to appear in the 24th UDMS Symposium, Venice, Italy.

Kifer, M., Lausen, G., & Wu, J. (1995, July). Logical foundations of object oriented and frame-based languages. *Journal of the ACM (JACM), 42*(4), 741-843.

Knox, R. E., & Logan, D. (2003, September 10). *What taxonomies do for the enterprise.* Gartner Research Articles, AV-20-8780.

Kripke, S. (1959). A completeness theorem in modal logic. *The Journal of Symbolic Logic, 24,* 1-14.

Lemmon, E. J. (with D. Scott) (1977). An introduction to modal logic. In K. Segerberg (Ed.), *American philosophical quarterly monograph series, no. 11.* Oxford: Basil Blackwell.

MacGregor, R. (1991, June). Inside the LOOM classifier. *ACM SIGART Bulletin, Special Issue on Implemented Knowledge Representation and Reasoning Systems, 2*(3), 88-92.

Copyright © 2006, Idea Group Inc. Copying or distributing in print or electronic forms without written permission of Idea Group Inc. is prohibited.

McCarthy, J. (1993). Notes on formalizing contexts. *Proceedings of the 13th International Joint Conference on Artificial Intelligence*, San Mateo, CA (pp. 555-560). San Francisco: Morgan Kaufmann.

Mena, E., Illarramendi, A., Kashyap, V., & Sheth, A. (2000). Observer: An approach for query processing in global information systems based on interoperability between pre-existing ontologies. *Distributed and Parallel Databases, 8*(2), 223-271.

Mitra, P., Wiederhold, G., & Kersten, M. L. (2000, March 27-31). A graph-oriented model for articulation of ontology interdependencies. *Proceedings of the 7th International Conference on Extending Database Technology,* Konstanz, Germany (pp. 86-100).

Mizoguchi, R., Ikeda, M., & Sinitsa, K. (1997, August). Roles of shared ontology in AI-ED research: Intelligence, conceptualization, standardization, and reusability. *Proceedings of the 8th World Conference On Artificial Intelligence In Education AIED-97*, Kobe, Japan (pp. 537-544).

Nuseibeh, B., Kramer, J. & Finkelstein, F. (1994). Expressing the relationships between multiple views in requirements specification. *Proceedings of the 15th ICSE* (pp. 760-773). Piscataway, NJ: IEEE Press.

Obrst, L. (2003). Ontologies for semantically interoperable systems. *Proceedings of the 12th International Conference on Information and Knowledge Management,* New Orleans, LA (pp. 366-369). New York: ACM Press.

Osterwalder, A. (2002, June). *An e-business model ontology for the creation of new management software tools and IS requirement engineering.* CAiSE 2002 Doctoral Consortium, Toronto. Retrieved August 8, 2004, from http://www.mics.ch/getDoc.php?docid=226&docnum=1

Ötztürk, P., & Amodt, A. (1997, February). Towards a model of context for case-based diagnostic problem solving. *Proceedings of the 1st International and Interdisciplinary Conference on Modeling and Using Context (Context-97),* Rio de Janeiro, Brazil (pp. 198-208).

Peerce, A., et al. (1999). The Kraft architecture for knowledge fusion and transformation. *Proceedings of the 19th SGES International Conference on Knowledge-Based Systems and Applied Artificial Intelligence (ES'99).* Springer.

Rifaieh, R. (2004, December). *Using contextual ontologies for semantic sharing within enterprise information systems.* PhD Thesis, National Institute of Applied Science of Lyon. Retrieved from http://csidoc.insa-lyon.fr/these/2004/rifaieh/these.pdf

Copyright © 2006, Idea Group Inc. Copying or distributing in print or electronic forms without written permission of Idea Group Inc. is prohibited.

Rifaieh, R., Arara, A., & Benharkat, N. (2004). *MuRO: A multi-representation ontology as a foundation of enterprise information systems.* Accepted to the 7th International Conference on Information Technology CIT'04 (forthcoming in LNCS-Springer).

Rifaieh, R., Arara, A., & Benharkat, A. N. (2004, August 6-8). *Multi-representation ontologies in the context of enterprise information systems.* American Conference on Information Systems, New York.

Rifaieh, R., & Benharkat, A. (2002, November). Query-based data warehousing tool. *Proceedings of the 5th ACM International Workshop on Data Warehousing and OLAP, DOLAP'02,* McLean (pp. 35-42). New York: ACM Press.

Rifaieh, R., & Benharkat, A. N. (2002, December). A mapping expression model used for a neta-data driven ETL tool. In K. Yétongnon, & M. Amin (Eds.), *Proceedings of the 2nd IEEE ISSPIT,* Marrakech, Morocco (pp. 288-293). IEEE.

Rifaieh, R., & Benharkat, N. (2003, June 5-7). An analysis of EDI's message translation and integration problems. *Proceedings of the International Conference on Computer Science, Software Engineering, Information Technology, e-Business, and Applications (CSITeA'03),* Rio de Janeiro, Brazil (pp. 254-260).

Rifaieh, R., & Benharkat, N. (2003, July). A mapping framework for EDI message translation. *Proceedings of ACS/IEEE Conference AICCSA'03,* Tunisia (p. 87). Piscataway, NJ: IEEE Press.

Roelofsen, F., & Serafini, L. (2004, July 25-29). Complexity of contextual reasoning. *Proceedings of AAAI-04 Conference,* San Jose, CA (pp. 118-123).

Sowa, J. F. (2000). *Knowledge representation: Logical, philosophical, and computational foundations.* Pacific Grove, CA: Brooks Cole Publishing.

Strang, T. (2003, October). *Service interoperability in ubiquitous computing environments.* PhD Thesis, Ludwig-Maximilians-University, Munich. Research Report. Berlin: VDE-Verlag.

Strang, R., & Linnhoff-Popien, C. (2004, September). *A context modeling survey.* Accepted for Workshop on Advanced Context Modelling, Reasoning and Management as part of The Sixth International Conference on Ubiquitous Computing, Nottingham, UK.

Stuckenschmidt, H., Wache, H., Vgele, T., & Visser, U. (2000). Enabling technologies for interoperability. *Proceedings of Workshop on the 14th International Symposium of Computer Science for Environmental Protection,* Bonn, Germany (pp. 35-46).

Copyright © 2006, Idea Group Inc. Copying or distributing in print or electronic forms without written permission of Idea Group Inc. is prohibited.

Uschold, M., King, M., Moralee, S., & Zorgios, Y. (1998). The enterprise ontology. In M. Uschold, & A. Tate (Eds.), *The Knowledge Engineering Review, Special Issue on Putting Ontologies to Use, 13.*

Wache, H. et al. (2001). Ontology-based integration of information — A survey of existing approaches. *Proceedings of the International Joint Conference on Artificial Intelligence IJCAI'01* (pp. 108-117).

Wang, X., Zhang, D., et al. (2004, March 14-17). Ontology-based context modeling and reasoning using OWL. *Proceedings of the 2nd IEEE International Conference on Pervasive Computing and Communications (PerCom 2004),* Orlando, FL (pp. 18-22). Washington, DC: IEEE Computer Society.

Welty, C. A. (2003). Ontology research. Guest Editorial. *AI Magazine, 24*(3), 11-12.

Wolter, F., & Zakharyaschev, M. (1998). On the decidability of description logics with modal operators. *Proceedings of the 6th International Conference on Principles of Knowledge Representation and Reasoning (KR'98),* Montreal, Canada (pp. 512-552). Morgan Kaufman.

Copyright © 2006, Idea Group Inc. Copying or distributing in print or electronic forms without written permission of Idea Group Inc. is prohibited.

Chapter VI

A Comparison of Semantic Annotation Systems for Text-Based Web Documents

Lawrence Reeve, Drexel University, USA
Hyoil Han, Drexel University, USA

Abstract

The Semantic Web promises new as well as extended applications, such as concept searching, custom Web page generation, and question-answering systems. Semantic annotation is a key component for the realization of the Semantic Web. The volume of existing and new documents on the Web makes manual annotation problematic. Semi-automatic semantic annotation systems, which we call platforms because of their extensibility and composability of services, have been designed to alleviate this burden for text-based Web documents. These semantic annotation platforms provide services supporting annotation, including ontology and knowledge base access and storage, information extraction, programming interfaces, and end-user interfaces. This chapter defines a framework for examining semantic annotation platform differences based on platform characteristics,

Copyright © 2006, Idea Group Inc. Copying or distributing in print or electronic forms without written permission of Idea Group Inc. is prohibited.

surveys several recent platform implementations, defines a classification scheme based on information extraction method used, and discusses general platform architecture.

Introduction

The Semantic Web, as described in Berners-Lee (1998), is the next generation of the Web providing machine-understandable information that is based on meaning. One way to provide meaning to Web information is by creating ontologies, and then linking information on a Web page to specifications contained in the ontology using a markup language (Berners-Lee et al., 2001). A key problem for the realization of the Semantic Web is providing these annotations for both existing and new documents on the Web. Semantic annotation is the process of mapping instance data to an ontology. Ontologies are conceptualizations of a domain that typically are represented using domain vocabulary (Chandrasekaran, Josephson, & Benjamins, 1999). Benefits of adding meaning to the Web include: query processing using concept-searching rather than keyword-searching (Berners-Lee et al., 2001); custom Web page generation for the visually-impaired (Yesilada, Harper, Goble, & Stevens, 2004); using information in different contexts, depending on the needs and viewpoint of the user (Dill et al., 2003); and question-answering (Kogut & Holmes, 2001).

It is not yet possible to automatically identify and classify all entities within source documents with complete accuracy (Popov et al., 2003). Manual annotation can be done using tools such as Semantic Word (Tallis, 2003), which provides a single interface for authoring as well as document markup. Manual approaches, however, suffer from several drawbacks. Human annotators can provide unreliable annotation for many reasons: complex ontology schemas, unfamiliarity with subject material, and motivation, to name a few (Bayerl, Lüngen, Gut, & Paul, 2003). It is expensive to have human annotators markup documents (Cimiano, Handschuh, & Staab, 2004), and the human annotator may not consider using multiple ontologies (Dingli, Ciravegna, & Wilks, 2003). Documents and ontologies can change, requiring new or modified markup, which leads to document markup maintenance issues (Dingli et al., 2003). Finally, the volume of existing documents on the Web can lead to an overwhelming task for humans to manually complete (Kosala & Blockeel, 2000). For all these reasons, manual efforts have been identified as a "knowledge acquisition bottleneck" (Maedche & Staab, 2001). Semi-automatic annotation platforms offer advantages over manual efforts, primarily document volume scalability through reduction of the human workload (Dill et al., 2003).

Copyright © 2006, Idea Group Inc. Copying or distributing in print or electronic forms without written permission of Idea Group Inc. is prohibited.

Semi-automatic semantic annotation systems, which we call Semantic Annotation Platforms (SAPs), have been developed to overcome the scalability issue of providing annotations for the large number of documents on the Web. Semi-automatic annotation of text documents can be seen as a typical information extraction for named-entity recognition process, but is different in that type information from a rich ontology is more specific and also the entities must be clearly identified and not just recognized as an entity of some type, as is the case with basic information extraction efforts (Popov et al., 2003). Semi-automatic annotation has been developed using research done in the areas of information extraction, information integration, wrapper induction, and machine learning (Dingli et al., 2003).

The remainder of this chapter defines a framework for viewing the differences between platforms, provides an overview of a few representative platforms, and briefly analyzes each platform using the platform characterization framework. This chapter expands on our initial work (Reeve & Han, 2005), which classified and surveyed several platforms, but did not fully identify the platform characteristics framework and platform classification scheme that the survey was based on, or identify a general semantic annotation platform architecture. The platform characteristics framework and classification system is useful for evaluating potential applications of a semantic annotation platform, where each has its own strengths based on domain, document structure, information extraction methods, ontology support, and manual effort required. The platform architecture is useful for understanding how these multiservice platforms are composed.

Development of Semantic Annotation Platforms

The semi-automatic annotation systems compared in this chapter provide semantic annotation of text-based Web documents. We call such systems platforms, largely due to their extensibility and composability; in addition, some of the literature also refers to them as platforms (Popov et al., 2003; Dill et al., 2003). There have been many efforts to build platforms for semantic annotation. Several of these platforms are briefly discussed in the *Platform Overviews* section, and were selected as a representative sample, rather than an exhaustive list, of the platform classifications discussed in the *Platform Classification* section. These semantic annotation platforms are diverse enough to show the distinguishing platform characteristics identified in the *Platform Characteristics* section.

Copyright © 2006, Idea Group Inc. Copying or distributing in print or electronic forms without written permission of Idea Group Inc. is prohibited.

Platform Architecture

Figure 1 shows the general architecture of a semantic annotation platform (SAP) as a composable system. Most SAPs are extensible, meaning various components can be replaced with alternate implementations. The advantage of an extensible annotation platform is that it can be adapted to serve many needs, such as changing domains, languages, or providing scalability.

The Application layer is responsible for providing an end-user interface to the annotation services provided by a SAP. Examples include facilities for annotating a document or document set and then potentially confirming the annotations before committing them, providing a query interface for searching annotations, and providing a user interface for configuring the information extraction component. The Application layer is the primary application programming interfaces (API) layer. A set of general programmatic interfaces designed to shield the applications from changes in the middle layer are defined in this layer. Applica-

Figure 1. General architecture of a semantic annotation platform

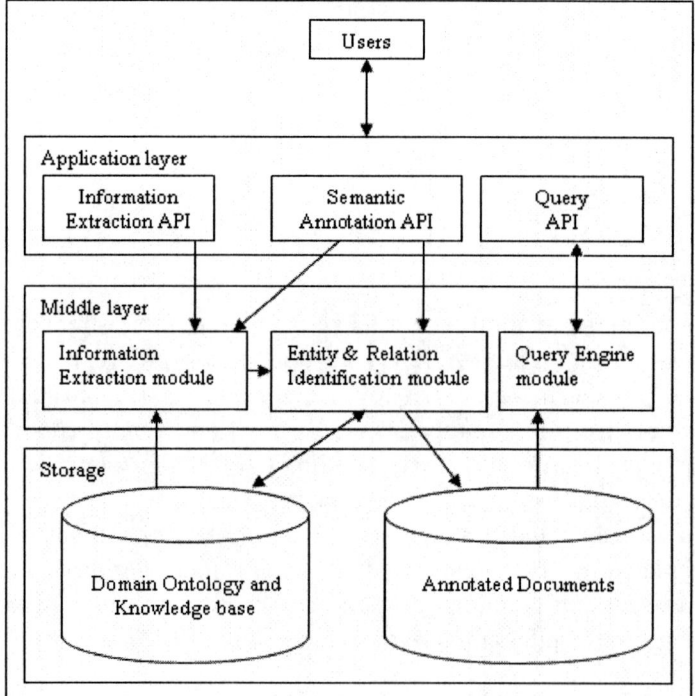

Copyright © 2006, Idea Group Inc. Copying or distributing in print or electronic forms without written permission of Idea Group Inc. is prohibited.

tions call the defined APIs in order to perform actions on behalf of an application and they can be quite numerous, covering annotation, information extraction, search, storage management, and many other provided services. The middle layer contains the actual components that perform work for an application, such as information extraction for concept (names and relationships) identification. The Application layer provides a consistent view to an application, but the middle layer is tied to an existing or adapted tool. For example, the information extraction component may switch from a pattern-based tool to a statistical tool, and it is unlikely the programmatic interface is the same for both. Finally, the Storage layer is designed to provide storage and storage management facilities for storing long-term data such as ontologies, document annotations and knowledge bases.

Platform Classification

Current semantic annotation platforms use several methods for information extraction (IE) from Web documents. Figure 2 shows a hierarchical classification of annotation platforms based on their IE component. This classification scheme can be used to organize the platforms performing semantic annotation. While semantic annotation platforms have many aspects, the information extraction approach currently used to find entities within text has the most impact on the effectiveness of the platform. For this reason, the information extraction (IE) approach of each platform is used to organize the platforms.

The top-level approach is multistrategy, which uses a combination of the lower level approaches. A platform using a multistrategy approach is able to adapt its IE methods based on the text it is processing in order to obtain the best results. The multistrategy approach uses a high-level identification of text genre, and then executes the appropriate IE methods (either pattern-based or machine learning based). No semantic annotation platform to date is using a complete multistrategy approach incorporating both pattern and machine learning approaches.

The two components levels composing the multistrategy approach are pattern-based methods and machine learning methods. Pattern-based methods are systems composed of manual rules. The rules are typically hand-crafted rules or seed patterns that define how entities can be found in text. A limiting factor on the scalability of such systems is that the manual rule generation process can be maintenance-intensive. Each time a data source changes, the pre-defined rules may also need to be changed. Machine learning approaches are mainly based on supervised learning and use pre-annotated examples to learn how to identify entities. Rules are learned from a pre-annotated corpus, and then later applied

Copyright © 2006, Idea Group Inc. Copying or distributing in print or electronic forms without written permission of Idea Group Inc. is prohibited.

Figure 2. Classification of semantic annotation platforms based on information extraction method used

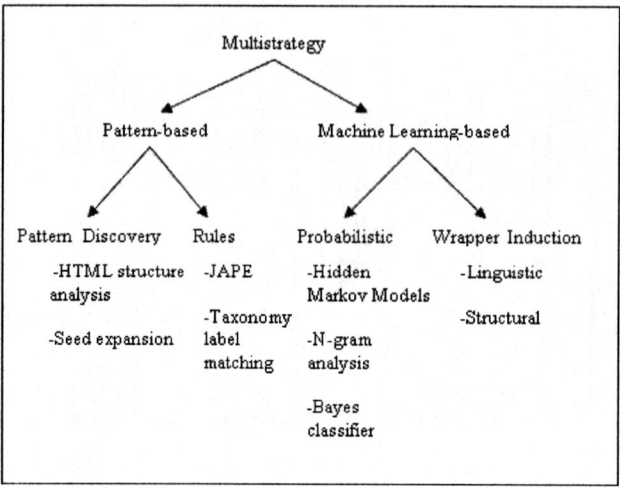

to perform annotation. The implication is that sufficient training data exists to adequately train the system. Hidden Markov Model (Rabiner, 1989) is an example of a machine learning approach that can be used. The Hidden Markov Model approach is not currently being used by any of the semantic annotation platforms as an information extraction method, but has been used successfully to extract attributes from text (Svab, Labsky, & Svatek, 2004).

Pattern-Based Methods

Patterns are widely used in semantic annotation platforms. **Pattern discovery** works by taking a few seed samples, finding entities based on the patterns, expanding the seed samples with patterns from the new entities found, and repeating the process until no more instances are found, or the user stops the iterative process (Brin, 1998). Patterns can exploit known linguistic patterns, such as Hearst patterns (Hearst, 1992), to find entities in text.

Rules can be manually generated or they can be learned using machine learning techniques. For classification purposes, manual rule generation is considered as part of the pattern-based classification, as shown in Figure 2. The reason is because the rules are initially manually specified by the user. Rules can take

Copyright © 2006, Idea Group Inc. Copying or distributing in print or electronic forms without written permission of Idea Group Inc. is prohibited.

many forms. They can be as simple as labels, as in a gazetteer, or they can be complex definitions expressed in a grammar using a tool such as the Java Annotations Pattern Engine (Maynard, 2003). In general, rule-based systems do not perform as well as machine learning based systems. However, adaptive rule platforms, such as the MUSE platform (see the *Platform Overviews* section), perform competitively with machine learning based platforms.

Machine Learning Based Methods

Platforms based on machine learning are divided using two approaches: probabilistic and wrapper induction. The more common of the two approaches is wrapper induction, most likely because wrapper induction excels at extraction of text from structured documents, and is a more quickly solvable problem than extraction of entities from unstructured text.

Probabilistic methods use algorithms such as Hidden Markov Models (Rabiner, 1989) to perform information extraction. Probabilistic approaches are not yet widespread in semantic annotation work.

Wrapper induction is a way to automatically construct a procedure for extracting content from a particular resource, such as a Web page. A wrapper is a function from a page to the set of tuples it contains (Kushmerick, Weld, & Doorenbos, 1997). Wrappers are used when there exists a repeatable structure to extract information from. Many Web sites have pages generated using from a back-end database, and the pages generated follow a common template. Wrappers reverse the page generation process to retrieve the original database tuples. Wrappers can be hand-crafted, or they can be learned. Manual wrappers require the user to mark areas of interest within a document. The machine can then extract entities from documents with a similar structural format as the manually marked-up document (Vargas-Vera et al., 2002). Kushmerick (1997) defined a method for performing wrapper induction, where the wrappers are automatically learned from example query responses from a data source. Wrappers are most effective when the data is presented in a structured format, such as product catalogs (Dingli et al., 2003).

Wrappers can also be linguistic-based, where the wrapper induction process discovers linguistic rules for identifying entities (Vargas-Vera et al., 2002). Amilcare (University of Sheffield, 2002) implements the LP^2 algorithm (Ciravegna, 2001), which performs rule induction using both linguistic and structural information.

Copyright © 2006, Idea Group Inc. Copying or distributing in print or electronic forms without written permission of Idea Group Inc. is prohibited.

Platform Overviews

This section provides an overview of each of the semantic annotation platforms surveyed. For further details of each system, readers are recommended to refer to our previous paper (Reeve & Han, 2005). The performance as reported by each platform author is then discussed and compared.

AeroDAML

AeroDAML (Kogut & Holmes, 2001) is designed to map proper nouns and common relationships to corresponding classes and properties in the DARPA Agent Markup Language (DAML) (DARPA, 2004) ontologies. AeroDAML uses AeroText (Lockheed Martin, 2005) for its information extraction component. The default ontology used by AeroDAML consists of two parts. The upper level ontology uses the WordNet noun synset hierarchy, while the lower level ontology uses the knowledge base provided by AeroText (Kogut & Holmes, 2001).

AeroText consists of four main components: (1) a Knowledge Base (KB) compiler for transforming linguistic data into a run-time knowledge base; (2) a Knowledge Base Engine for applying the KB to source documents; (3) a development environment for building and testing KBs; and (4) a Common Knowledge Base containing domain independent rules for extracting proper nouns and relations.

Armadillo

Armadillo (Dingli et al., 2003) is used to mine home pages of computer science faculty to find personal contact information, such as name, position, home page, and e-mail address. Armadillo uses a pattern-based approach to find entities, but is unique in that it finds its own initial set of seed patterns, rather than requiring an initial set of seeds, as described in Brin (1998). Once the seeds are found in the corpus, pattern expansion is then used to discover additional entities. The seed discovery and expansion finds faculty names in Web pages. Since many names may be discovered, Web services, such as Google and CiteSeer, are queried to provide evidence a person actually works in the computer science department. The names are then used to discover home pages, where detailed information about a person can often be found and extracted. The idea of information redundancy is exploited to verify entities that have been extracted.

Copyright © 2006, Idea Group Inc. Copying or distributing in print or electronic forms without written permission of Idea Group Inc. is prohibited.

KIM

The Knowledge and Information Management (KIM) platform (Popov et al., 2003) contains an ontology, knowledge base, a semantic annotation, indexing and retrieval server, as well as front-ends for interfacing with the server. For ontology and knowledge base storage it uses the SESAME RDF repository (openRDF.org, 2005), and for search it uses a modified version of the Lucene (Cutting, 2004) search engine. The semantic annotation process relies on a prebuilt lightweight ontology called KIMO as well as an interdomain knowledge base. KIMO defines a base set of entity classes, relationships, and attribute restrictions. The knowledge base is populated with 80,000 entities consisting of locations and organizations, gathered from a general news corpus. Named-entities found during the annotation process are matched to their type in the ontology and also to a reference in the knowledge base. The dual mapping allows the information extraction process to be improved by providing disambiguation clues based on attributes and relations.

The information extraction component of semantic annotation is performed using components of the GATE (Cunningham, Maynard, Bontcheva, & Tablan, 2002) toolkit. GATE provides IE implementations of tokenizers, part-of-speech taggers, gazetters, pattern-matching grammars (JAPE), and coreference resolution (Popov et al., 2003). Some components of GATE have been modified to support the KIM server. The gazetteer, for example, performs entity alias lookups using the knowledge base rather than an external file.

MnM

MnM (Vargas-Vera et al., 2002) provides an environment to manually annotate a training corpus, and then feed the corpus into a wrapper induction system based on Amilcare (University of Sheffield, 2002). Once the platform has been trained and rules have been induced from a training corpus, the system annotates text documents based on user defined semantic tags. Each document is annotated, presented to the user for approval, and sent to the ontology server to populate the ontology with instance data. Populating the ontology with instance data is done by taking the data from the semantic annotations and populating each ontology entry with its attribute values. The population phase attempts to provide values for as many attributes as possible. If the semantic annotation process does not provide all attribute values, the user will need to manually provide attribute values. This can occur, for example, if the attribute value is not mentioned in the text, or if the annotation rule set is incomplete and needs more training.

Copyright © 2006, Idea Group Inc. Copying or distributing in print or electronic forms without written permission of Idea Group Inc. is prohibited.

MUSE

MUSE (Maynard, 2003) uses an adaptive rule-based approach to perform annotation. Text attributes are used to conditionally run various processing resources, such as different gazetteers, over a document. Semantic tagging is accomplished using the Java Annotations Pattern Engine, also known as JAPE (Cunningham, Maynard, & Tablan, 2000). MUSE has been used to perform annotation in languages other than English to demonstrate its adaptability and compare its performance with machine learning systems, which typically require large training data. Since MUSE is modular, only the language-dependent parts needed to be converted between languages. For example, the part-of-speech tagger, gazetteer lists, and potentially language-dependent parts of the semantic tagger may need to be modified. The multiple language adaptation projects demonstrated that MUSE provides an advantage over machine learning based approaches because it requires a smaller amount of training data.

Ont-O-MAT Using Amilcare

Ont-O-Mat is an implementation of the S-CREAM (Semi-automatic CREAtion of Metadata) semantic annotation framework (Handschuh, Staab, & Ciravegna, 2002). The information extraction component is based on Amilcare. Amilcare is machine learning based and requires a training corpus of manually annotated documents. Amilcare uses the ANNIE (A Nearly-New IE system) part of the GATE toolkit to perform information extraction tasks such as tokenization, part-of-speech tagging, gazetteer lookup, and named entity recognition (Handschuh et al., 2002). The result of ANNIE processing is passed to Amilcare, which then induces rules for information extraction using a variant of the LP^2 algorithm. The wrapper induction process uses linguistic information, and is the same Amilcare wrapper induction process that MnM (Vargas-Vera et al., 2002) uses, generating tagging and correction rules.

Ont-O-MAT Using PANKOW

Ont-O-Mat provides an extensible architecture that allows replacement of selected components. The original Ont-O-Mat implementation was done using Amilcare. In this case, Ont-O-Mat replaces the annotation portion with a implementation of the PANKOW (Pattern-based Annotation through Knowledge On the Web) algorithm (Cimiano et al., 2004). The PANKOW process takes proper nouns from the information extraction phase and generates

Copyright © 2006, Idea Group Inc. Copying or distributing in print or electronic forms without written permission of Idea Group Inc. is prohibited.

hypothesis phrases based on linguistic patterns and the specified ontology. For example, a sports ontology may generate hypothesis phrases from the proper noun "Pete Rose" using patterns such as "Pete Rose *is a* Player" and "Pete Rose *is a* Team," where "Player" and "Team" are ontology concepts. The hypothesis phrases are then presented to the Google Web service. The phrase with the highest query result count is then used to annotate the text with the appropriate concept. The core principle is called "disambiguation by maximal evidence" (Cimiano et al., 2004). This principle is similar to the approach used by Armadillo (Dingli et al., 2003), which uses multiple Web services to find maximal evidence.

SemTag

SemTag is the semantic annotation component of Seeker, a comprehensive platform for performing large-scale annotation of Web pages (Dill et al., 2003). SemTag has been applied as a non-domain specific semantic annotation tool. It has annotated 264 million Web pages, generating 434 million automatically disambiguated semantic annotations. The taxonomy used by SemTag is TAP. TAP is shallow and covers a range of lexical and taxonomic information about popular items such as music, movies, authors, sports, health, and so forth. The annotations generated by SemTag are stored separate from the source document. It is assumed that the source document is read-only, since the annotator in this case is not the original author. The intent of the SemTag/Seeker design is to provide a public repository with an API that will allow agents to retrieve the Web page from its source and then request the annotations separately from a Semantic Label Bureau. SemTag performs annotation by finding terms in the text corresponding to entries in the knowledge base, and then using a novel algorithm called Taxonomy-based Disambiguation to determine a term's position in the taxonomy.

Platform Performance

Table 1 shows the author-reported performance of various platforms, with the exception of AeroDAML, Ont-O-Mat using Amilcare, and SemTag, whose authors provided incomplete performance information. The systems were evaluated by the platform authors, using different corpora in sometimes different domains, so direct comparisons are not possible, but the results should give some idea of the performance of each system. The standard measures of Precision and Recall, taken from the information retrieval field, were used by the remaining SAP authors in determining annotation effectiveness. In the general definition of recall and precision shown below, "accurate" and "inaccurate" refer to annota-

Copyright © 2006, Idea Group Inc. Copying or distributing in print or electronic forms without written permission of Idea Group Inc. is prohibited.

Table 1. Information extraction methods and performance measurements of semantic annotation platforms, as reported by the platform authors

Platform	IE Method	Precision	Recall	F-Measure
AeroDAML	Rule	Not reported	Not reported	Not reported
Armadillo	Pattern Discovery	91	74	87
KIM	Manual Rules	86	82	84
MnM	Wrapper Induction	95	90	Not reported
MUSE	Manual Rules	93	92	93
Ont-O-Mat using Amilcare	Wrapper Induction	Not reported	Not reported	Not reported
Ont-O-Mat using PANKOW	Pattern Discovery	65	28	25
SemTag	Semi-automatic Rules	82	Not reported	Not reported

tions generated semi-automatically by a SAP, while "all" refers to all annotations generated by a human annotator.

$$Annotation\ \mathrm{Re}call = \frac{accurate}{all}$$

$$Annotation\ \mathrm{Pr}ecision = \frac{accurate}{accurate + inaccurate}$$

The highest performing machine learning based platform is MnM, which uses Amilcare to take advantage of both structural and linguistic information within a document. For pattern-based platforms, MUSE performs best, most likely because it is adaptive to different text types and its rules have been hand-tuned for the domain it is annotating. The worst performing is Ont-O-Mat using PANKOW, which is a recent effort to use unsupervised learning with linguistic patterns.

Copyright © 2006, Idea Group Inc. Copying or distributing in print or electronic forms without written permission of Idea Group Inc. is prohibited.

Platform Characteristics

Figure 1 shows the various layers of abstraction in the design of a Semantic Annotation Platform (SAP). At each layer, the implementer must make decisions which impact the performance and effectiveness of the level based on a set of design goals. These goals are considerations platforms must take into account. As shown in Table 2 and briefly discussed in the *Platform Overviews* section, several SAPs have been developed, and each SAP was designed to address a slightly different annotation need. Table 2 shows some key characteristics for each platform. These characteristics were identified by examining the literature for each platform and noting what characteristics helped distinguish each SAP. We consider the characteristics shown in Table 2 to be a minimum of elements to consider when gaining a sense of a SAP's strengths and weaknesses. In this section, the primary SAP characteristics and their application to the platforms are discussed.

Table 2. Characteristics of semantic annotation platforms

Platform	Doc. Type	IE Method	MR	External Input	X	IE Tools	Initial Ontology
AeroDAML	HTML	Rule	Y	Rule	N	AeroText	WordNet; AeroText KB
Armadillo	HTML	PD	Y	Seed	N	Amilcare ANNIE	Address Book; Paper Citation
KIM	HTML	PM	Y	Gazetteer KB population	N	GATE	KIMO
MnM	HTML, Plain Text	WI using ML-LP^2	N	Annotated corpus	Y	Amilcare	KMi
MUSE	Plain Text	Rule	Y	Gazetteer Rules	Y	GATE JAPE	User constructed
Ont-O-Mat: Amilcare	HTML	WI using ML-LP^2	N	Annotated corpus	Y	Amilcare	User constructed
Ont-O-Mat: PANKOW	HTML	PD	N	Web pages	Y	PANKOW	User constructed
SemTag	HTML	TLM	N	Taxonomy with labels	Y	Seeker platform	TAP with 72K labels

(Doc: Document; IE: Information Extraction; JAPE: Java Annotations Pattern Engine; KB: Knowledge Base; ML: Machine Learning; MR: Manual Rules; N: No; PD: Pattern Discovery; PM: Pattern Matching; TLM: Taxonomy Label Matching; WI: Wrapper Induction; X: Extensible; Y: Yes)

Copyright © 2006, Idea Group Inc. Copying or distributing in print or electronic forms without written permission of Idea Group Inc. is prohibited.

Document Type

The document type shows the type of input typically presented to the platform. Some platforms support many document types, while others only a single or several. Most efforts at Semantic Web annotation are focused primarily on HTML because of its ubiquity, although other document formats such as XHTML, SGML, XML, RTF and others are also supported. HTML presents a challenging environment for entity extraction because HTML markup is designed for presentation and not data manipulation. Unless documents are annotated by customized XML tags with ontological concepts, it is hard to interpret semantics on the documents. While machines can effectively render the format, they cannot easily provide semantic interpretation of it. Applying information extraction methods to the goal of identifying entities and relationships in Web documents requires that information extraction systems be adapted to this environment. Document structure is classified as unstructured, semistructured or structured. Unstructured documents consist of natural language text without any intended structure, such as magazine and journal articles. An information extraction method using natural language processing techniques is needed to parse such documents. Structured and semistructured documents are usually generated documents from back-end content management systems using templates (Mukherjee, Yang, & Ramakrishnan, 2003). Wrapper induction techniques can be used to deduce the location of entities within a document effectively if the structure is well-defined (Kushmerick et al., 1997). All surveyed platforms except MUSE work with HTML documents. MUSE works with plain text documents, as does MnM.

Information Extraction Method

The information extraction method used is a key characteristic in the performance of a SAP, as it is the component which identifies entities in text. Most platforms use a pattern-based approach. Armadillo uses pattern discovery with an initial seed set that is expanded by analyzing HTML structures, such as lists and tables, to find entities. Onto-O-Mat using PANKOW uses linguistic patterns to find maximal evidence to confirm the type of an entity. KIM uses natural language processing to locate entities and then match them against entries in a knowledge base. AeroDAML also uses natural language processing to locate entities within text. SemTag uses a simple approach of tagging entities in the text which match entries in the knowledge base, and in a later step disambiguating the entity tags which may fall into multiple places in the ontology. MUSE uses an adaptive approach, where rules are applied during processing based on attributes found in the text, such as language, domain, and so forth. Both MnM and Onto-

Copyright © 2006, Idea Group Inc. Copying or distributing in print or electronic forms without written permission of Idea Group Inc. is prohibited.

O-Mat use Amilcare to perform entity identification. Amilcare uses a machine learning, rule induction approach. Table 1 shows MnM achieves very good performance among machine learning approaches, while MUSE achieves good performance among pattern-based approaches.

External Input

All systems require some type of manual effort in order to begin the process of semantic annotation. Rules, gazetteers and knowledge bases are common methods of implementing semantic components. Rules can be specified using a grammar such as JAPE (Cunningham et al., 2000), or can be a small set of initial seed patterns to facilitate pattern discovery (Brin, 1998). Gazetteers are lookup lists that map specific literal strings into a semantic concept. The string mappings are predefined by a user for a specific domain. Knowledge bases contain additional entity information than can be stored in the ontology alone. For example, the City of New York entity can have several literal expressions, such as [NYC], [N.Y.C], and [New York] (Kiryakov et al., 2003). Alternatively, knowledge bases can store a large number of additional entity details. For example, in the KIM platform which targets news article annotation, the knowledge base has been prepopulated with about 80,000 entities (Popov et al., 2003).

There exist several problems common to manual external input, such as manual rules, gazetteers, and knowledge bases. First, manual effort is still required to develop the external input to semantic components, although this effort can be spread over many thousands of documents, which is still substantially less than annotating each document manually. Second, the rules and gazetteers must be changed to accommodate different domains and languages (Maynard, 2003). To overcome some of these problems, semantic annotation platforms often incorporate approaches to alleviate this manual bottleneck by automating the construction of the external input. For example, the KIM system incorporates a mechanism where the knowledge base is continually expanded with new instance data, which is verified against a manually annotated corpus and a manually constructed smaller knowledge base (Popov et al., 2003). To alleviate manual construction of rules, several systems use rule induction in order to automatically learn rules to process document text. For example, both the Ont-O-Mat (Handschuh et al., 2002) and MnM (Vargas-Vera et al., 2002) platforms use the LP^2 algorithm from the natural language processing community to learn rules. The algorithm is a wrapper induction algorithm that uses linguistic data as well as structural data, and is based on facilities within the Amilcare toolkit (Vargas-Vera et al., 2002). There has also been work done in the information extraction community using wrapper induction for structural data in order to

Copyright © 2006, Idea Group Inc. Copying or distributing in print or electronic forms without written permission of Idea Group Inc. is prohibited.

extract entities from structured or semistructured sources (Kushmerick et al., 1997). A drawback with the wrapper induction approaches, as with any machine learning approach, is that enough training samples must be provided to achieve the desired accuracy. Interestingly, the developers of the MUSE system report that they can move their rule-based system to different languages more quickly than machine learning based systems (Maynard, 2003). The MUSE system requires that language-specific components, such as gazetteers, be changed to support new languages, but in contrast to machine learning based approaches, does not require a substantial amount of training data with new languages to achieve high accuracy. The existing semantic rules can be re-used, greatly easing the support of new languages.

Extensibility

Extensible platforms allow for various components to be exchanged. For example, the middle layer of a SAP in Figure 1 contains the critical components for semantic annotation performance. In an extensible SAP, a rule induction for information extraction (IE) component can be substituted with a statistical one. The benefits of the rest of the platform components continued to be used while newer IE components are evaluated and integrated. Another advantage is allowing the SAP to adapt to different domains where the information extraction component may perform differently based on the domain document input, as demonstrated by the MUSE system (Maynard, 2003).

Information Extraction Tools

There are number of information extraction (IE) technique implementations available as a result of work done in the information extraction and natural language processing communities. Semantic annotation platforms usually take advantage of the work that in some cases has a long history. The most common toolkits are GATE (University of Sheffield, 2004) and Amilcare (University of Sheffield, 2002). GATE is produced by the University of Sheffield's Natural Language Processing Group and is a language process environment for building human language processing systems. GATE is divided into three parts: (1) an architecture which defines a goal component for componentizing a language processing system; (2) a concrete framework implementation; and (3) a graphical development environment. GATE includes an information extraction system called ANNIE (A Nearly-New IE system) that is composed of a tokenizer, gazetteer, sentence splitter, part-of-speech tagger, semantic tagger, coreferencing (OrthoMatcher), and other related IE components.

Copyright © 2006, Idea Group Inc. Copying or distributing in print or electronic forms without written permission of Idea Group Inc. is prohibited.

Amilcare is produced by the University of Sheffield's Computer Science Department and is specifically designed to perform semantic annotation (University of Sheffield, 2002). Amilcare is a rule-based system, where the rules are induced with a learning algorithm run against a training corpus pre-annotated with XML tags. The transformation-based rule algorithm is called LP^2 which induces two types of rules: (1) initial rules for annotating text, and (2) rules that correct mistakes generated by the first type of rules. Interestingly, GATE is used within Amilcare for tokenization, sentence identification, part of speech tagging, gazetteer lookup and named entity recognition.

Initial Ontology

Semantic annotation requires the use of an ontology in order to perform concept instance mapping. Ontologies are usually architected using levels, such as upper and lower. The upper ontology consists of general concepts, while the lower ontology has a deeper specialization of the upper ontology concepts (Missikoff, Navigli, & Velardi, 2002). Some semantic annotation platforms place the responsibility on the user for constructing an initial ontology. Examples include MUSE (Maynard, 2003) and Ont-O-Mat (Handschuh et al., 2002). Other platforms provide an initial ontology as part of their development. The KIM platform provides an ontology called KIMO that is designed to provide a minimal open-domain ontology, and is based on OpenCyc (OpenCyc.org, 2005), WordNet (Princeton University, Cognitive Science Laboratory, 2005), DOLCE (Italian National Resource Council - Institute of Cognitive Science and Technology, 2005) and other upper-level resources (Popov et al., 2003). KIMO is composed of approximately 250 classes, 100 attributes and relations, and the specialization of classes is derived from an analysis of a corpus of general news (Popov et al., 2003). The Seeker component of SemTag uses TAP, which is a shallow knowledge base that contains information about a broad range of popular culture subjects, such as movies, sports, and so forth (Dill et al., 2003). The TAP knowledge base has about 72,000 labels that are used to tag instances found in documents. The MnM platform uses a hand-crafted ontology called KMi (Knowledge Management Institute) (Vargas-Vera et al., 2002). The AeroDAML platform uses the commercial product Aerotext, and utilizes an upper-level ontology based on Wordnet, while the lower-level ontology uses the common knowledge base of AeroText (Kogut & Holmes, 2001). Armadillo provides an example of a platform where the initial ontology is very lightweight, consisting of an address-book type of ontology where members of a computer science department are discovered and populate address information, such as name, phone number, address, and so forth (Dingli et al., 2003).

Copyright © 2006, Idea Group Inc. Copying or distributing in print or electronic forms without written permission of Idea Group Inc. is prohibited.

Table 2 shows some key characteristics for each platform. In this section, the key characteristics among SAPs were compared and investigated to show a sense of each SAP's strengths and weaknesses. With semantic annotation platforms shown in Table 2, we can see the advantages of the SAP architecture in Figure 1 to integrate new external sources or new information extraction tools and how the existing SAPs utilized such an architecture. For example, the Ont-O-Mat system (Handschuh et al., 2002) was originally designed using Amilcare to learn rules about linguistic patterns in order to identify entities. Later the Ont-O-Mat platform was extended using the PANKOW (Pattern-based Annotation through Knowledge on the Web) algorithm (Cimiano et al., 2004). SAPs designed with extensible architectures can adapt to evolving technology. Information extraction components can be replaced as different approaches are developed. The most common toolkits used for entity identification are GATE (University of Sheffield, 2004) and Amilcare (University of Sheffield, 2002).

Future Trends

Semantic annotation platforms for the Web have only recently been developed, and they are not complete in their accuracy and elimination of manual effort. The precision and recall still vary widely depending on the platform used, information extraction (IE) methods, and data source type (unstructured, semistructured, or structured), as shown in Table 2. There still exists an opportunity to improve the performance of SAPs and reduce their required manual effort. A future trend in semantic annotation platforms, then, is the continued integration of technologies originally developed in the field of information extraction.

With the number of available semantic annotation platforms currently available, as shown in Table 2, it is possible to extend existing SAPs with newer annotation implementations that may lead to improved annotation accuracy beyond what current platforms are producing. As an example, the integration of other methods into an existing SAP has been done. The Ont-O-Mat system (Handschuh et al., 2002) was originally designed using Amilcare to learn rules about linguistic patterns in order to identify entities. Further work done by separate researchers developed a method called PANKOW (Pattern-based Annotation through Knowledge on the Web) and integrated it into the Ont-O-Mat platform (Cimiano et al., 2004). The advantage of the integration is that the work could focus on the method rather than on constructing supporting services provided by a SAP. In particular, PANKOW utilizes the ontology and document management facilities provided by Ont-O-Mat (Cimiano et al., 2004).

Copyright © 2006, Idea Group Inc. Copying or distributing in print or electronic forms without written permission of Idea Group Inc. is prohibited.

Conclusion

The Semantic Web requires the widespread availability of document annotations in order to be realized. Benefits of adding meaning to the Web include: query processing using concept-searching rather than keyword-searching (Berners-Lee et al., 2001); custom Web page generation for the visually-impaired (Yesilada et al., 2004); using information in different contexts, depending on the needs and viewpoint of the user (Dill et al., 2003); and question-answering (Kogut & Holmes, 2001). Annotations are currently problematic for several reasons. Manual annotation does not scale to the volume of documents on the Web, and suffers from problems such as annotator motivation and domain knowledge (Bayerl et al., 2003). There are additional problems with manual annotation, such as changing ontologies, and having a document annotated using multiple ontologies, providing multiple perspectives (Dill et al., 2003).

Semantic Annotation Platforms (SAPs) were developed to provide a level of automation to the semantic labeling process, and overcome the limitations of manual annotation. Several semantic annotation platforms currently exist, distinguished primarily by their information extraction method, as that component has the largest impact on the effectiveness of semantic annotation. The two primary approaches are pattern-based and machine learning based. Machine learning algorithms often perform more effectively than pattern-based methods, but the MUSE system shows that a rule-based system using conditional processing can perform as well as a machine learning system (Maynard, 2003). SAPs designed with extensible architectures can adapt to evolving technology. Information extraction components can be replaced as different approaches are developed. The most common toolkits used within SAPs are GATE (University of Sheffield, 2004) and Amilcare (University of Sheffield, 2002).

Future work on semantic annotation platforms for text-based Web documents will likely focus on integrating other entity identification approaches from the information extraction and natural language processing communities. In addition, ways to bootstrap the semantic annotation process and reduce the amount of manual effort will also evolve. This includes ontology learning and knowledge base population. Extensible platforms will enable the rapid development of new advances by allowing individual components of a semantic annotation platform to be replaced or extended. The continuing evolution of semantic annotation platforms to provide better annotation and new features while extending existing ones is vital to the realization of the Semantic Web.

Copyright © 2006, Idea Group Inc. Copying or distributing in print or electronic forms without written permission of Idea Group Inc. is prohibited.

References

Bayerl, P. S., Lüngen, H., Gut, U., & Paul, K. I. (2003). Methodology for reliable schema development and evaluation of manual annotations. *Proceedings of the Workshop on Knowledge Markup and Semantic Annotation at the Second International Conference on Knowledge Capture (K-CAP 2003),* Florida.

Berners-Lee, T. (1998). *Semantic Web road map.* Retrieved April 24, 2005, from http://www.w3.org/DesignIssues/Semantic.html

Berners-Lee, T., Hendler, J., & Lassila, O. (2001). The Semantic Web. *Scientific American, 284*(5), 34-43.

Brin, S. (1998). Extracting patterns and relations from the World Wide Web. *Proceedings of the WebDB Workshop at 6th International Conference on Extending Database Technology,* Valencia, Spain (Vol. 1590, pp. 172-183).

Chandrasekaran, B., Josephson, J. R., & Benjamins, V. R. (1999). What are ontologies and why do we need them? *IEEE Intelligent Systems, 14*(1), 20-26.

Cimiano, P., Handschuh, S., & Staab, S. (2004). Towards the self-annotating Web. *The 13th International Conference on World Wide Web,* New York (pp. 462-471).

Ciravegna, F. (2001). Adaptive information extraction from text by rule induction and generalisation. *Proceedings of 17th International Joint Conference on Artificial Intelligence (IJCAI 2001),* Seattle, WA (pp. 1251-1256).

Cunningham, H., Maynard, D., Bontcheva, K., & Tablan, V. (2002). *GATE: A framework and graphical development environment for robust NLP tools and applications.* The 40th Anniversary Meeting of the Association for Computational Linguistics (ACL'02).

Cunningham, H., Maynard, D., & Tablan, V. (2000). *JAPE: A Java annotation patterns engine* (2nd ed.). Technical report CS—00—10, University of Sheffield, Department of Computer Science. Retrieved April 24, 2005, from ftp://ftp.dcs.shef.ac.uk/home/hamish/auto_papers/Cun00e.ps.gz

Cutting, D. (2004). *Apache Jakarta Lucene.* Retrieved April 24, 2005, from http://lucene.apache.org/java/docs/index.html

DARPA (2005). *DARPA agent markup language.* Retrieved June 6, 2005, from http://www.daml.org

Dill, S., Eiron, N., Gibson, D., Gruhl, D., Guha, R., Jhingran, A., et al. (2003). *SemTag and seeker: Bootstrapping the Semantic Web via automated*

Copyright © 2006, Idea Group Inc. Copying or distributing in print or electronic forms without written permission of Idea Group Inc. is prohibited.

semantic annotation. The 12th International World Wide Web Conference, Budapest, Hungary (pp. 178-186).

Dingli, A., Ciravegna, F., & Wilks, Y. (2003). Automatic semantic annotation using unsupervised information extraction and integration. *Proceedings of the Workshop on Knowledge Markup and Semantic Annotation at the Second International Conference on Knowledge Capture (K-CAP 2003),* Florida.

Handschuh, S., Staab, S., & Ciravegna, F. (2002). S-CREAM: Semi-automatic CREAtion of metadata. *Proceedings of the 13th International Conference on Knowledge Engineering and Knowledge Management (EKAW02).*

Hearst, M. (1992). Automatic acquisition of hyponyms from large text corpora. *Proceedings of the 14th International Conference on Computational Linguistics,* Nantes, France (Vol. 2, pp. 539-545).

Italian National Resource Council - Institute of Cognitive Science and Technology. (2005). *DOLCE: A descriptive ontology for linguistic and cognitive engineering.* Retrieved April 24, 2005, from http://www.loa-cnr.it/DOLCE.html

Kiryakov, A., Popov, B., Ognyanoff, D., Manov, D., Korilov, A., & Goranov, M. (2003). Semantic annotation, indexing, and retrieval. *ISWC 2003, 2nd International Semantic Web Conference,* Sanibel Island, FL (pp. 834-849).

Kogut, P., & Holmes, W. (2001). AeroDAML: Applying information extraction to generate DAML annotations from Web pages. *Proceedings of the Workshop on Knowledge Markup and Semantic Annotation at the First International Conference on Knowledge Capture (K-CAP 2001),* Victoria, BC.

Kosala, R., & Blockeel, H. (2000). Web mining research: A survey. *SIGKDD Explorations, 2*(1), 1-15.

Kushmerick, N. (1997). *Wrapper induction for information extraction.* Doctoral Dissertation, University of Washington.

Kushmerick, N., Weld, D. S., & Doorenbos, R. (1997). Wrapper induction for information extraction. *Proceedings of the 15th International Joint Conference on Artificial Intelligence (IJCAI '97),* Nagoya, Japan (pp. 729-737).

Lockheed Martin. (2005). AeroText. Retrieved June 6, 2005, from http://www.lockheedmartin.com/wms/findPage.do?dsp=fec&ci=11255&rsbci=0&fti=126&ti=0&sc=400

Maedche, A., & Staab, S. (2001). Ontology learning for the Semantic Web. *IEEE Intelligent Systems, 16*(2), 72-79.

Copyright © 2006, Idea Group Inc. Copying or distributing in print or electronic forms without written permission of Idea Group Inc. is prohibited.

Maynard, D. (2003). Multi-source and multilingual information extraction. *Expert Update.*

Missikoff, M., Navigli, R., & Velardi, P. (2002). *The usable ontology: An environment for building and assessing a domain ontology.* The 1st International Semantic Web Conference (ISWC2002) (pp. 39-53).

Mukherjee, S., Yang, G., & Ramakrishnan, I. V. (2003). Automatic annotation of content-rich HTML documents: Structural and semantic analysis. *Proceedings of the 2nd International Semantic Web Conference,* Sanibel Island, FL.

OpenCyc.org. (2005). *OpenCyc.* Retrieved April 24, 2005, from http://www.opencyc.org/

openRDF.org. (2005). Sesame RDF repository. Retrieved June 6, 2005, from http://www.openrdf.org/

Popov, B., Kiryakov, A., Kirilov, A., Manov, D., Ognyanoff, D., & Goranov, M. (2003). KIM: Semantic annotation platform. *The 2nd International Semantic Web Conference (ISWC2003)* (Vol. 2870, pp. 834-849).

Princeton University, Cognitive Science Laboratory. (2005). *WordNet: A lexical database for the English language.* Retrieved April 24, 2005, from http://wordnet.princeton.edu/

Rabiner, L. R. (1989). A tutorial on hidden Markov Models and selected applications in speech recognition. *Proceedings of the IEEE, 77*(2), 257-285.

Reeve, L., & Han, H. (2005). Survey of semantic annotation platforms. *Proceedings of the 20th Annual ACM Symposium on Applied Computing, Web Technologies and Applications track,* Santa Fe, NM.

Ondrej, S., Labsky, M., & Svatek, V. (2004). RDF-based retrieval of information extracted from Web product catalogues. *Proceedings of the 27th Annual ACM SIGIR Conference on Research and Development in Information Retrieval, Semantic Web Workshop,* Sheffield, UK.

Tallis, M. (2003). *Semantic word processing for content authors.* The 2nd International Conference on Knowledge Capture, Sanibel, Florida.

University of Sheffield. (2002). *Amilcare — Adaptive IE tool.* Retrieved December 28, 2004, from http://nlp.shef.ac.uk/amilcare/amilcare.html

University of Sheffield. (2004). *GATE — A general architecture for text engineering.* Retrieved December 28, 2004, from http://gate.ac.uk/

Vargas-Vera, M., Motta, E., Domingue, J., Lanzoni, M., Stutt, A., & Ciravegna, F. (2002). MnM: Ontology driven semi-automatic and automatic support for semantic markup. *The 13th International Conference on Knowledge Engineering and Management (EKAW 2002)* (pp. 379-391).

Copyright © 2006, Idea Group Inc. Copying or distributing in print or electronic forms without written permission of Idea Group Inc. is prohibited.

Yesilada, Y., Harper, S., Goble, C., & Stevens, R. (2004). Ontology based semantic annotation for visually impaired Web travellers. *Proceedings of the 4th International Conference on Web Engineering (ICWE 2004),* Munich, Germany (Vol. 3140, pp, 445-458).

Copyright © 2006, Idea Group Inc. Copying or distributing in print or electronic forms without written permission of Idea Group Inc. is prohibited.

Section III

Ontologies-Based Querying and Knowledge Discovery

Copyright © 2006, Idea Group Inc. Copying or distributing in print or electronic forms without written permission of Idea Group Inc. is prohibited.

Chapter VII

Ontology Enhancement for Including Newly Acquired Knowledge About Concept Descriptions and Answering Imprecise Queries

Lipika Dey, Indian Institute of Technology, Delhi, India

Muhammad Abulaish, Jamia Millia Islamia, India

Abstract

This chapter presents a text-mining-based ontology enhancement and query-processing system. The key ideas introduced here are that of learning and including imprecise concept descriptions into ontology structures. This is essential for ontology-based text information extraction since it is not necessary that text description of the concepts or user-specified descriptions will exactly match stored concept descriptions. The traditional property-value framework for concept description has been extended to a property-

Copyright © 2006, Idea Group Inc. Copying or distributing in print or electronic forms without written permission of Idea Group Inc. is prohibited.

value-qualifier framework for this purpose. The system also supports ontology enhancement by identifying, defining, and adding new precise and imprecise concept descriptions mined from text documents. The acquired knowledge is stored in a structured knowledge base for answering user queries. Since user queries may contain concept descriptions, which do not exactly match stored or known concepts, the query processor uses fuzzy reasoning for query processing. Each answer is accompanied by a confidence value that reflects its similarity to the original query concept.

Introduction

Ontological structures used for describing concepts and their interrelationships are gaining popularity for designing efficient Web-based information processing systems. The Semantic Web visualized by Berners-Lee (2001) aims at integrating knowledge from heterogeneous Web resources and ontology can be a key enabling technology to realize it. Ontologies have found their use in a wide range of applications like search engines, e-commerce, biomedical knowledge processing, knowledge engineering, information retrieval and extraction from unstructured texts, natural language processing, multi-agent systems, qualitative modeling of physical systems, database design, geographic information science and digital libraries (Baader et al., 2003; Bodner & Song, 1996).

An ontology can be viewed as a model of a domain that defines the concepts existing in that domain, their attributes and the relationships between them and is typically represented as a knowledge base. For example, plant ontology specifies structural organization of plants in terms of parts and subparts like stems, leaves, cells, etc.; the various categories of plants like algae, legumes, ferns, and so on. Thus an ontology can be viewed as an explicit formal specification of how to represent the objects, concepts, and other entities that are assumed to exist in some area of interest and the relationships that hold among the specified concepts.

Even as the use of ontology for domain-specific applications are fast gaining popularity, researchers are actively engaged in tackling some of the chief bottlenecks that still hinders the use of ontology for general-purpose applications. Though ontology plays a key role by defining concepts and relationships in an unambiguous way, it is unreal to expect that there exists a unique, unambiguous way of defining every concept which all authors and users will adhere to. Hence, for all large, complex applications it is imperative that facilities for learning and updating of underlying ontological structures are integrated into the system.

Copyright © 2006, Idea Group Inc. Copying or distributing in print or electronic forms without written permission of Idea Group Inc. is prohibited.

The use of ontologies has been mostly restricted to building Web Applications, which adhere to the conceptualization specified in the ontology. It is still not possible to build information retrieval systems like a general-purpose question-answering system, which can extract exact information from free-form Web documents about a concept present in the user query. The reason being that concept descriptions need not be exhaustive and may be even imprecise at times. We say that a concept definition is not exhaustive since new or modified definitions of concepts may emerge. Hence, any agent or system designed for reasoning with these concepts should be able to adapt to changes. In an ontological setup where one accepts a basic structure to describe a concept, this may mean addition of new properties or values or adding new interconceptual relations. The set of properties and values used to describe a concept should be adapted to accommodate new descriptors that an agent comes across during the course of its use. For example, though it may be a well-established truth that wines of other colors exist, the wine ontology[1] uses only three values to describe color of wine. However, this should not be a severe deterrent in utilizing this ontology for document processing tasks, provided it is possible to enhance the basic structure with new values. New values for colors may be learned through knowledge discovery process as new information is encountered while mining texts. Again, assumption that a concept description is always precise may not be correct. At times, a user may be unclear about his or her preferences and so a fuzzy concept description comes as a good and natural way to represent them. Finally, it is also possible that user-given concept descriptions, ontology-stored descriptions and the concept descriptions appearing in the documents do not match exactly, rather have varying degrees of overlap. In that case, an uncertainty based reasoning scheme is needed to integrate the information collected from the Web documents and the ontology specifications to determine the relevance of the document to the user query. These notions will be further clarified with examples later in the chapter, where we will highlight some of the problems encountered during information extraction from text documents and thereby build up on our proposed methodology to handle them.

Thus we see that the major problem in designing ontology-based applications stems from the bottleneck of knowledge acquisition and the time-consuming activity of building exhaustive knowledge repositories for all perceivable domains. Besides, it is also not possible to converge to unambiguous descriptions of all concepts. Hence ontology construction has to be a dynamic process. In this chapter, we will be presenting an overview of some of the key activities going on in the area of ontology learning and knowledge-integration from multiple sources. We will also present our work on enriching property descriptors of ontological concepts with new descriptors mined from Web documents. The expanded ontology structure supports definition of *imprecise concepts* through

Copyright © 2006, Idea Group Inc. Copying or distributing in print or electronic forms without written permission of Idea Group Inc. is prohibited.

the use of embedded qualifiers. To the best of our knowledge, though fuzzification of relations and properties have been considered earlier in several studies like that of Wallace and Avrithis (2004), Quan, Hui, and Cao (2004), and Widyantoro and Yen (2001), using qualifiers for property description of objects has not been explored earlier in ontology studies. Our system is also integrated with the capability to build a structured knowledge repository with the help of background ontology and new information extracted from documents. The structured knowledge base supports imprecise query processing through a fuzzy reasoner.

The remaining chapter is structured as follows. We shall first present formal definitions of ontology and mechanisms for ontology based reasoning. Thereafter we shall review related work on ontology based text information extraction and ontology enhancement. We then establish through examples why imprecise concept descriptions are required for ontology-based information extraction systems. This discussion is followed by the schematic representation of our system and functional details of the various modules. Results of knowledge acquisition are then presented along with a performance evaluation of the proposed system on two different domains. We round off with a discussion on the future trends of ontology based information extraction from texts. A brief summary of the proposed system is given again at the end.

Background:
Ontology Specification and Reasoning

The word ontology has been borrowed from philosophy, where it means a systematic explanation of being. The knowledge engineering community has adopted ontology as a key enabling technology since the nineties. One of the first definitions of ontology given by Neches et al. (1991), is as follows: "an ontology defines the basic terms and relations comprising the vocabulary of a topic area as well as the rules for combining terms and relations to define extensions to the vocabulary". According to the above definition, an ontology includes not only the terms that are explicitly defined in it, but also the knowledge that can be inferred from it. Gruber (1993) defined an ontology as "an explicit specification of a conceptualization", which has become one of the most acceptable definitions to the ontology community. Guarino et al. (1995) collected and analyzed seven definitions of ontology and provided their corresponding syntactic and semantic interpretations. They proposed to consider an ontology as "a logical theory which gives an explicit, partial account of a conceptualization", where conceptualization is basically an idea of the world that a person or a group of people can have.

Copyright © 2006, Idea Group Inc. Copying or distributing in print or electronic forms without written permission of Idea Group Inc. is prohibited.

Ontology provides an abstract view of a set of world objects. Ontology specifies the key concepts in a domain and their interrelationships to provide an abstract view of an application domain (Fensel et al., 2001). Usually a concept in ontology is defined in terms of its mandatory and optional properties along with the value restrictions on those properties. Along with concept descriptions, it provides a taxonomic classification of concepts in the world to be used as semantic primitives. An ontology is machine-interpretable and represented in one of the many languages like Ontology Inference Language (OIL) (Horrocks, 2002), Web Ontology Language[2] (OWL), eXtensive Markup Language[3] (XML), and more.

With the support of ontology, both user and system can communicate with each other by the shared and common understanding of a domain. Ontologies are built by knowledge engineers with inputs from domain experts. Since ontologies provide a framework for unambiguous representation of a set of domain concepts and their interrelations, they can help in intelligent Web-based information retrieval wherein it is possible to overcome the heterogeneity of Web resources through the use of a shared conceptualization.

The conceptual structuring of domain knowledge can be exploited to develop logical reasoning schemes for drawing inferences with domain knowledge. Description Logics is one such reasoning scheme that employs a structuring of domain knowledge similar to ontologies. Description Logics are subsets of first-order logic. They serve to disambiguate imprecise representations that these formalisms permit. Description Logics is the most expressive decidable logics with classical semantics, and "good" reasoning procedures suitable for reasoning over large bodies of knowledge. It ensures the latter by considering the trade-off between the expressivity of first order logic formalisms and the tractability of implementing such formalisms for practical and usable representation of, and reasoning over large bodies of domain knowledge like ontologies. For a complete treatise on Description Logics one can refer to Baader et al. (2003). Ontology representation languages like DAML+OIL and OWL are based on Description Logics (DLs) thus enabling the knowledge representation systems to provide reasoning support for Web Applications (Horrocks & Sattler, 2001).

Ontology Learning:
A Review of Related Work

In this section we will review some of the recent research efforts that have been directed towards the problems of automatic ontology enhancement and design of ontology-based text processing systems. As novel uses of the domain concepts

Copyright © 2006, Idea Group Inc. Copying or distributing in print or electronic forms without written permission of Idea Group Inc. is prohibited.

are encountered while gathering information, ontology learning helps in induction of new concepts and concept descriptors into the existing ontology. The use of ontological models to access and integrate large knowledge repositories in a principled way has an enormous potential to enrich and make accessible unprecedented amounts of knowledge for reasoning (Crow & Shadbolt, 2001).

Luke et al. (1997) have proposed SHOE, a set of Simple HTML Ontology Extensions, which allow World Wide Web authors to annotate their pages with semantic knowledge. The annotations are expressed in terms of ontological knowledge, which can be generated by using or extending standard ontologies. The annotator has to choose an appropriate ontology for tagging a Web page correctly. SHOE provides the ability to define and enhance ontologies using HTML, which lay out classifications and entity-relationship rules, declare entity attributes and relationships among entities. In order to annotate a Web page, the annotator must select an appropriate ontology and then use that ontology's vocabulary to tag the document concepts appropriately. A Web crawler performs a graph traversal to discover related Web pages. The performance depends heavily on the quality of annotation.

Velardi, Fabriani, and Missikof (2001) suggest a scheme for enhancing existing ontological structures with new information extracted from texts. Their work is based on identification of three primary kinds of concepts: *actor,* which defines a relevant entity of the domain and is able to activate or perform process; *object,* which is a passive entity on which a process operates; *process,* which is an activity aimed at the satisfaction of an actor's goal. Secondary concepts include *information components,* which are clusters of information pertaining to the information structure of an actor or an object, *information elements* which are atomic information elements that are parts of an information component and *elementary action* which denote activities that constitute process components and are not further decomposable. Based on the above definition of primary and secondary concepts, candidate terminological expressions are usually captured with more or less shallow techniques, ranging from stochastic methods to more sophisticated syntactic approaches. The proposed system generates concept forest which is then manually integrated into the hand-crafted upper level ontology. This paper does not suggest any formal reasoning mechanism to match query concepts and document concept descriptions.

Hahn and Marko (2002) introduce a dual-use methodology for learning both grammar and ontologies. Their system automates the maintenance and growth of two types of knowledge sources, which are crucial for natural language text understanding: background knowledge of the underlying domain and linguistic knowledge about the lexicon, and the grammar of the underlying natural language. Learning occurs simultaneously with the ongoing text understanding process. The knowledge assimilation process is centred around the linguistic and

Copyright © 2006, Idea Group Inc. Copying or distributing in print or electronic forms without written permission of Idea Group Inc. is prohibited.

conceptual *quality* of various forms of evidence underlying the generation, assessment, and on-going refinement of lexical and concept hypotheses. On the basis of the strength of evidence, hypotheses are ranked according to qualitative plausibility criteria, and the most reasonable ones are selected for assimilation into the already given lexical class hierarchy and domain ontology.

Liddle, Hewett, and Embley (2003) have proposed an ontology-based data extraction system which uses an application ontology that describes a data-rich, ontologically narrow domain in a conceptual fashion. From the underlying application the system automatically generates a single wrapper that can be applied to any page relevant to the application domain. They have also developed a Java-based tool that helps domain experts by providing a graphical interface to create domain ontology, which is then used to extract data from Web documents and to store them in structured form. The main disadvantage of the system is the requirement of domain knowledge facilitator who can provide the knowledge for creating the appropriate application ontology in an appropriate format.

Shamsfard and Barforoush (2004) have suggested an ontology building approach in which the system starts from a small ontology kernel and constructs the ontology through text understanding automatically. The kernel contains the primitive concepts, relations and operators to build an ontology. This model uses dynamic categories to handle changes and floating categories to handle multiple viewpoints. Presently, the system has been implemented to extract information from natural language texts comprised of simple Persian (Farsi) sentences.

Li and Zhong (2004) have described a methodology for ontology learning over an XML ontology scheme. Here the original ontology is in XML and a collection of data from users is represented as a list. The formal model proposed in this work comprises of a list of facts and the frequency of their occurrences, which are provided by the users. Each fact supplies an individual opinion that specifies which class in the ontology the fact belongs to. For inclusion of an information object into the ontology, a mass distribution of user profiles on the ontology is used to incorporate the object appropriately.

Ontologies have assumed a very prominent role in Bioinformatics since much of Biology works by applying prior knowledge to an unknown entity. Stevens et al. (2002) have described how a Bioinformatics ontology can be built using OIL. Transparent Access to Multiple Bioinformatics Information Sources (TAMBIS) is a mediation system that uses an ontology to enable biologists to ask questions over multiple external databases using a common query interface. The TAMBIS ontology (TaO) contains description of the principal concepts of molecular biology and bioinformatics: macromolecules; their motifs, their structure, function, cellular location, and the processes in which they act. For each portion of TaO the onotologist asserts a basic framework consisting of concepts and their

Copyright © 2006, Idea Group Inc. Copying or distributing in print or electronic forms without written permission of Idea Group Inc. is prohibited.

properties. Description Logic based inference mechanism is then employed to reclassify or infer further relationships. OIL can support a cyclical ontology development process where starting with a primitive taxonomy, concept descriptions can be further enriched.

Muller, Kenny, and Strenber (2004) have developed an ontology-based text-mining system TEXTPRESSO for scientific literature that splits documents into sentences, and sentences into words or phrases. Each word or phrase is then labeled according to concepts defined in the Gene Ontology (GO). The labeling is based on identification of a set of terms associated to a concept. These terms and the set of entities to be looked for were handcrafted in consultation with Biological databases. An index on all sentences with respect to labels and words is created to allow a rapid search for sentences that have a desired label and/or keyword. Textpresso also serves as a curation tool, in addition to being a search engine for researchers.

Creation of Fuzzy ontology structures have also received a lot of attention in recent times. Widyantoro and Yen (2001) have shown how fuzzy membership values associated to ontology concepts, along with a concept hierarchy, can be used for intelligent text retrieval. In this work, they have used abstracts of papers from several IEEE Transactions that have been manually typed and tagged based on their title, authors, publication date, abstract body, and author supplied keywords. In addition, their system extracts some keywords from the paper abstract and then a fuzzy ontology is built on the collection of keywords. The hierarchical arrangement of the terms in the newly generated ontology is dependent on their co-occurrence measures. However, the system is dependent on the user judgment about the relevance of articles to user queries.

Wallace and Avrithis (2004) have extended the idea of ontology-based knowledge representation to include fuzzy degrees of membership for a set of interconcept relations defined in an ontology. They have also utilized the membership of these relations to judge the context of a set of entities, the context of a user and the context of the query for the purpose of intelligent information retrieval. A fixed set of commonly encountered semantic relations have been identified and their combinations are used to generate fuzzy, quasi-taxonomic relations.

Quan, Hui, and Cao (2004) have proposed an automatic fuzzy ontology generation framework — FOGA. They have incorporated fuzzy logic into formal concept analysis to handle uncertainty information for conceptual clustering and concept hierarchy generation. However, the assignment of meaningful labels on initial class names, attributes, and its relations is required, which is manual and requires domain expertise. This system is also not designed to work on the fuzzy relational concepts present in unstructured or semistructured text documents.

Copyright © 2006, Idea Group Inc. Copying or distributing in print or electronic forms without written permission of Idea Group Inc. is prohibited.

The Problem of Handling
Imprecise Concept Descriptions

It has been already stated earlier in our discussions that one of the primary concerns in designing ontology-based information extraction applications stems from the fact that stored domain descriptions are not exhaustive or unambiguous. Ontology enhancement provides a way to circumscribe this problem. However, another problem that has not gained quite as much attention is that ontology descriptions may not exactly match the concept descriptions appearing in documents or those specified in user queries for retrieving relevant information. We will illustrate this point further with a few examples.

The wine ontology built by World Wide Web Consortium[4] (W3C) is one of the most exhaustive domain ontologies. Besides a thorough taxonomic classification, it contains property descriptors like color, flavor, body, taste, region of brewing, kind of grape used, and so forth for describing a wine. Each property has its value restrictions. For example color has value restrictions (red, rose, white), taste has value restrictions (sweet, bitter, dry) and so on. The Web abounds in documents describing various kinds of wine. Hence an ontology-based query processing system can help users find information about any kind of wine from the Web. However, let us consider a few Web documents on wine.

- Oaky Pinot Gris: This is a very serious, very sweet wine, with classy oak on the finish, and toasty, crystallised apples, pears and pineapples on the aftertaste.

- Alsace Grand Cru Altenberg De Bergheim Tokay Pinot Gris SGN 2001: Wonderfully clean, crisp, concentrated fruit. Intensely sweet.

- Alsace Pinot Gris Clos Windsbuhl 2002 Zind-Humbrecht: Beautiful crystallised fruits nose and palate. Very pure and fresh. Quite sweet.

- Moscato d'Oro and Sauvignon Blanc Botrytis are both sweet wines, crafted by different methods.

- Niagara: A medium-sweet, white varietal table wine with the fresh fruity flavor and luscious aroma of the famed Niagara grape. Serve well chilled with fish, roasted chicken, or after dinner.

- Chateau Du Monte Pink: Pink, bubbly, and slightly sweet.

Our first observation is that property names may or may not occur explicitly in a wine description. Secondly, new values like *toasty, pure, bubbly*, etc. appear as descriptions. Last, but not the least, descriptors are associated with adverbs

Copyright © 2006, Idea Group Inc. Copying or distributing in print or electronic forms without written permission of Idea Group Inc. is prohibited.

like intensely, slightly, etc. which clearly alter the values of the descriptors themselves. For example, if a user wants to know the name of a *very sweet wine*, the answers should be arranged in decreasing order of relevance starting with "Oaky Pinot Gris" and ending with "Chateau Du Monte Pink".

Let us consider another set of documents, this time excerpts from weather reports of various places.

- The sky is expected to be at least half covered with clouds, with a very slight chance of some showers; thunder and lightning may be associated with the showers. The showers may be heavy, but will be short-lived.

- (Sydney) The sky is expected to be at least half covered with clouds, with some showers possible; thunder and lightning may be associated with the showers. The showers may be very heavy and last for a long period of time.

- Hong Kong experiences both hot and cold seasons. The Observatory maintains a close watch on the local temperature changes. It issues warnings whenever Hong Kong is threatened by cold or very hot weather, to alert members of the public to the danger of low body temperature or the risk of heatstroke and sunburn due to cold weather or very hot weather respectively.

- Possible showers or thunderstorms in far southeast ahead and near the trough. Dry in the remainder. Warm to hot. Northwest to northeast winds ahead of the trough and southwest to southeast winds behind.

Obviously, the second document contains prediction of *heavier showers* than the first document. Similarly, if one wants to know the name of a *tolerably hot city*, Hong Kong should be avoided with a high probability.

Thus a critical issue that should be addressed for ontology-based Web information retrieval is how to handle the task of identifying, defining, and entering imprecise concepts into domain ontology specifications and thereby reason with these imprecise definitions. Though imprecise query processing from databases have been addressed earlier (Nambiar & Kambhampati, 2003), we observe that works on ontology-based text processing and knowledge acquisition have not addressed this.

We suggest the extension of ontological structures to use property qualifiers along with property values to tackle imprecision in concept descriptions. Again since there are virtually infinite number of qualifiers that one encounters in any domain, facility for incorporation of new qualifiers into an existing ontology is an integral part of our system. Our system is also equipped to perform fuzzy reasoning based on these linguistic qualifiers to determine the degree of relevance between a target query concept and document contents.

Copyright © 2006, Idea Group Inc. Copying or distributing in print or electronic forms without written permission of Idea Group Inc. is prohibited.

Proposed Extension to Ontology Structures for Including Imprecise Concept Descriptions

It has been established in the earlier section that in some cases, concepts are best described through the use of imprecise property descriptors rather than through a <concept-property-value> kind of structure. Incorporating imprecision into the ontology structure itself can help in resolving ambiguities arising due to differences in user requirement specification and concept descriptions embedded in text documents.

Our work, a preliminary version of which was reported in Abulaish and Dey (2004), uses qualifiers along with values for defining and learning the description of a concept in an enhanced <concept-property-value-qualifier> framework. Starting with a base ontology, our system uses a Part-Of-Speech (POS) tagger to identify relevant portions of documents through the identification of property names or their synonyms or known property values. It then uses a two-way decision-making system to identify concept descriptor values and qualifiers from relevant segments. New values and qualifiers are used to enhance the ontology structure. The information extracted from the documents is stored thereafter in a structured knowledge base to enable user query processing. The newly acquired qualifiers are presented to a knowledge engineer, who describes a partial ordering relation among the qualifiers. The engineer may also suggest some additional qualifiers or removal of some of these qualifiers from the domain knowledge base. The partial ordering relation is used while reasoning for similarity between user-specified requirements and ontology-based concept descriptions. The query processing mechanism employs a fuzzy reasoning principle to compute the relevance between a user-given query concept and the concept descriptions extracted.

A System to Enhance Ontology Structures to Include Imprecise Concept Descriptions

Figure 1 presents the schematic diagram of our system. The system has two major components — *Ontology Enriching Module* and *Fuzzy Query Processor*. The *Ontology Enriching Module* enhances an existing domain ontology structure to a fuzzy ontology structure by learning imprecise concept descriptors and property values from Web documents. This module also creates the framework for a structured knowledge base to store the information extracted

Copyright © 2006, Idea Group Inc. Copying or distributing in print or electronic forms without written permission of Idea Group Inc. is prohibited.

Figure 1. Schematic diagram of the system

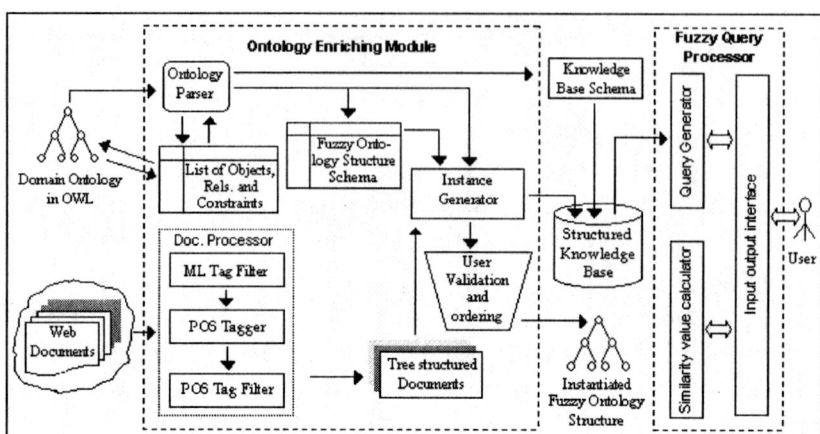

from Web documents. The *Fuzzy Query Processor* extracts information from the structured knowledge base to answer user queries. The following subsections provide the functional details of each module.

Ontology Enriching Module

The *Ontology Enriching Module* consists of following submodules.

(i) Ontology Parser. This module is responsible for parsing the ontology structures to perform the following tasks:

a. Initially an existing ontology structure is parsed to analyze the parent domain-concepts and their relationships for ontology-aided document processing.

b. The existing ontology structures are thereby enhanced to fuzzy ontology structures by incorporating fuzzy classes into them. Fuzzy classes are explained later.

c. A knowledge base schema is generated by parsing the fuzzy ontology structure. This acts as the structured storage for the information extracted from documents.

Copyright © 2006, Idea Group Inc. Copying or distributing in print or electronic forms without written permission of Idea Group Inc. is prohibited.

(ii) Document Processor. Since our aim is to extract *qualifiers* — as well as *property-values* from unstructured documents and instantiate the fuzzy ontology structure, the task of the document processor is to extract relevant portions of a text document. The processor accepts free-form text documents and identifies information components to instantiate the fuzzy ontology structure appropriately. Document processing has three phases.

a.   **Markup Language (ML) Tags Filter.** This divides an unstructured Web document into individual record-size chunks, which in our case are paragraphs. These paragraphs are stripped of the ML tags, and presented as individual unstructured records for further processing.

b.   **Part-of-Speech (POS) Tagger.** This module assigns Part-Of-Speech tags to all words in a sentence. A sentence is a part of the unstructured record (paragraph) generated by the ML Tags filter.

c.   **POS Tags Filter.** While a free-form text may contain all types of words, the information extraction process uses only a subset of these words in an application guided way or a domain specified way. For example, while dealing with imprecise concept descriptions adverbs play an important role in conjunction with the adjectives, since adverbs denote qualifiers for properties which appear as adjectives. However for biomedical documents verbs play an important role to designate actions/functions of different biological entities and adverbs are of no special significance. To speed up the document processing activity for extracting information, we have incorporated the POS tags filtering module which takes in as input the kinds of unwanted POS tags that are to be filtered out.

(iii) Instance Generator. This module is responsible for extracting relevant concepts from the records generated by the document processor. These concepts fill up the structured knowledge base and the fuzzy ontology structure. The module uses a bidirectional inferencing mechanism for identifying relevant concepts. On invocation, this module uses the parent domain-concepts present in the ontology structure and then scans the record components guided by this knowledge. It populates the fuzzy ontology structure as well as a structured knowledge base appropriately.

Further details about each component are provided in the following subsections.

Copyright © 2006, Idea Group Inc. Copying or distributing in print or electronic forms without written permission of Idea Group Inc. is prohibited.

Ontology Parser

To accommodate imprecise concept descriptions, the fuzzy ontology structure uses modified concept descriptor classes. The fuzzy ontology structure contains two generic classes: a "Value" class and a "Qualifier" class. For each property descriptor class in the original ontology structure, two subclasses are included in the fuzzy ontology structure: a "PropertyValue" class and a "PropertyQualifier" class, which are subclasses of the "value" and "qualifier" classes respectively. A qualifier class is constrained to have a collection of linguistic qualifiers. A set of linguistic qualifiers can be modeled as a graded set. The qualifier class along with its value is used to describe the property of a concept with varying degree of precisions. A property value can also be associated with a NULL qualifier. In the fuzzy ontology structure a FuzzyProperty class is created through multiple inheritance from value class and qualifier class. This is illustrated with Figures 2 and 3.

Figure 2 shows a partial view of property-descriptors as defined in the wine ontology developed by W3C. Using this structure either the taste of both "Oaky Pinot Gris" and "Alsace Grand Cru Altenberg De Bergheim Tokay Pinot Gris SGN 2001" can be defined as "sweet" or the first one can be defined as "very sweet" and the second one can be defined as "intensely sweet". With the first representation, the difference in the degree of sweetness is missed. With the second representation, their similarity of being sweet but to varying degrees is not captured. However, using the structure of Figure 3, the value of color for both the wines will be stored as "sweet", though the color-qualifier values will be different. For the first wine it will be "very" and for the second one it will be "intensely". This representation captures the essence of both wines being sweet quite unambiguously. Using a fuzzy reasoner, the degree of similarity can be computed based on the similarity between "very" and "intensely". We have used Protégé 2.1.2 to develop the new fuzzy ontology structure. Protégé[5] is an integrated software tool used by system developers and domain experts to develop knowledge-based systems. Protégé is currently being used in all those

Figure 2. Wine ontology (partial)

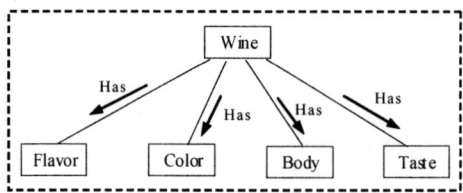

Copyright © 2006, Idea Group Inc. Copying or distributing in print or electronic forms without written permission of Idea Group Inc. is prohibited.

Figure 3. Fuzzy properties for some wine properties shown in Figure 2

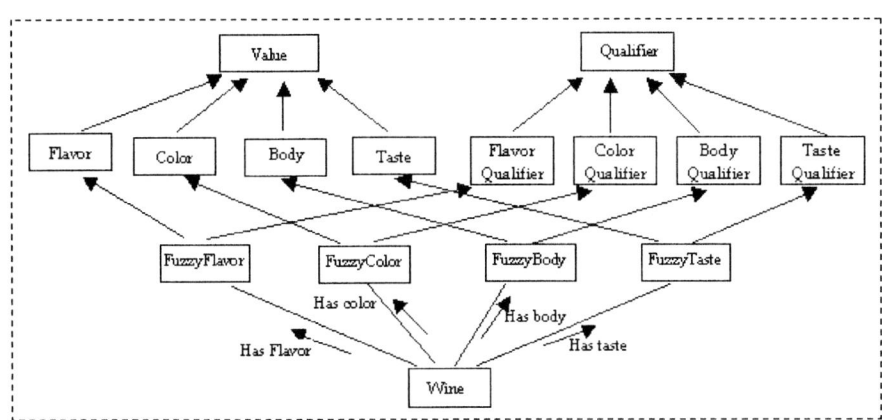

Table 1. Structured knowledge base schema

Entity Name	Property-1 Qualifier	Property-1 Value	Property-2 Qualifier	Property-2 Value	....	Property-n Qualifier	Property-n Value

fields including clinical medicine and the biomedical sciences in which concepts can be modeled as a class hierarchy.

The ontology parser thereafter parses the fuzzy ontology structure and creates an SQL schema for the structured knowledge base. This is accomplished through embedded SQL statements. Fuzzy property values and fuzzy qualifiers are attributes in this table. The ontology parser also generates the list of objects, relationships, and constraints, which provide a basis for mapping the relationships in the ontology into the table declarations in SQL schema. It also provides the cardinality constraints on the relationships like one-one, one-many, and many-many. The general layout of a knowledge base structure that is generated from the fuzzy ontology structure is shown in Table 1.

Document Processor

The main function of the document processor is to extract segments from unstructured text documents, which can contain information. The document processor consists of a ML tags filter which divides the Web documents into individual record-size chunks, cleans them by removing the unwanted ML tags,

Copyright © 2006, Idea Group Inc. Copying or distributing in print or electronic forms without written permission of Idea Group Inc. is prohibited.

and stores them as individual unstructured records for further processing. Each record consists of a collection of sentences which are a part of the same paragraph, where a sentence termination is identified by the occurrence of a full-stop.

Parts-Of-Speech play an important role in information extraction. Concept names are usually nouns, concept descriptors are adjectives and description qualifiers mostly consist of adverbs. Thus words with these parts-of-speech are to be extracted from sentences while generating imprecise concept descriptions. For this purpose a POS tagger is employed to assign a POS tag to each word in a sentence.

To extract imprecise concept descriptions, a POS tagged sentence is converted into a ternary tree structure. The entire sentence is first divided into segments on the basis of commas (,), semicolons (;), conjunctions (c), pronouns (which), and prepositions (with). Other than the first segment in a sentence, wherein a noun is searched for, a segment is incorporated into the tree provided it has at least one adjective tag (likely to contain a property value). If the first segment has at least one adjective tag it is added as a subtree in the document tree otherwise it is merged with the next segment having adjective tag(s). The ternary tree structure is defined as follows:

Structure Tree
Begin

String	**Value;*
Tree	**Lchild;*
Tree	**Mchild;*
Tree	**Rchild;*

End

Each sentence of a document and thereby the whole document is converted into an instance of the tree by distributing the tags in the following way:

Root (**R**): A node that contains the right most adjective tag of a segment.

Lchild (**L**): A node that contains all tags that are to the left of the tag considered at R.

Mchild (**M**): A node that contains all tags that are to the right of the tag considered at R

Rchild: points to the root of the subtree constructed from the next segment.

Copyright © 2006, Idea Group Inc. Copying or distributing in print or electronic forms without written permission of Idea Group Inc. is prohibited.

The equivalent context-free grammar for this is given as follows:

Document (D) → LRMD | ∈

L → (N+P+A+V+J)*

R → J

M → (N+P+V)*, where N, P, A, J, and V denote nouns, pronouns, adverbs, adjectives, and verbs, respectively.

Figure 4 illustrates sample tree structures created from the following sentences: "Dolcetto is a red table wine, which is quite dry and has a slightly fruity flavor" and "Syrah has a full body, less fruity flavor and deep red color that grow originally in California's coastal areas". Multiple sentences of the same paragraph are linked through the lowest, rightmost child.

Instance Generator

This module parses the ternary tree representation of the documents in conjunction with parent domain-concepts and constraints present in the fuzzy ontology structure to instantiate the fuzzy ontology structure as well as populate the structured knowledge base with information extracted from Web documents. We now state some of the key observations about extracting information from text documents in an ontology-guided fashion.

1. A particular object may have been described in a document by using some or all properties present in the ontology structure or by some other values, which are not present in the ontology structure.

Figure 4. Sample ternary tree structures created from text documents

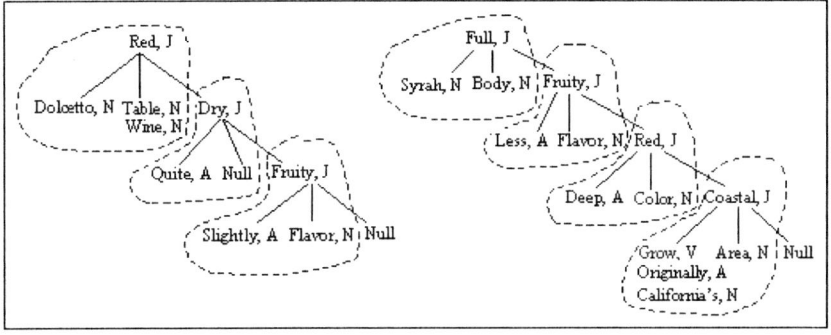

Copyright © 2006, Idea Group Inc. Copying or distributing in print or electronic forms without written permission of Idea Group Inc. is prohibited.

Figure 5. Extracted instances of fuzzy classes of wine ontology structure

Instances of *fuzzy_flavor* class

FlavorQualifier	FlavorValue
Slightly	*Bitter*
Null	Delicate
Zesty	Delicate
Null	*Dry*
Null	*Exquisite*
Null	*Favorite*
Fresh	*Fruity*
Null	*Fruity*
Usually	*Fruity*
Mellow	*Nutty*
Null	*Nutty*
Smooth	*Nutty*
Null	*Robust*
Null	*Spicy*
Harmonious	*Velvety*

Instances of *fuzzy_color* class

ColorQualifier	ColorValue
Pale	*Cherry-red*
Clear	*Lively*
Null	Red
Ruby	Red
Orange	Red
Indigenous	Red
Everyday	Red
Light	Red
Dry	Red
Robust	Red
Premium	Red
Null	Red
Fruity	Red
Bright	*Ruby-red*
Tenuous	*Straw*
Null	*Straw*
Softly	White
Null	White
Pale	White
Renowned	White
Festive	White
Bright	Yellow

Instances of *fuzzy_taste* class

TasteQualifier	TasteValue
Null	*Appealing*
Slightly	Bitter
Deliciously	Dry
Fine	Dry
Light	Dry
Medium	Dry
Null	Dry
Quite	Dry
Slightly	*Flinty*
Null	*Fresh*
Full	*Mellow*
Medium	Sweet
Null	Sweet
Slightly	Sweet
Very	Sweet

Instances of *fuzzy_body* class

BodyQualifier	BodyValue
Null	Full
Null	Light
Null	Medium
Very	Light

Legend:
1. All values in the qualifier columns are new and learned by the system from text documents.
2. All *ITALIC* values in the property value columns are new and learned by the system.

2. A document may or may not use property name in conjunction with the property values for describing a concept. For example, in the document — "Roussanne is a light bodied, light red and very sweet wine from France's Loire Valley, often blended with Merlot" — though the property name body is mentioned explicitly, the property descriptors taste and color.

Guided by these observations, our instance-generator employs a two-way approach to enrich the Fuzzy ontology structure and populate the knowledge base. On encountering a property name — the instance generator looks for values to fill up the object description through POS tags. This method allows learning of object descriptions with property values that are not present in the underlying ontology. In absence of a property name, a property value from the underlying ontology is used as a pointer to fill up the particular property slot.

Algorithm Instance_Generator implements the two-way approach. This algorithm accepts as input the ontological structures and entities, the ternary tree structure generated from the parsed documents, and the structured knowledge base schema as input. The output of this algorithm consists of values which are used to fill up the slots of the Fuzzy ontology structure as well as the structured knowledge base.

Copyright © 2006, Idea Group Inc. Copying or distributing in print or electronic forms without written permission of Idea Group Inc. is prohibited.

Algorithm Instance_Generator

- **Input:** Root node **R** of the tree structure generated from document; fuzzy ontology structure schema; list of ontological concepts and their relationships, structured knowledge base schema.

- **Step1:** Search the left child **Lchild** of **R** for the entity name and put it in the column entity_name of the knowledge base.

- **Step2:** Search the middle child **Mchild** for a property name. When a property name explicitly appears in the document, it occurs either as a noun tag or a verb tag. Since a property name may or may not be explicitly present, the following two cases have to be considered.

    - **Case 1:** *If Property name is found:* In this case, the value can be used directly.

      Extract the value appearing in the value field of the root node and store it in the corresponding property_value column of the knowledge base as well as in the corresponding slot of the Fuzzy ontology structure and go to step 4.

    - **Case 2:** *If Property name is not found:* In this case, the value appearing in the value field of the root node can be used to identify the property.

      Search the ontology structure to determine whether the value is present in the property-value set of a property.

        - **Case2.1:** *The value is present in the property-value set of some property:* In this case extract the value appearing in the value field of the root node and store it in the corresponding property_value column of the knowledge base as well as in the corresponding slot of the fuzzy ontology structure and go to step 4.

        - **Case 2.2:** *The value is not present:*In this case go to step 3.

- **Step 3:** Search the left child **Lchild** for a property-value.

    - **Case 1:** *A property-value is found:* Extract the found value and store it in the property_value column of the knowledge base as well as in the corresponding slot of the Fuzzy ontology structure and go to step 4.

Copyright © 2006, Idea Group Inc. Copying or distributing in print or electronic forms without written permission of Idea Group Inc. is prohibited.

- **Case 2:** *No property-value found:* In this case the subtree under consideration is assumed not to contribute any value for the fuzzy ontology structure and we go to step 6.

- **Step 4:** Search the left child **Lchild** of **R** for a qualifier. The qualifier for a value is likely to appear in this as an adverb or adjective tag. This node may have more than one qualifier so, the associativity of the search is from right-to-left and we associate the very first qualifier for the value found in the above steps. If a valid qualifier is found, extract and store it in the corresponding Qualifier_value column of the knowledge base as well as in the corresponding slot of the Fuzzy ontology structure.

- **Step 5:** If any of the above steps have yielded a match, block the matched property-value set for further search in the tree under consideration.

- **Step 6:** Follow the right child **Rchild** and replace **R** by it to consider the next subtree and repeat steps 2-6 until the value of the right child pointer is NULL.

Figure 5 illustrates some property and qualifier values retrieved from Web documents starting with the wine ontology. Figure 6 shows a partial view of the knowledge base that is created from this information. In the next section we present the details of the query processing mechanism that is employed to answer user queries using this knowledge base.

Fuzzy Query Processor

Query processing is a two-step process of acceptance and conversion of the user query into a corresponding SQL query and then finding the relevant answer from the structured knowledge base.

- **Accepting user query:** The query processor interface, shown in Figure 7 uses the underlying fuzzy ontology structure to guide the user to enter his/her queries, which is then converted into an SQL query. A user can specify a property by using only a qualifier, or only a value or both. Complex queries can be formulated by selecting one or more of the logical operators AND/OR/NOT.

- **Relevant answer generation from structured knowledge base:** During this phase the SQL query is first constructed from user given parameters. It may be noted that while all user-given property names, property values, property-value qualifiers, and logical operators (if any) are to be

Copyright © 2006, Idea Group Inc. Copying or distributing in print or electronic forms without written permission of Idea Group Inc. is prohibited.

Figure 6. Sample of knowledge base created from Web documents on wine

WineName	ColorQualifier	ColorValue	TasteQualifier	TasteValue	BodyQualifier	BodyValue	FlavorQualifier	FlavorValue
Fereisa	Pale	Cherry-red	Null	Null	Null	Null	Null	Null
Dolcetto	Null	Red	Quite	Dry	Null	Null	Slightly	Bitter
Carema	Null	Red	Null	Null	Null	Full	Null	Null
Brachetto	Ruby	Red	Null	Sweet	Null	Null	Null	Delicate
Barolo	Orange	Red	Null	Dry	Null	Full	Harmonious	Velvety
Barbaresco	Null	Null	Null	Dry	Null	Null	Null	Spicy
Barbera	Ruby	Red	Null	Null	Null	Null	Null	Null
Asti Spumante	Null	Null	Slightly	Sweet	Null	Null	Null	Null
Grignolino	Indigenous	Red	Slightly	Bitter	Null	Null	Null	Null
Gavi Docg	Teneous	Straw	Null	Dry	Null	Null	Usually	Fruity
Bardolino	Clear	Lively	Null	Fresh	Null	Null	Null	Null
Valpolicella	Everyday	Red	Null	Dry	Null	Null	Null	Null
Soave	Null	Straw	Null	Dry	Null	Null	Null	Null
Recioto di Soave	Bright	Yellow	Null	Null	Null	Full	Null	Null
Prosecco	Softly	White	Null	Null	Null	Null	Null	Null
Chianti	Bright	Ruby-red	Null	Null	Null	Null	Null	Dry
Verdlet	Null	White	Slightly	Flinty	Null	Null	Null	Null
Muscat Di Tanta Maria	Pale	White	Null	Sweet	Null	Null	Null	Null
Vidal Blanc	Null	Null	Medium	Dry	Null	Light	Null	Fruity
Rhine	Null	Null	Light	Dry	Null	Null	Null	Null
Chablis Blanc	Null	Null	Null	Appealing	Null	Light	Zesty	delicate
Edelweiss	Null	Null	Medium	Dry	Null	Null	Null	Fruity
Niagara	Null	White	Medium	Sweet	Null	Null	Fresh	Fruity
Seyval Blanc	Null	White	Medium	Dry	Null	Null	Null	Null
White Muscadine	Null	Null	Null	Sweet	Null	Null	Null	Fruity
Chardonnay	Renowned	White	Deliciously	Dry	Null	Null	Null	Exquisite
Vin Rose Sec	Null	Null	Fine	Dry	Null	Null	Null	Null
Pink Catawba	Null	Null	Null	Null	Null	Medium	Null	Favorite
Alpine Rose	Null	Null	Null	Null	Null	Null	Null	Null
Blush Chablis	Null	Null	Null	Null	Null	Light	Zesty	Delicate
Blush Niagara	Null	Null	Medium	Sweet	Null	Null	Fresh	Fruity
White Zinfandel	Light	Red	Null	Dry	Null	Null	Null	Null
Blush Muscadine	Null	Null	Null	Sweet	Null	Null	Null	Fruity
Cynthiana	Dry	Red	Null	Null	Null	Null	Null	Null
Burgundy	Robust	Red	Null	Dry	Null	Null	Null	Null
Sangria	Premium	Red	Null	Null	Null	Null	Null	Null
Beau Noir	Null	Red	Null	Null	Null	Null	Null	Null
Altus Spumante	Festive	White	Null	Null	Null	Null	Null	Null
Red Sparkler	Null	Null	Slightly	Sweet	Null	Null	Null	Null
White Sparkler	Null	Null	Slightly	Sweet	Null	Null	Null	Null
Cabernet Franc	Fruity	Red	Null	Null	Null	Null	Null	Null
Gewurztraminer	Null	White	Slightly	Sweet	Null	Null	Null	Fruity
Johannisberg Reisling	Null	White	Deliciously	Dry	Null	Null	Null	Null
Alpine Burgundy	Null	Null	Null	Sweet	Null	Null	Null	Null
Red Muscadine	Null	Null	Null	Sweet	Null	Null	Null	Fruity
Cabernet Sauvignon	Dry	Red	Null	Null	Null	Null	Null	Robust
Merlot	Null	Null	Full	Mellow	Null	Null	Null	Null
Sherry	Null	Null	Medium	Dry	Null	Null	Mellow	Nutty
Cocktail Sherry	Null	Null	Null	Null	Null	Null	Smooth	Nutty
Cream Sherry	Null	Null	Null	Null	Null	Full	Null	Nutty

selected, only the property names, values and logical operators are used to formulate the SQL query predicates. For example if the user enters the query — "List the wines with light red color or slightly bitter taste" the corresponding SQL query is generated as:

Select WineName, ColorQualifier, ColorValue, TasteQualifier, TasteValue
From WineKnowledgebase
Where ColorValue = "red" or TasteValue = "Bitter";

Copyright © 2006, Idea Group Inc. Copying or distributing in print or electronic forms without written permission of Idea Group Inc. is prohibited.

Figure 7. Query interface

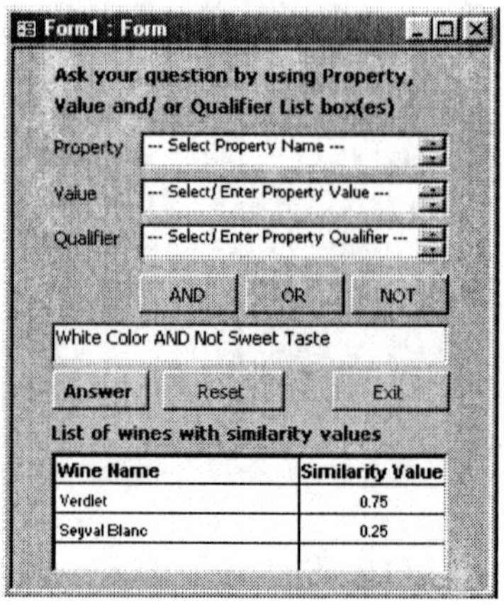

Table 2. Partial listing of entries retrieved

WineName	ColorQualifier	ColorValue	TasteQualifier	TasteValue
Dolcetto	Null	Red	Quiet	Dry
Carema	Null	Red	Null	Null
Brachetto	Ruby	Red	Null	Sweet
Grignolino	Indigenous	Red	Slightly	Bitter

This will retrieve all entries from the table where the property values match. Similarity computation will follow with the qualifier column values. Table 2 shows a partial list of entries retrieved for the above-mentioned SQL query.

The only exception in formulating SQL queries is implemented for the "not" qualifier. For queries including a not qualifier, SQL queries are constructed using a separate principle. In this case, the SQL query has two ORed predicates. The first predicate retrieves all entries with values other than the negated property value. The second predicate retrieves all entries that do not have both the negated qualifier and the value together. This is illustrated through an example in query 3 later.

The answer generation mechanism first processes the SQL query for finding possible relevant entries from the knowledge base. It then applies fuzzy reasoning to find the relevance of each entity and lists them in decreasing order of relevance.

Copyright © 2006, Idea Group Inc. Copying or distributing in print or electronic forms without written permission of Idea Group Inc. is prohibited.

To calculate the relevance of each retrieved answer with respect to the original user query, we have used similarity measures. If u and v are two objects in a given universe of discourse U, a similarity measure is a function: SIM (u, v) → [0, 1]. In our framework since the qualifier set for different properties is modeled as a graded set, the distance between two qualifier values associated to two retrieved entries, reflects their degree of dissimilarity. The distance between the value v_i at position i and the value v_j at position j within a set is defined as:

$$d (v_i, v_j) = | i - j | \tag{1}$$

The similarity between two qualifiers u and v is computed as:

$$SIM (u, v) = 1 - d (u, v) / (MAX + 1),$$

where MAX is the maximum distance between two concepts in the graded set. This satisfies the condition $0 < SIM(u,v) \leq 1$. The relevance of an extracted instance to the user given query is computed in terms of the degree of match between user-specified property values and qualifiers to those appearing in the knowledge base.

Let the user given description for each property P be represented by $C = <P, V, Q_1>$, where V denotes the property value and Q_1 denotes the qualifier. Let x be a retrieved instance which has value V for property P and qualifier Q_2. Let $\mu_C(x)$ denote the similarity between the instance x and the user given description C for property P. $\mu_C(x)$ is computed as follows:

- **Case 1:** When Q_1 = NULL or $Q_1 = Q_2$: The first situation arises when user does not select/enter a qualifier with a property value. In this case, $\mu_C(x)$ = 1. That is, all instances with matching values are accepted as exact matches. Similarly, if there is an exact match between the user given concept descriptions and those present in the retrieved instances the similarity value $\mu_C(x) = 1$.

- **Case 2:** Otherwise, if both Q_1 and Q_2 are present in the qualifier set of P in the fuzzy ontology, $\mu_C(x) = SIM(Q_1, Q_2)$.

- **Case 3:** Else; if either Q_1 or Q_2 or both are not present in the qualifier set and they are not NULL — in this case, it is judged that the query processor is not able to resolve the difference in the descriptions and so $\mu_C(x) = 0$.

Copyright © 2006, Idea Group Inc. Copying or distributing in print or electronic forms without written permission of Idea Group Inc. is prohibited.

For complex queries with multiple property descriptions $C_1, C_2, ..., C_n$, combined with logical AND, OR, and NOT operators, the final similarity value is computed by using fuzzy intersection, union and negation operations respectively. Traditional fuzzy membership computation function for intersection, union, and negation of fuzzy concepts C_1, C_2, ..., C_n are used (Zadeh, 1965):

$$\mu_{C1 \cup C2 \cup ... Cn}(x) = MAX \; [\mu_{C1}(x), \mu_{C2}(x), ..., \mu_{Cn}(x)]$$
$$\mu_{C1 \cap C2 \cap ... Cn}(x) = MIN \; [\mu_{C1}(x), \mu_{C2}(x), ..., \mu_{Cn}(x)]$$
$$\mu_{\neg C1}(x) = 1 - \mu_{C1}(x)$$

Following are some sample queries and the retrieved answers for them.

Query 1. List the wines that are white, fruity and very sweet.

SQL>   Select WineName, ColorQualifier, ColorValue, FlavorQualifier, FlavorValue, TasteQualifier, TasteValue

From WineKnowledgebase

Where ColorValue = 'White' AND FlavorValue = 'Fruity' AND TasteValue = 'Sweet';

Result: The first entry is judged to have more relevance to the query since the qualifier *medium* is more close to *very* than *slightly* is.

WineName	Color Qualifier	Color Value	Flavor Qualifier	Flavor Value	Taste Qualifier	Taste Value	Similarity Value
Niagara	Null	White	Fresh	Fruity	Medium	Sweet	0.56
Gewurztraminer	Null	White	Null	Fruity	Slightly	Sweet	0.22

Query 2: List the wines with light yellow color or very bitter taste

SQL> Select WineName, ColorQualifier, ColorValue, TasteQualifier, TasteValue

From WineKnowledgebase

Where ColorValue = 'yellow' or TasteValue = 'Bitter';

Copyright © 2006, Idea Group Inc. Copying or distributing in print or electronic forms without written permission of Idea Group Inc. is prohibited.

Result: The first entry is judged more relevant since qualifiers "Bright" and "light" are more similar than qualifiers "very" and "slightly". Since none of them satisfy all properties, so none of them are judged absolutely relevant.

WineName	ColorQualifier	ColorValue	TasteQualifier	TasteValue	Similarity Value
Racioto di soave	Bright	Yellow	Null	Null	0.67
Grignolino	Null	Null	Slightly	Bitter	0.40

Query 3: List the wines that are not medium sweet.

SQL> Select WineName, BodyQualifier, BodyValue, FlavorQualifier, FlavorValue

  From WineKnowledgebase

  Where TasteValue ≠ 'Sweet' OR (TasteQualifier ≠ 'medium' AND TasteValue = 'Sweet');

Result: It may be noted that all entries with tastevalue anything but sweet are retrieved along with those entries which do not have "medium" and "sweet" for colorqualifier and colorvalue respectively. It may also be noted that all non-sweet wines have more relevance to the user query than the wines which are slightly sweet or sweet, etc.

WineName	TasteQualifier	TasteValue	Similarity Value
Dolcetto	Quite	Dry	1.00
Barolo	Null	Dry	1.00
Barbaresco	Null	Dry	1.00
Grignolino	Slightly	Bitter	1.00
Gavi Docg	Null	Dry	1.00
Bardolino	Null	Fresh	1.00
Valpolicella	Null	Dry	1.00
Soave	Null	Dry	1.00
Verdlet	Slightly	Flinty	1.00
Vidal Blanc	Medium	Dry	1.00
Rhine	Light	Dry	1.00
Chablis Blanc	Null	Appealing	1.00

table continued on following page

Copyright © 2006, Idea Group Inc. Copying or distributing in print or electronic forms without written permission of Idea Group Inc. is prohibited.

WineName	TasteQualifier	TasteValue	Similarity Value
Edelweiss	Medium	Dry	1.00
Seyval Blanc	Medium	Dry	1.00
Chardonnay	Deliciously	Dry	1.00
Vin Rose Sec	Fine	Dry	1.00
White Zinfandel	Null	Dry	1.00
Burgundy	Null	Dry	1.00
Johannisberg Reisling	Deliciously	Dry	1.00
Merlot	Full	Mellow	1.00
Sherry	Medium	Dry	1.00
Brachetto	Null	Sweet	0.60
Muscat Di Tanta Maria	Null	Sweet	0.60
White Muscadine	Null	Sweet	0.60
Blush Muscadine	Null	Sweet	0.60
Alpine Burgundy	Null	Sweet	0.60
Red Muscadine	Null	Sweet	0.60
Asti Spumante	Slightly	Sweet	0.40
Red Sparkler	Slightly	Sweet	0.40
White Sparkler	Slightly	Sweet	0.40
Gewurztraminer	Slightly	Sweet	0.40

Query 4: List the wines that are clear with any color.

SQL>   Select WineName, ColorQualifier, ColorValue
From WineKnowledgebase
Where ColorValue <> 'Null';

Result: Only one wine matches the description exactly. The last few entries are judged to have no relevance since the relationship of these qualifiers with others are not known a priori.

WineName	ColorQualifier	ColorValue	Similarity Value
Bardolino	Clear	Lively	1.00
Fereisa	Pale	Cherry-red	0.92
Brachetto	Ruby	Red	0.92
Barbera	Ruby	Red	0.92
Muscat Di Tanta Maria	Pale	White	0.92

table continued on following page

Copyright © 2006, Idea Group Inc. Copying or distributing in print or electronic forms without written permission of Idea Group Inc. is prohibited.

WineName	ColorQualifier	ColorValue	Similarity Value
Prosecco	Softly	White	0.85
Sangria	Premium	Red	0.85
Grignolino	Indigenous	Red	0.77
Gavi Docg	Teneous	Straw	0.77
White Zinfandel	Light	Red	0.70
Burgundy	Robust	Red	0.70
Dolcetto	Null	Red	0.62
Carema	Null	Red	0.62
Soave	Null	Straw	0.62
Recioto di Soave	Bright	Yellow	0.62
Chianti	Bright	Ruby-red	0.62
Verdlet	Null	White	0.62
Niagara	Null	White	0.62
Seyval Blanc	Null	White	0.62
Beau Noir	Null	Red	0.62
Gewurztraminer	Null	White	0.62
Johannisberg Reisling	Null	White	0.62
Barolo	Orange	Red	0.00
Valpolicella	Everyday	Red	0.00
Chardonnay	Renowned	White	0.00
Cynthiana	Dry	Red	0.00
Altus Spumante	Festive	White	0.00
Cabernet Franc	Fruity	Red	0.00
Cabernet Sauvignon	Dry	Red	0.00

Thus it can be seen that imprecision in user queries can be tackled within a structured format with information extracted from the unstructured sources.

Performance Evaluation of the Proposed System

The overall system performance is dependent on the performance of two independent units — the *Ontology Enriching Module* and the *Fuzzy Query Processor*. We have evaluated the two components separately and provide here the experimental details of evaluation. First we shall present performance analysis for the *Ontology Enriching Module* and then discuss the efficacy of the *Fuzzy Query Processor*.

Evaluation of the Ontology Enriching Module

A system for automatic extraction of information from unstructured or semi-structured text documents is only useful if it is as accurate and reliable as human

Copyright © 2006, Idea Group Inc. Copying or distributing in print or electronic forms without written permission of Idea Group Inc. is prohibited.

curation. Since terminology and complex proper names are not found in Dictionaries it is a problematic task to provide objective performance evaluation for any automatic method for concept extraction. Velardi et al. (2001) have proposed three possible ways of formally evaluating a terminology.

1. Using the extracted terms for a Natural Language application like document classification and measure the performance of the application with and without the component. However such an evaluation strategy may not produce clear-cut results, especially when the influence of the component on the overall system performance is not predominant.

2. Using some existing thesaurus as a "golden standard", and to measure the precision and recall of the method at extracting the terms included in the available thesaurus. This approach is sufficiently assessed for named entities, since large gazetteers of proper names do exist. But, evaluation of not-named terminology is far more difficult, since no method would detect terms that are absent or appear rarely in the corpus used for term extraction. Moreover, the notion of "term" is too vague to consider available terminological database as "closed" sets, unless the domain is extremely specific.

3. Evaluating the performance of entity extractor by a team of experts: Since the notion of Named Entity is quite precise, therefore manual judgement of extracted names is a relatively reliable approach, but as far as not-named terms (e.g., property name and qualifier) are concerned, reaching the consensus about the introduction of a new concept description is more difficult.

Our intention in this work was to study the applicability of ontology-guided precise and imprecise concept description extraction from general-purpose unstructured texts. We considered documents on wine as our test-bed, since a well-defined ontology exists for this domain. For another set of documents we have used the domain of "wood" since we found concept descriptions in that domain also tended to be similar. For the second domain, since no standard ontology existed, we started with a small, seed ontology, and we have been able to obtain a more complete ontology, enhanced with more information extracted from documents. The Wood Ontology[1] is available in OWL. Since no standard tagged corpora were available for this purpose, we have selected the third approach to evaluate the performance of our system.

To build our corpora we have downloaded a large collection of wine and wood documents from the World Wide Web. Each of these documents describes one type of wine or wood and contains around 10 sentences each. Presently, the wine

Copyright © 2006, Idea Group Inc. Copying or distributing in print or electronic forms without written permission of Idea Group Inc. is prohibited.

corpus has 500 documents consisting of 52,320 words, and the wood corpus has 475 documents consisting of 55,590 words. These documents were manually inspected and a complete compilation of all possible elements to be extracted was made under different categories (shown in column1 of Table 3): wine or wood names, property names, property values and property value qualifiers. The elements extracted by the Ontology Enriching Module are automatically stored in different columns of the database depending on their types. It may be noted that property names are not stored explicitly in the database since the database attributes are created from the ontology structure itself. However, the recognition of property names is important since it affects the property-value recognition process and thereby the overall system performance.

We evaluated the performance of this module using recall and precision values defined as follows:

Precision = Number of relevant elements extracted by OEmodule/Total number of elements extracted

Recall = Number of relevant elements extracted by OEmodule/Total number of relevant elements actually present in the corpus.

Table 3 summarizes these values for the Ontology Enriching Module. The table shows the relevance values for three different, but not mutually exclusive, subtasks performed by this module: entity name extraction, property value extraction, and property qualifier extraction. We now discuss each type of recognition result separately.

Table 3. Performance metrics for the OEmodule in terms of precision and recall

Domain	Type of Extracted Information	# Elements in Source	# Extracted Elements Recognized Correctly	#Incorrect Elements Extracted	Precision	Recall
Wine	Named Entities (Wine names)	1092	971	42	95.85	88.92
	Property Values	2172	1591	31	98.09	73.25
	Qualifier Values	1492	1126	185	85.89	75.47
Wood	Named Entities (Wood Names)	965	856	49	93.65	88.70
	Property Values	1962	1177	109	91.52	59.99
	Qualifier Values	650	486	82	85.56	74.77

Copyright © 2006, Idea Group Inc. Copying or distributing in print or electronic forms without written permission of Idea Group Inc. is prohibited.

- **Named-entity recognition.** It may be observed from Table 3 that the precision of recognized names in both the cases is quite high, though the recall value is slightly lower. This indicates that most of the names were correctly recognized as wine or wood names, though a few other nouns were also identified as names. These other names were mostly names of places where a specific wine is found or a wood is grown. The reason for the recall value to be lower was identified as follows. In the case of named-entity recognition, we observed that the system sometimes extracts partial names, even when the complete name was present in the document. Some other times a single name is split into two. This is due to the lack of a consistent naming convention for these domains. For example, while the name *Verdlet* is easy to identify as a noun, it is not so for the name *Blush Niagara*. The tagger that is used to identify Parts-Of-Speech patterns, generally influences this. This problem can be overcome by employing a set of additional Entity-Recognition rules, which encompass the peculiar naming conventions of the domain. This is the usual approach adopted for Biological document processors, since there one has to deal with a large entity collection with rather inconsistent naming standards (Zhou & Su, 2002; Velardi et al., 2001).

- **Property value recognition.** According to the current design, the OEM recognizes a word as a property value if either it is a member of a property-value set in the ontology structure or it is associated with a property name in the Web document. For example, let us consider the following document along with its generated tags:

Verdlet is a delicate, fruity, white table wine, with the pleasingly crisp, slightly
N   XT N   J   J   N   N   R   T   A   V   A
flinty taste, from the rare French developed grape variety Verdlet Blanc.
J   N   R T J   N   V   N   N   N   N

In the above example, though the property *color* does not explicitly appear in the description, the knowledge base is filled up correctly, since *white* is recognized as an adjective by the tagger, and it is present in the property value set of *color* in the domain ontology. The description "flinty" is also correctly recognized as a *taste value* since it occurs explicitly in association to property name — *taste*. However, the word *fruity*, though assigned the correct parts-of-speech by the tagger, it is not recognized as a *property value* for flavor by our system, because it is neither present in any *property value set* in the underlying ontology structure nor is it associated to any specific *property name* in the document. However, *fruity* does not introduce any error since it is not considered as a property value at all.

Copyright © 2006, Idea Group Inc. Copying or distributing in print or electronic forms without written permission of Idea Group Inc. is prohibited.

- **Property value qualifier recognition.** This task has relatively less precision and recall values than other tasks. There are two reasons that we have identified for this. Firstly, in many cases a property value may be associated with more than one qualifier. In such cases our module extracts only the one that is closest to the property value. For example, consider the following sentence picked-up from the wine domain:

Chianti: a very bright ruby red color; a dry flavor, which becomes delicate ...

Here we found that the *color value* of "Chianti" is associated with two different qualifiers *bright* and *very*. Our system extracts only *bright* as a qualifier value and associates it with the *color value* of wine. In order to reflect the true performance of the module, during performance evaluation we have counted *very* as missed relevant element. This problem can be easily tackled by considering adverb chains rather than single adverbs as qualifiers. However, since that will add to the complexity of reasoning during query processing, we have not implemented this yet.

The second problem creeps in due to the fact that many adverbs occur in the document which are not really qualifiers though they occur sufficiently close to some property values. An example of such a sentence is given below in which *occasionally* is wrongly judged as a qualifier for *taste* whose value is *slight*.

Valpolicella, an everyday red wine, ... and has plenty of body, and occasionally a slight
N   T   J   J N......C X   N   R N   C   A   T J
taste of bitter almonds.
N   R   J   N

Evaluation of the Fuzzy Query Processor

It is difficult to provide a performance analysis of the query-processing module since no benchmark set of queries exists for judging the performance of such a system. Since the concept descriptions are finally stored in a database, the system can obviously retrieve all exact matches correctly. When it comes to judging the relevance of answers to fuzzy query, the quality of retrieval is dependent on the similarity computation procedure. For example, it can be seen from the examples cited above that in some cases, the fuzzy Min-Max function seems to be too restrictive, though we have chosen it since this provides a

Copyright © 2006, Idea Group Inc. Copying or distributing in print or electronic forms without written permission of Idea Group Inc. is prohibited.

standard way of interpreting ANDing and ORing of entities. The similarity between two fuzzy qualifiers also depends on the expert's judgment and fixing the orders of the qualifiers in the corresponding qualifier sets. We refrain from giving any relevance figure for this module, since right now correctness of an answer generated is largely dependent on the user's perspective.

Future Trends in Text Information Extraction

Text information processing systems are gradually shifting towards more semantic interpretation of document contents and thereby integration of domain ontologies into these systems is becoming a necessity. As a result, a significant amount of research is being directed towards automatic ontology enrichment with new concepts encountered. Semantic analysis of contents can provide better answers to user queries since they will be exploiting domain knowledge.

One of the domains in which tremendous amount of work on ontology-based information processing is already in progress is the biomedical domain (Ohta et al., 2001). The availability of structured domain knowledge has led to the creation of a number of ontologies like GENIA ontology, Gene Ontology (GO), etc. Efforts are on to harness the ever-growing document repository in this area into a structured collection using these ontologies. The collection is also contemplated to be used for precise question-answering for users. The aim is to provide the users with specific information to questions like "Which gene causes color-blindness" rather than the current practice of providing a list of URLs from which he/she is expected to search the answer. Combining and collating relevant information from multiple sources is the obvious next step to this.

Text understanding for nontechnical documents is yet more challenging. Ontology based domain knowledge representation plays a crucial role in building intelligent information extraction systems which can save the user from the problem of information overload. Since text is unstructured, therefore Natural Language Processing principles are playing an increasingly important role in the design of such systems. Simple methods like POS tagging, etc., have already been explored for domains like text summarizing and abstraction. However, for most of these systems it is found that though they work with a high precision value, recall rate is not as exciting. This can be attributed to the various nuances of natural language like polysemy, synonymy and the inexact nature of the grammar itself.

A lot of effort is directed towards processing the unstructured information off-line and thereby create structured knowledge repositories out of it. Annotation

Copyright © 2006, Idea Group Inc. Copying or distributing in print or electronic forms without written permission of Idea Group Inc. is prohibited.

based repositories serve the purpose to some extent (Katz & Lin, 2002; Katz et al., 2003). However, it is difficult to manually annotate the tremendously huge collection that is also expanding at an alarming rate. One of the solutions to this problem is to consider systems for automatic curation of documents. Curation refers to the process of building and maintaining structured repositories which contain information about contents of documents. Online query processing can thereby access the structured repositories only and not the free-form text document. This can definitely reduce the problem of information overload that users of the Web face today.

Future Work in Text Mining and Ontology Enhancement

The work presented in this chapter can be categorized as ontology-guided text mining for automatic curation. Our aim is to build a structured knowledge base which can use an underlying domain ontology to extract specific information from text documents and store them in a structured way for query processing. Presently the system uses shallow NLP techniques to extract this information and populate the database. Concept descriptions can be either precise or accompanied by fuzzy qualifiers which allows for fuzzy property values to be specified for concepts. The design of the database is also ontology guided. The system has an inbuilt query-processing mechanism, which can handle inexact matches between user-given concept descriptions and document concept descriptions.

Our intention is to strengthen the knowledge acquisition and ontology enhancement processes in future by amalgamating NLP techniques with validity, reliability and relevance parameters. Any knowledge curation process should ideally take into account the source of information provider in order to judge its validity. Reliability of information can be computed from association analysis of information components. For example, when a certain set of property-value pairs consistently co-occur, these may be accepted to enhance the ontology set with high degree of belief or certainty associated with them. On the other hand, pairs which occur with low co-occurrence values may have a lower degree of belief associated with them. A similar belief association is being currently explored for inferring inter-concept relations from documents. Relevance computation for concepts is a slightly more tricky issue since relevance is by and large user dependent. For example, even though the price of an article may appear to be "very high" for one user, it may be "just right" for another. Thus, combining the fuzzy reasoning module with user profiling is a very promising research area which has applications in building Web-based recommender systems with high commercial and scholastic implications. In general, combination of soft-computing techniques like fuzzy reasoning with information extraction principles prom-

Copyright © 2006, Idea Group Inc. Copying or distributing in print or electronic forms without written permission of Idea Group Inc. is prohibited.

ises to be very exciting, since it can help in the design of intelligent question answering systems.

Conclusion

In this work, we have presented a text-mining system, which has twin objectives of enriching an existing domain ontology with new and/or imprecise concept descriptions and answer user queries. The ontology enrichment module extends an existing ontology structure to introduce fuzzy classes and creates a fuzzy ontology structure. The fuzzy ontology structure uses a property-value-qualifier framework to describe concepts. This enables concept-descriptors to have a degree of qualification to be attached to them. These descriptors may be learnt as new information is encountered through mining of Web documents.

A structured knowledge base is also created to store concepts mined from Web documents. The knowledge-base schema for this is prepared by the system using the fuzzy ontology structure. Concept descriptions may consist of values and qualifiers. The extracted qualifiers are arranged into a graded set with the help of domain experts. The system uses the structured knowledge base along with the fuzzy ontology structure to answer user queries which may not have precise matches in any document.

Imprecise query processing proceeds in two-stages. In the first stage, exact partial matches for the user query are extracted from the knowledge base. In the second stage, fuzzy similarity computation functions are applied to filter out the most relevant instances from this set. The answers are ranked according to their relevance to the user query.

Presently, the concept extraction process is dependent on the underlying grammar used for parsing sentences. Though initial experiments suggest that this grammar can be made domain independent since it can be aligned with English grammar as such, more work is required to establish this. We are currently working on techniques to generate the requisite grammar for a domain through statistical analysis of documents from that domain. Secondly, our system does not consider the reliability of information to be added and considers all new information encountered as candidates to be inserted. Integration of fuzzy relations with membership values based on reliability of source and frequency of domain concepts is being experimented with. This system is also being extended to extract predicates from domain documents rather than simple entities. Our aim is to extend logical reasoning for these predicates through the use of extended Description Logics.

Copyright © 2006, Idea Group Inc. Copying or distributing in print or electronic forms without written permission of Idea Group Inc. is prohibited.

References

Abulaish, M., & Dey, L. (2004). Using part-of-speech patterns and domain ontology to mine imprecise concepts from text documents. *Proceedings of the 6th International Conference on Information Integration and Web based Applications & Services,* Indonesia (pp. 91-100).

Baader, F., Calvanese, D., McGuinness, D., Nardi, D., & Patel-Schneider, P. (Eds.). (2003). *The description logic handbook, theory, implementation and applications.* Cambridge University Press.

Berners-Lee, T., Hendler, J., & Lassila, O. (2001, May). The Semantic Web. *Scientific American,* 28-31.

Bodner, R. C., & Song, F. (1996). Knowledge-based approaches to query expansion in information retrieval. In G. McCalla (Ed.), *Advances in artificial intelligence* (pp. 146-158). New York: Springer.

Crow, L., & Shadbolt, N. (2001). Extracting focused knowledge from the Semantic Web. *International Journal Human-Computer Studies, 54,* 155-184.

Fensel, D., Horrocks, I., van Harmelen, F., McGuinness, D. L., Patel-Schneider, P. (2001, March/April). OIL: Ontology infrastructure to enable the Semantic Web. *IEEE Intelligent Systems, 16*(2), 38-45.

Gruber, T. R. (1993). A translation approach to portable ontology specification. *Knowledge Acquisition,* 199-220.

Guarino, N., Carrara, M., & Giaretta, P. (1995). Ontologies and knowledge bases: Towards a terminological clarification. In N. Mars (Ed.), *Towards very large knowledge bases, knowledge building and knowledge sharing* (pp. 25-32). Amsterdam: IOS Press.

Hahn, U., & Marko, K. G. (2002). An integrated, dual learner for grammars and ontologies. *Data & Knowledge Engineering, 42*(3), 273-291.

Heflin, J., Hendler, J. A., & Luke, S. (2003). SHOE: A blueprint for the Semantic Web. *Spinning the Semantic Web,* 29-63.

Horrocks, I. (2002). DAML+OIL: A description logic for the Semantic Web. *Bulletin of the IEEE Computer Society Technical Committee on Data Engineering, 25*(1), 4-9.

Horrocks, I., & Sattler, U. (2001). Ontology reasoning in the SHOQ(D) description logic. *Proceedings of IJCAI'01.* Morgan Kaufmann.

Katz, B., & Lin, J. (2002). Annotating the Semantic Web using natural language. *Proceedings of the 2nd Workshop on NLP and XML (NLPXML'02) at COLING,* Taipei, Taiwan.

Copyright © 2006, Idea Group Inc. Copying or distributing in print or electronic forms without written permission of Idea Group Inc. is prohibited.

Katz, B., Lin, J., Loreto, D., Hildebrandt, W., Bilotti, M., Felshin, S., et al. (2003). Integrating Web-based and corpus-based techniques for question answering. *Proceedings of the 12th Text REtrieval Conference (TREC 2003)*, Gaithersburg, MD.

Li, Y., & Zhong, N. (2004). Web mining model and its applications for information gathering. *Knowledge Based Systems, 17*, 207-217.

Liddle, S. W., Hewett, K., & Embley, D. W. (2003). An integrated ontology development environment for data extraction. *Proceedings of ISTA'03* (pp. 21-33).

Luke, S., Spector, L., Rager, D., & Hendler, J. (1997). Ontlogy-based Web agents. *Proceedings of the 1st International Conference on Autonomous Agents* (pp. 59-66).

Nambiar, U., & Kambhampati, S. (2003). Answering imprecise database queries: A novel approach. *Proceedings of the 5th ACM-CIKM Workshop on Web Information and Data Management (WIDM'03)*, New Orleans (pp. 126-133).

Neches, R., Fikes, R. E., Finin, T., Gruber, T. R., Senator, T., & Swartout, W. R. (1991). Enabling technology or knowledge sharing. *AI Magazine, 12*(3), 36-56.

Muller, H. M., Kenny, E. E., & Strenber, P. W. (2004). Textpresso: An ontology-based information retrieval and extraction system for biological literature. *PloS Biol, 2*(11): e309. Retrieved from http://www.plosbiology.org

Ohta, T., Tateisi Y., & Kim, J. (2001). Ontology based corpus annotation and tool. *Genome Informatics, 12*, 469-470.

Quan, T. T., Hui, S. C., & Cao, T. H. (2004). FOGA: A fuzzy ontology generation framework for scholarly Semantic Web. *Proceedings of the 2004 Knowledge Discovery and Ontologies Workshop (KDO'04)*, Pisa, Italy.

Shamsfard, M., & Barforoush, A. A. (2004). Learning ontologies from natural language texts, *International Journal of Human-Computer Studies, 60*(1), 17-63.

Stevens, R., Goble, C. A., Horrocks, I., & Bechhofer, S (2002). Building a bioinformatics ontology using OIL. *IEEE Transactions on Information Technology in Biomedicine, 6*(2), 135-141.

Velardi, P., Fabriani, P., & Missikof, M. (2001). Using text processing techniques to automatically enrich a domain ontology. *Proceedings of the ACM Conference on Formal Ontologies and Information Systems (FOIS'01)*, Ogunquit, ME (pp. 270-284).

Velardi, P., Missikoff, M., & Basili, R. (2001). Identification of relevant terms to support the construction of domain ontologies. *Proceedings of ACL*

Copyright © 2006, Idea Group Inc. Copying or distributing in print or electronic forms without written permission of Idea Group Inc. is prohibited.

Conference on Human Language Technology and Knowledge Management, Toulouse, France.

Wallace, M., & Avrithis, Y. (2004). Fuzzy relational knowledge representation and context in the service of semantic information retrieval. *Proceedings of the IEEE International Conference on Fuzzy Systems (FUZZ-IEEE),* Budapest, Hungary.

Widyantoro, D. H., & Yen, J. (2001). A fuzzy ontology-based abstract search engine and its user studies. *Proceedings of the 10th IEEE International Conference on Fuzzy Systems,* Melbourne, Australia (pp. 1291-1294).

Zadeh, L. A. (1965). Fuzzy sets. *Journal of Information and Control, 8,* 338-353.

Zhou, G. D., & Su, J. (2002). Named entity recognition using an HMM-based chunk tagger. *Proceedings of the 40th Annual Meeting of the Association for Computational Linguistics (ACL)* (pp. 473-480).

Endnotes

[1] www.w3.org/TR/owl-guide/wine.owl

[2] http://www.w3.org/TR/owl-ref/

[3] http://www.w3.org/XML/

[4] http://www.w3.org/Consortium/

[5] http://protégé.stanford.edu

[6] http://protege.stanford.edu/ontologies/ontologies.html

Copyright © 2006, Idea Group Inc. Copying or distributing in print or electronic forms without written permission of Idea Group Inc. is prohibited.

Chapter VIII

Dynamic Knowledge Discovery in Open, Distributed and Multi-Ontology Systems: Techniques and Applications

Silvana Castano, Università degli Studi di Milano, Italy

Alfio Ferrara, Università degli Studi di Milano, Italy

Stefano Montanelli, Università degli Studi di Milano, Italy

Abstract

In open distributed systems like peer-to-peer networks and Grids, many independent peers, possibly spanned across multiple organizations, need to share information resources (e.g., data, documents, services) provided by other nodes. By dynamic knowledge discovery we mean the capability of each node of finding knowledge in the system about information resources that, at a given moment, best match the requirements of a request for given target resource(s). The chapter will focus on describing models and

Copyright © 2006, Idea Group Inc. Copying or distributing in print or electronic forms without written permission of Idea Group Inc. is prohibited.

techniques for ontology metadata management and ontology-based dynamic knowledge discovery in open distributed systems, by describing the architecture of a toolkit for information resource discovery and sharing developed in the HELIOS peer-based system.

Introduction

In open distributed systems like peer-to-peer networks and Grids, many indepen-dent peers, possibly spanned across multiple organizations, need to share information resources (e.g., data, documents, services) provided by other nodes. In such a distributed context, a key problem is related to *dynamic knowledge discovery*. By dynamic knowledge discovery we mean the capability of each node of finding knowledge in the system about information resources that, at a given moment, best match the requirements of a request for given target resource(s). As shown in Figure 1, we can define an open distributed system as a network of many independent peers with equal role and capabilities. Each peer exposes to the system the shared information resources together with a set of corresponding metadata, and interacts with the other parties with the intention to (1) discover peers containing relevant knowledge with respect to a target request, and (2) acquire the information resources of interest provided by the other peers. To this end, the knowledge discovery process is based on a query/ answer paradigm in which each peer in the system acts both as a client and as a server interacting with other nodes directly, by submitting queries containing a request for one or more concepts of interest and by replying to queries with concepts relevant to (i.e., matching) the target.

The following features affect knowledge discovery and sharing in open distrib-uted systems: (1) *dynamism of the system*, regards the fact that peers are allowed to join and leave the network at any moment; (2) *autonomy of nodes*, in that each peer is responsible for its own knowledge management and representation; (3) *absence of a-priori agreement*, about ontology specifica-tion vocabulary and language to be used for knowledge specification; (4) *equality of node responsibilities*, no centralized nodes with coordinating tasks are recognized and each peer enforces interaction facilities with other nodes for knowledge sharing. In this context, the following main requirements need to be addressed by providing appropriate techniques:

- **Ontology metadata management.** In order to provide a semantically rich representation of exposed information resources in terms of metadata descriptions, each peer defines a *peer ontology*. The peer ontology

Copyright © 2006, Idea Group Inc. Copying or distributing in print or electronic forms without written permission of Idea Group Inc. is prohibited.

Figure 1. Reference architecture for open distributed systems

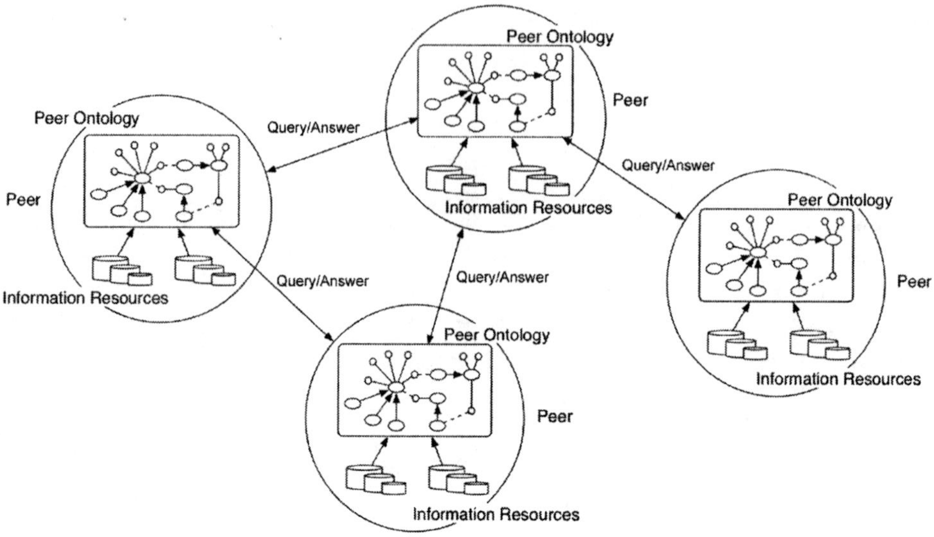

contains the knowledge the peer brings to the network and the knowledge a peer has of the network contents and it is represented using some Semantic Web compatible language (e.g., RDF(S) [W3C RDF Core Working Group, 2004], DAML+OIL [Connolly et al., 2001], OWL [W3C Web Ontology Working Group, 2004]). Since each peer can decide to adopt a different ontology language, models, and techniques for ontology metadata management are required to ensure independence from the specific language adopted by each peer for its peer ontology representation.

- **Knowledge discovery support.** In open distributed systems, no centralized authority is responsible of providing an integrated view of the overall knowledge available in the network. For this reason, each peer has to dynamically identify relevant parties with respect to a target concept of interest by means of a knowledge discovery process. A key role in open distributed systems is related to the capability of providing models and techniques capable of exploiting ontology metadata contained in peer ontologies to enforce effective dynamic knowledge discovery. Appropriate queries, called *probe queries*, are used to formulate knowledge requests to the peers of the system. A peer answers to a probe query by sending back concept descriptions in form of metadata extracted from its peer ontology. Furthermore, in distributed contexts with multiple independent peer ontologies, *ontology matching techniques* are required to compare the incoming probe queries against a peer ontology in order to identify whether there are concepts matching the request. To this end, the matching techniques have

Copyright © 2006, Idea Group Inc. Copying or distributing in print or electronic forms without written permission of Idea Group Inc. is prohibited.

to ensure flexibility by providing a wide spectrum of metrics and models to be used in correspondence of different level of richness in target concept specification.

- **Data acquisition support.** Once relevant partners are identified in the knowledge discovery process, a peer can interact with such nodes in order to acquire data/information related to one or more target resources. Conventional *search queries* are sent to access the information resource repositories stored at each peer. In order to support search query processing, appropriate techniques are required by each peer to provide the access to its information resource repositories.

The chapter will focus on describing models and techniques for ontology metadata management and ontology-based dynamic knowledge discovery, by describing the architecture of a toolkit for information resource discovery and sharing developed in the HELIOS peer-based system.

Background

The problem of dynamic knowledge discovery in open distributed systems has been addressed in the context of the Semantic Web and P2P computing (Bernstein et al., 2002; Halevy, Ives, Suciu, & Tatarinov, 2003). To enable knowledge discovery in open distributed contexts with a multitude of autonomous ontologies, appropriate semantic interoperability and matching techniques are required to determine semantic affinity between concepts of different ontologies that are semantically related. The problem of ontology matching in open distributed systems has been addressed in Bouquet, Magnini, Serafini, and Zanobini (2003) and Doan, Madhavan, Domingos, and Halevy (2002), where intelligent techniques based on a Description Logics approach are described, which compare the knowledge contained in different concept ontologies, by looking for semantic mappings denoting similar concepts. Recent research in P2P systems focuses on providing techniques for evolving from basic P2P networks supporting only file exchanges using simple filenames as metadata, to more complex systems like schema-based P2P networks, capable of supporting the exchange of structured contents (e.g., documents, relational data) by exploiting explicit schemas to describe knowledge, usually using RDF and thematic ontologies as metadata.

- **Piazza** (Halevy et al., 2003). The Piazza system proposes a solution for the semantic integration of heterogeneous information sources in a distributed

Copyright © 2006, Idea Group Inc. Copying or distributing in print or electronic forms without written permission of Idea Group Inc. is prohibited.

framework. Network nodes develop different functionalities according to their capabilities. In particular, nodes with high resource capabilities play the role of mediators in the network. The system implements a hybrid P2P solution: a mediator node receives a set of information sources schemas and executes the semantic integration step to derive an ontology view of the acquired information. A set of mediators can be organized in a hierarchy, unifying their ontologies in a global view. When a mediator receives a query from any host, it consults its own ontology and returns a list of sources eligible to offer an answer to the query. A query can be received and analyzed by more mediators, without source clusterization.

- **Edutella** (Nejdl et al., 2002). In the Edutella project, the P2P model is applied by using the JXTA protocol. The network is segmented into thematic clusters. In each cluster, a mediator semantically integrates source metadata. This approach is an example of hybrid P2P architecture, in that each source sends queries to the mediator of its own cluster, and the mediator returns a list of nodes eligible to offer semantically related information. The effective data access holds in direct network connections among peers. The mediator handles a request either directly or indirectly: directly, by answering queries using its own integrated schema; indirectly, by querying other cluster mediators.

- **Swap** (Broekstra et al., 2003). The Swap system aims at overcoming the lack of semantics in current Peer-to-Peer systems. To this purpose, an RDF(S) metadata model for encoding semantic information is introduced, allowing peers to handle heterogeneous and even contradictory views on the domain of interest. Each peer implements an ontology extraction method to extract from its different information sources an RDF(S) description (ontology) compatible with the metadata model. Such ontologies are used to perform query processing by means of the SeRQL Query Language: peers storing knowledge semantically related to a target concept are localized through SeRQL views defined on specific similarity measures. Views from external peers are integrated through an ontology merging method to extend the knowledge of the receiving peer according to a rating model.

- **Edamok** (Bouquet et al., 2003). Edamok is a P2P system aiming at realizing knowledge sharing among peer communities of interest (federations). The system is based on the concept of context of a peer, to represent the interests of the peer. All peers are equal in functionalities, and every peer can act on the network as a Seeker (when looking for documents and information) or as a Provider (when answering to incoming queries). Peers that agree to appear as a unique entity to the other peers (e.g., if they provide homogeneous contents) can form a federation. In order to point out

Copyright © 2006, Idea Group Inc. Copying or distributing in print or electronic forms without written permission of Idea Group Inc. is prohibited.

semantic mapping between concepts stored in distinct peers, the system exploits the Ctx-Match algorithm. This algorithm compares the knowledge contained in different contexts looking for semantic mappings denoting peers interested in similar concepts. These mappings are stored in order to assist the query resolution components to direct queries to peers that store relevant information.

- **Chatty Web** (Aberer, Cudrè-Mauroux, & Hauswirth, 2003). The Chatty Web project presents an approach that applies to any system which provides a communication infrastructure (e.g., networked systems, P2P systems) and offers the opportunity to study semantic interoperability as a global phenomenon in a network of information sharing communities. Each peer offers data that are organized according to some schema expressed in a data model (e.g., relational, XML, RDF). Semantic interoperability is accomplished by assuming the existence of local agreements provided as mappings between different schemas. Peers introduce their own schemas and exchanging translations between them; then peers can incrementally come up with an implicit *consensus schema* that gradually improves the global search capabilities of the system. The authors identify different methods that can be applied to establish global forms of agreement starting from a graph of local mappings among schemas and present the gossiping algorithm which is used to identify the sufficiently large set of peers capable of rendering meaningful results on a specified query.

- **Hyperion** (Arenas et al., 2003). The Hyperion project proposes an architecture for peer database management systems. These systems build a network of peers that coordinate most of the typical DBMS tasks such as querying, updating, and sharing of data. Such a network works in a way similar to conventional distributed databases. The proposed approach assumes total absence of any central authority, the absence of a global schema, transient participation of peer databases, and evolving coordination rules among different databases, but is not based on ontological description of the information resources.

With respect to these approaches, we present models and techniques that work in open distributed context without super-peer nodes and integrated schemas, in the presence of autonomous peer ontologies designed without local agreements on the knowledge representation language and the vocabulary, and without predefined mappings among peer ontologies.

Copyright © 2006, Idea Group Inc. Copying or distributing in print or electronic forms without written permission of Idea Group Inc. is prohibited.

Ontology Metadata Management

In a typical open distributed system, peer ontologies can be specified by means of different Semantic Web languages (e.g., RDF(S), DAML+OIL, OWL). Furthermore, different ontologies can describe the same resource using different descriptions, also using the same language. An important requirement for knowledge sharing and discovery is to provide a suitable model for representing and managing the ontological description of resources in a language independent, Semantic Web-compatible manner. To this purpose, we describe a reference ontology model, which is subsequently used as the basis for knowledge discovery techniques.

H-MODEL for Ontology Metadata Representation

H-MODEL provides a representation of ontology metadata in terms of concepts, properties, and semantic relations, according with the following definition. Given a set N of names and a set T of predefined basic data types (e.g., basic data types of XML schema or RDF (W3C XML Schema Working Group, 2004; W3C RDF Core Working Group, 2004)), an ontology O is defined as a 4-tuple of the form $O = (C, P, PC, SR)$, where:

- C is a set of concepts of O, where a concept $c \in C$ is defined as a pair of the form $c = (n_c, P_c)$, where $n_c \in N$ is the concept name, and $P_c \in P$ is a set, possibly empty, of concept properties.

- P is a set of properties. A property p in H-MODEL is defined as a pair of the form $p = (n_p, PC_p)$, where n_p is the property name, and $PC_p \in PC$ is a set of property constraints.

- PC is a set of property constraints. Each property constraint associates a property p with a concept c, by specifying the minimal cardinality and the property value v_p of p in c. A property constraint $pc_p \in PC_p$ is a 3-tuple of the form $pc_p = (c, k_p, v_p)$, where c is a concept, $k_p \in \{0, 1\}$ is the minimal cardinality associated with p when applied to c, and v_p is the value associated with p when applied to c, and can be a data-type $d_t \in T$ or a reference name $n \in N$. We call *strong properties* the properties with $k = 1$, and *weak properties* the ones with $k = 0$.

- SR is a set of semantic relations. A semantic relation $sr \in SR$ is defined as a binary relation of the form $sr(c,c')$, where c and $c' \in C$ are concepts and sr is the relation holding between them. Semantic relations state the kind of semantic dependency or semantic connection that holds between two

Copyright © 2006, Idea Group Inc. Copying or distributing in print or electronic forms without written permission of Idea Group Inc. is prohibited.

concepts in the ontology. To obtain a semantically rich and expressive representation of the knowledge in an ontology, H-MODEL supports the following semantic relations:

> **same-as.** The same-as relation is defined between two concepts c and c' which are considered semantically equivalent, that is, which denote the same real world entity or have the same meaning;

> **kind-of.** The kind-of relation defined between two concepts c and c' states that the concept c is a specialization of the concept c';

> **part-of.** The part-of relation defined between two concepts c and c' states that the concept c represents a component of the concept c';

> **associates.** The associates relation defined between two concepts c and c' states that a generic positive association is defined between c and c'.

A graphical representation of H-MODEL constructs is given in Table 1.

In order to show how Semantic Web ontology languages are mapped onto H-MODEL constructs, we focus on the OWL language, which provides three increasingly expressive sublanguages designed for use by specific communities of implementers and users, with different needs in terms of resource description

Table 1. Graphical representation of H-MODEL constructs

H-MODEL construct	Graphical representation
Concept (c)	◯ c
Strong property (p with $k_p = 1$)	——◉ p
Weak property (p with $k_p = 0$)	——◯ p
Property value (p with v_p)	p ◯- - - - v_p
Same-as semantic relation	◄——►
Kind-of semantic relation	——►
Part-of semantic relation	——▷
Associates semantic relation	◄——»

Copyright © 2006, Idea Group Inc. Copying or distributing in print or electronic forms without written permission of Idea Group Inc. is prohibited.

Table 2. Correspondences between OWL statements and H-MODEL constructs

OWL	H-MODEL
`<owl:Class rdf:ID="Book" />`	◯ Book

`<owl:Restriction>`
`<owl:onProperty rdf:resource="#author" />`
`<owl:minCardinality>1</owl:minCardinality>`
`</owl:Restriction>`
	——● author

`<owl:Restriction>`
`<owl:onProperty rdf:resource="#price" />`
`<owl:minCardinality>0</owl:minCardinality>`
`</owl:Restriction>`
	——◯ price

`<owl:Restriction>`
`<owl:onProperty rdf:resource="#author" />`
`<owl:allValuesFrom rdf:resource="#Person" />`
`</owl:Restriction>`
	——● - - - ◯ Person
	author

`<owl:Class rdf:about="Book">`
`<owl:equivalentClass rdf:resource="#Volume" />`
`</owl:Class>`
	Book ◯◀▶◯ Volume

`<owl:Class rdf:about="Book">`
`<owl:subClassOf rdf:resource="#Publication" />`
`</owl:Class>`
	Book ◯▶◯ Volume

`<owl:Class rdf:ID="Technical_Journal">`
`<owl:intersectionOf rdf:parseType="Collection" />`
`<owl:Class rdf:about="#Journal />`
`<owl:Class rdf:about="#Technical_Publication />`
`</owl:intersectionOf>`
`</owl:Class>`
	Technical_Publication ◯
	
	◯ Journal
	◯ Technical_Journal

`<owl:Class rdf:ID="Conference_Publication">`
`<owl:unionOf rdf:parseType="Collection" />`
`<owl:Class rdf:about=`
`        "#Conference_Proceedings />`
`<owl:Class rdf:about=`
`        "#Workshop_Proceedings />`
`</owl:unionOf>`
`</owl:Class>`
	Conference_Publication ◯
	
	◯ Conference_Proceedings
	◯ Workshop_Proceedings

accuracy (W3C Web Ontology Working Group, 2004). We show how a subset of the OWL constructs is mapped onto H-MODEL (see Table 2), in order to capture the ontology elements that are used for knowledge discovery and to represent them in terms of concepts, properties and semantic relations.

Each concept class declaration in OWL is represented by a concept c in H-MODEL. The concept name n_c is set by referring to the RDF ID associated with the class declaration in OWL. Each property declaration in OWL is represented by a property p in H-MODEL. The property name n_p is set by referring to the RDF ID associated with the property declaration. The property cardinality k_p as well as the property value v_p are enforced by exploiting OWL property restrictions. In particular, OWL cardinality restrictions are exploited for setting the cardinality of p, while the AllValuesFrom, SomeValuesFrom, and hasValue OWL

Copyright © 2006, Idea Group Inc. Copying or distributing in print or electronic forms without written permission of Idea Group Inc. is prohibited.

clauses are exploited for setting the value of the property, which can be a data type, or a reference name that represents the name of a concept or of an individual in OWL. For representing the semantic relations and the class operators used in OWL, same-as and kind-of semantic relations of H-MODEL are used. In particular, the same-as and the kind-of relations correspond to the equivalentClass and the subClassOf relations in OWL, respectively. Moreover, the kind-of relation is also used for mapping the intersectionOf, and the unionOf OWL operators. For an intersection clause of the form $A = B \cap C$ we set two kind-of relations of the form A kind-of B and A kind-of C. For an union clause of the form $A = B \cup C$ we set two kind-of relations of the form B kind-of A and C kind-of A. The part-of and the associates relations are not used for OWL constructs. Rather, they are used for those ontology languages and formalisms that support the definition of aggregations and associations between classes (e.g., UML, semantic database models). In fact, even if UML has been designed for building systems in object-oriented programming languages, and extended to the design of database schemas, there have been proposals for using UML also for ontology representation (Cranefield & Purvis, 1999).

Models and Techniques
for Knowledge Discovery

Knowledge discovery in open, distributed, and multi-ontology systems requires appropriate matching models and techniques in order to find concepts of a peer ontology that have a semantic affinity with (i.e., that match) given target concept(s). The problem of ontology matching in multi-ontology contexts introduces a number of challenging requirements to be addressed.

- A key requirement of ontology matching is the capability of coping with different levels of detail in specifying target concept(s) in probe queries, ranging from simply concept names to more structured concept specifications, with properties and/or semantic relations.

- Ontology matching must consider both the linguistic features and the contextual features of concept specifications in order to suitably capture concept meaning. Linguistic features refer to names of ontology elements and their meaning. Contextual features refer to the structure of a concept in terms of properties and concepts directly related to the considered concept.

Copyright © 2006, Idea Group Inc. Copying or distributing in print or electronic forms without written permission of Idea Group Inc. is prohibited.

- Ontology matching algorithm should be dynamically configurable in order to cope with different levels of detail of target concept specifications, with different combinations of linguistic and contextual features.

We address the problem of ontology matching with reference to the H-MATCH algorithm developed to address the requirements of ontology matching in multi-ontology contexts. Furthermore, we describe the role of H-MATCH in processing probe queries for dynamic knowledge discovery.

Models for Ontology Matching

The goal of ontology matching is to evaluate the level of affinity between target concept(s) (e.g., concepts specified in a probe query) and the concepts specified in a peer ontology. The meaning of ontology concepts depends basically on the names chosen for their definition (linguistic features) and on their properties and their relations with other concepts (contextual features).

- **Linguistic features.** To capture the meaning of names of concept specification, we exploit a thesaurus *Th* of terms and terminological relationships between them. *Th* is automatically derived from the lexical system WORDNET (Miller, 1995) where the different senses of English words are grouped by synonymy. The sets of synonyms (*synsets*) are organized hierarchically (i.e., each synset is connected to more general and more specific concepts by hypernymy and hyponymy relationships) and other semantic relations (e.g., meronymy) are available so as to build a semantic net. In the thesaurus construction, the WORDNET relationships are mapped onto the three terminological relationships supported by H-MATCH: (1) SYN (synonym): it is defined between two terms t_i and t_j that can be indifferently used to denote a certain concept. The SYN relationship is derived from synsets in WORDNET; (2) BT/NT (Broader/Narrower Term): BT relationship is defined between two terms t_i and t_j such has t_i has a broader, more general meaning than t_j; NT is the opposite of BT. BT/NT relationships correspond to hypernymy and hyponymy relationships in WORDNET, respectively; (3) RT (Related Term): it is defined between two terms t_i and t_j that are generally used together in the same context, either because t_i denotes a part of t_j or because t_i and t_j are specifications of a common term t_k. RT corresponds to meronymy relationship and coordinate terms in WORDNET, respectively.
- **Contextual features.** The context of an ontology concept c is composed by the properties of c (denoted by P_c) together with their corresponding

Copyright © 2006, Idea Group Inc. Copying or distributing in print or electronic forms without written permission of Idea Group Inc. is prohibited.

property constraints (denoted by PC_c) and by the concepts directly related to c, in the following called *adjacents* (denoted by C_c). The context Ctx_c of a concept c is thus defined as the union of the properties, the property constraints and the adjacents of c, that is, $Ctx_c = P_c \cup PC_c \cup C_c$. The context features of a concept express the level of detail of a concept description. In particular, different levels of detail in target concept description can characterize probe queries. The H-MATCH algorithm provides four different matching models that span from surface to intensive matching in order to deal with different levels of richness of target concept(s) description in a probe query. The H-MATCH matching models are defined as follows.

- ➤ **Surface Model:** it is defined to consider only the names of concepts in the matching process. Surface matching is suited for dealing with probe queries that provide high-level, poorly structured concept descriptions (e.g., only names of target concepts).

- ➤ **Shallow Model:** it is defined to consider both concept names and concept properties. With this model, we process probe queries that describe not only the concept names but also information about the presence of properties and about their cardinality constraints.

- ➤ **Deep Model:** it is defined to consider concept names and the whole context of concepts, by considering also semantic relations. The deep model is suited for probe queries that describe a target concept by specifying also the relations it has with other concepts.

- ➤ **Intensive Model:** it is defined to consider, in addition to the features of the deep model, also property values, for processing probe queries with the highest level of detail in concept description. In fact, by adopting the intensive model not only the presence and cardinality of properties, but also their values have an impact on the resulting semantic affinity value.

The H-MATCH Algorithm

The H-MATCH algorithm performs ontology matching by adopting a metric-based approach (Castano, Ferrara, & Montanelli, 2003b). To this end, a weight is associated with linguistic and contextual features in order to express the implication of each feature for semantic affinity evaluation. In particular, a weight W_{tr} is associated with each terminological relationship $tr \in \{$SYN, BT/NT, RT$\}$ in Th, with $W_{SYN} \geq W_{BT/NT} \geq W_{RT}$. In fact, synonymy is generally considered a more precise indicator of affinity than other relationships, consequently $W_{SYN} \geq W_{BT/NT}$. The lowest weight is associated with RT since it denotes a more

Copyright © 2006, Idea Group Inc. Copying or distributing in print or electronic forms without written permission of Idea Group Inc. is prohibited.

generic relationship than BT/NT [1]. Regarding contextual features, a weight W_{sr} is associated with each semantic relation to express the strength of the connection expressed by the relation on the involved concepts for semantic affinity evaluation purposes. In particular, we have $W_{same-as} \geq W_{kind-of} \geq W_{part-of} \geq W_{associates}$, where the greater the weight associated with a semantic relation, the higher the strength of the semantic connection between concepts. Finally, we associate a weight W_{sp} with strong properties, and a weight W_{wp} with weak properties, respectively, with $W_{sp} \geq W_{wp}$ to capture the presence of the property in characterizing the concept for matching purposes. In fact, strong properties are mandatory related to a concept and are thus relevant to characterize its structural description. Weak properties are optional for the concept in characterize its structure, and, as such, are less important in featuring the concept than strong properties.

The H-MATCH algorithm exploits the linguistic and contextual features of two concepts c and c' for evaluating a semantic affinity coefficient $SA(c,c')$ in the range $[0,1]$ that express their level of matching according to one of the matching models described above. The semantic affinity $SA(c,c')$ is obtained as the weighted sum of two affinity coefficients $LA(c,c')$ and $CA(c,c')$ that provide a measure of concept matching based on their linguistic affinity and their contextual affinity, respectively. H-MATCH is described in Algorithm 1.

The input of the H-MATCH algorithm is constituted by two ontological concepts to be matched and by three parameters. The first parameter specifies the matching model to be used. In open distributed systems, it is specified in the incoming probe query by taking into account the level of detail of the target concept specification. The second parameter is the threshold value set for specifying a minimum semantic affinity value required to consider the considered concepts as matching concepts. The third parameter is the weight associated with the linguistic affinity, which expresses the relevance that linguistic features have in the semantic affinity evaluation. In order to calculate the comprehensive semantic affinity value between the concepts, the algorithm dynamically composes the context of each concept according to the matching model specified. In the case of the surface model, the weight associated with the linguistic affinity is automatically set to 1 (and the contextual affinity is set to 0) because the surface matching takes into account only the concept names. In the remaining cases, the contextual affinity is evaluated by considering the contextual features involved in the corresponding matching model. The linguistic and contextual affinity coefficients are evaluated as follows.

- **Linguistic affinity coefficient.** The linguistic affinity coefficient $LA(c,c')$ $\rightarrow [0, 1]$ between two concepts c and c' is equal to the value of the highest-strength path of terminological relationships between their names n_c and $n_{c'}$.

Copyright © 2006, Idea Group Inc. Copying or distributing in print or electronic forms without written permission of Idea Group Inc. is prohibited.

Algorithm 1. The H-MATCH algorithm

Input:

    c, c': two ontological concepts.

    model: a matching model {surface, shallow, deep, intensive}

    threshold: a threshold t (0,1] specifying the minimum semantic affinity value required to consider c and c' as matching concepts

    W_{LA}: a weight [0,1] expressing the relevance of the linguistic affinity in the semantic affinity evaluation

Output:

    $SA(c,c')$: the semantic affinity value of c and c'

Method:

(1)   Evaluate the linguistic affinity $LA(c,c')$ between c and c' by taking into account their names n_c and $n_{c'}$, respectively;

(2)   **Switch** *model* //chooses the matching model

(3)   **Case** surface:

(4) $W_{LA} = 1$;

(5) $CA(c,c') = 0$;

(6)   **Case** shallow: // The context considers only concept properties

(7) $Ctx_c = P_c$;

(8) $Ctx_{c'} = P_{c'}$;

(9) $CA(c,c') = CA_{shallow}(c,c')$;

(10)   **Case** deep: // The context considers both properties and adjacents

(11) $Ctx_c = P_c \cup C_c$;

(12) $Ctx_{c'} = P_{c'} \cup C_{c'}$;

(13) $CA(c,c') = CA_{deep}(c,c')$;

(14)   **Case** intensive: // The context considers properties, adjacents, and property constraints

(15) $Ctx_c = P_c \cup C_c \cup PC_c$;

(16) $Ctx_{c'} = P_{c'} \cup C_{c'} \cup PC_{c'}$;

(17) $CA(c,c') = CA_{intensive}(c,c')$;

(18)   **EndSwitch**;

(19)   **If** $W_{LA} \cdot LA(c,c') + (1 - W_{LA}) \cdot CA(c,c') \geq t$ **Then**:

(20) $SA(c,c') = W_{LA} \cdot LA(c,c') + (1 - W_{LA}) \cdot CA(c,c')$; /*The semantic affinity balances the impact of the linguistic features and contextual features*/

(21)   **Else**:

(22) $SA(c,c') = 0$;

(23)   **EndIf**;

in *Th* if at least one path exists, and is zero otherwise. Path strength is computed by multiplying the weights associated with each terminological relationship involved in the path.

Copyright © 2006, Idea Group Inc. Copying or distributing in print or electronic forms without written permission of Idea Group Inc. is prohibited.

- **Contextual affinity coefficient.** The contextual affinity coefficient $CA(c,c')$ → [0, 1] between two concepts c and c' is proportional to the number of matching elements (i.e., properties and adjacents) they have in their contexts and to their level of matching. The level of matching of two context elements is computed by considering their linguistic affinity and the kind of relation they have with c and c', respectively. Depending on the matching model, the considered context elements can be either only properties (shallow) or properties and semantic relations (deep). Furthermore, when the intensive matching is adopted, the level of matching of property values is also considered in the computation of the contextual affinity coefficient. In the computation of $CA(c,c')$, each context element $e \in Ctx_c$ is compared with each context element $e' \in Ctx_{c'}$ in order to evaluate their level of matching. For a given element e, if only one matching element e' is identified, the corresponding matching value between e and e' is considered for the computation of CA. In case that more than one matching element e' is identified, the best matching pair (e,e') is considered for the evaluation of $CA(c,c')$, that is, the pair with the highest matching value. If no matching element e' is found, the best matching value for e is set to zero (i.e., e is not considered for CA computation). Finally, CA is computed as the ratio of the sum of the best matching value for each $e \in Ctx_c$ to the number of elements considered in the context of c (i.e., the cardinality of Ctx_c).

The H-Match algorithm evaluates a comprehensive measure of semantic affinity $SA(c,c')$ between the considered concepts as the weighted sum of their linguistic and contextual affinity, which is returned by the algorithm if this value is greater than or equal to the specified threshold. A detailed description of the H-Match algorithm and related matching techniques can be found in (Castano, Ferrara, Montanelli, & Racca, 2004b). We have analyzed the performance and the effectiveness of H-Match on real ontology matching case studies and experimental results on this topic can be found in (Castano, Ferrara, Montanelli, & Racca, 2005).

Example 1. As an example of matching, we consider a target concept Publication and a concept Book and we evaluate their semantic affinity by using H-Match with the deep matching model by setting the linguistic affinity weight $W_{LA} = 0.5$ and the threshold $t = 0.5$ as input. Moreover, the following predefined weights are used for H-Match configuration: $W_{SYN} = 1$, $W_{BT/NT} = 0.8$ $W_{RT} = 0.5^2$ and $W_{same-as} = 1$, $W_{kind-of} = 0.8$, $W_{part-of} = 0.7$, $W_{associates} = 0.3$, $W_{sp} = 1$, $W_{wp} = 0.5^3$. According to the H-Model representation of Figure 2, the context of Publication is composed of two weak properties (i.e., Publication.title and Publication.author) and the context of Book is composed of two weak properties (i.e., Book.author

Copyright © 2006, Idea Group Inc. Copying or distributing in print or electronic forms without written permission of Idea Group Inc. is prohibited.

Figure 2. H-MODEL representation of Publication and Book concepts

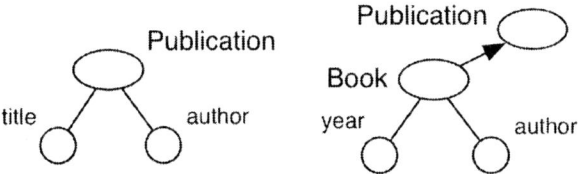

and Book.year) and a kind-of relation with the concept Publication. For what concern the linguistic affinity coefficient, in the thesaurus we have a NT relationship between Book and Publication, because in WORDNET the term Book is a hyponymy of the term Publication. Consequently, LA(Publication,Book) = 0.8, according to the weight defined for the NT terminological relationship. Regarding the contextual affinity coefficient, the deep matching model requires that both properties and adjacents have to be taken into consideration. Considering the Publication context, Publication.author is compared with the properties Book.author and Book.year, and with the adjacent concept Publication in the Book context. We find that Publication.author matches only Book.author with a matching value 1, due to the fact that their names coincides and they are both weak properties. No matching elements are identified in the Book context for the property Publication.title, then this latter property is not considered in CA evaluation. On the basis of the identified matching elements, CA(Publication,Book) $= (1 + 0) / 2 = 0.5$. Moreover, SA(Publication, Book) $= W_{LA} \cdot LA$(Publication,Book) $+ (1 - W_{LA}) \cdot CA$(Publication,Book)$=0.65$. Since $SA = 0.65 > t = 0.5$, Publication and Book are returned as matching concepts by H-MATCH.

Probe Query for Knowledge Discovery

The probe query type provides an expressive representation of target resources in terms of ontology concept specification (target concept[s]). A probe query template for information resource discovery is provided in Figure 3 and it is composed of the following clauses:

- **Find:** list of target concept(s) names.
- **With:** (optional) list of properties of the target concept(s).
- **Where:** (optional) list of conditions to be verified by the property values, and/or (optional) list of concepts related to the target by a semantic relation.

Copyright © 2006, Idea Group Inc. Copying or distributing in print or electronic forms without written permission of Idea Group Inc. is prohibited.

Figure 3. Probe query template

Probe query template	
Find	Target concept name [, ...]
[With	<Property name> [, ...]]
[Where	Condition, <related concept, semantic relation name> [, ...]]
[Matching model	Matching model to be used (Surface|Shallow|Deep|Intensive)]
[Matching threshold	t ∈ (0, 1]]

- **Matching model:** (optional) specification of the matching model asked by the requesting peer to process the query (i.e., surface, shallow, deep, intensive).
- **Matching threshold:** (optional) specification of the threshold value t, with $t \in (0, 1]$ to be used for the selection of matching concepts based on the semantic affinity value determined by the matching process. If a matching threshold is not specified in the query, the answering peer adopts its own default threshold.

The answer to a probe query is list of concepts that match the target. As described in Figure 4, the answer template contains the following clauses:

- **Concept:** name of the matching concept.
- **Properties:** (optional) list of properties of the matching concept.

Figure 4. Probe answer template

Probe answer template	
{Concept	Concept name
[Properties	<Property name> [, ...]]
[Adjacents	<related concept, semantic relation name> [, ...]]
Matching	<target concept, affinity value>[, ...]
[Matching model	Matching model used (Surface|Shallow|Deep|Intensive)]
[Matching threshold	t ∈ (0, 1]]}

Copyright © 2006, Idea Group Inc. Copying or distributing in print or electronic forms without written permission of Idea Group Inc. is prohibited.

- **Adjacents:** (optional) list of concepts related to the matching concept by a semantic relation.

- **Matching:** set of pairs <target concept, affinity value>, specifying the target concept with which the matching concept matches, together with the corresponding affinity value.

- **Matching model:** specification of the matching model applied to process the query.

- **Matching threshold:** (optional) specification of the threshold value t, with $t \in (0, 1]$ used for the selection of matching concepts based on the semantic affinity value determined by the matching process.

A peer interested in discovering nodes capable to provide information resources semantically related to a given target, composes and submits to the system a probe query according to the query template of Figure 3. Receiving a probe query, a peer uses the H-MATCH algorithm in order to identify if there are concepts matching the target request. In particular, an ontological description of the target concept(s) (extracted from the Find, With, and Where clauses), as well as the matching model and the threshold to apply (derived from the Matching model and the Threshold clauses, respectively) are provided to the matching algorithm. As a result, H-MATCH returns a (possibly empty) ranked list of concepts semantically related to the target, and, for each entry, the corresponding semantic affinity value. Finally, the results of the matching process are organized according to the answer template in Figure 4, and such an answer is replied back to the requesting peer. Collecting query replies from answering nodes, the requesting peer evaluates the results and decides whether to further interact with the nodes found to be relevant in order to access their information resources by means of appropriate search queries for information acquisition, as described in the next section.

Example 2. As an example of knowledge discovery, we consider an open distributed system composed of four peers, namely P_1, P_2, P_3, and P_4. As shown in Figure 5, the peer ontology of each peer contains concepts related to the publishing domain.

We suppose that P_1 intends to discover whether peers in the system can provide relevant information resources about the Publication concept. To this end, P_1 composes and submits to the system the query in Figure 6.

Exploiting its query propagation strategy[4], P_1 sends the query to P_2 and P_3. P_2 and P_3 catches the incoming query and invokes the H-MATCH algorithm which performs ontology matching using the deep model to evaluate the semantic affinity between the incoming query and the respective peer ontology. According

Copyright © 2006, Idea Group Inc. Copying or distributing in print or electronic forms without written permission of Idea Group Inc. is prohibited.

Figure 5. An example of knowledge discovery in an open distributed system

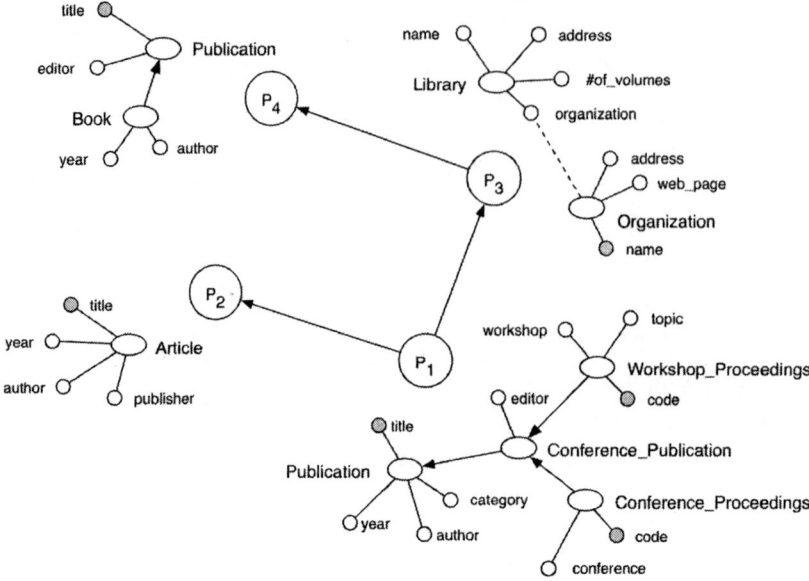

Figure 6. The probe query submitted by P_1

Find	Publication
With	title, author
Matching model	Deep
Matching threshold	0.5

to the H-MATCH results, P_2 replies to P_1 with the answer shown in Figure 7(a). Following the same procedure, P_3 does not identify matching concepts over the specified threshold in its peer ontology, since the matching results between the concepts Library and Organization and the target concept Publication are *0* and *0.125,* respectively.

Nevertheless, according to its query propagation strategy, P_3 forwards the query to P_4. According to its H-MATCH results, P_4 replies directly to P_1 with the answer shown in Figure 7(b).

Copyright © 2006, Idea Group Inc. Copying or distributing in print or electronic forms without written permission of Idea Group Inc. is prohibited.

Figure 7. The probe answer provided by (a) P_2 and (b) P_4

Concept	Article
Properties	title, author, year, publisher
Matching	<Publication, 0.675>
Matching model	Deep
Matching threshold	0.5
(a)	
Concept	Book
Properties	author, year
Adjacents	<Publication, kind-of>
Matching	<Publication, 0.65>
Matching model	Deep
Matching threshold	0.5
Concept	Publication
Properties	title, editor
Adjacents	<Book, kind-of>
Matching	<Publication, 0.75>
Matching model	Deep
Matching threshold	0.5
(b)	

Models and Techniques for Information Resource Data Acquisition

Search queries are used by a peer in order to acquire resource data from another node of the network. When a requesting peer has identified a partner containing relevant knowledge using probe queries, it can subsequently send a search query in order to access and acquire data about the target information resources. Search queries contain the Find, With and Where clauses as in the probe query template. The Find clause is used for selecting the concept of interest. The With clause is used for specifying the property that have to be returned in the answer. Finally, the Where clause is used for specifying the conditions to apply to one or more concept properties. In order to support search query processing, each peer provides appropriate techniques to access its repositories containing resource data.

- **Web Service-Based Techniques.** A peer provides a standard access to its shared information resources by means of a Web service. Standard protocols (e.g., SOAP, WSDL) can be adopted in order to interact with the Webservice and provide a seamless access to the underlying information resources. The requesting peer interacts with the Web Service by means

Copyright © 2006, Idea Group Inc. Copying or distributing in print or electronic forms without written permission of Idea Group Inc. is prohibited.

Figure 8. The search query submitted by P₁

Find	Article
With	title, author, year, publisher
Where	year = 2004

of the SOAP protocol which supports well-defined XML-based message communications. The WSDL document provides the specification of the set of public methods as well as the structure of the returned data extracted from the information resources.

- **Wrapper-Based Techniques.** The access to the shared information resources is provided by means of a wrapper service. The requesting peer interacts with the wrapper service by submitting target queries for information resource access. The wrapper service manages mapping rules defining the concepts of the peer ontology and the underlying information resources in order to reformulate the target query in terms of queries over specific structures of the repository where the information resources are stored (e.g., relational database structure). Finally, the answer to a search query is sent back to the requesting peer as an XML document containing the information about the shared resources and an additional contentURI element that specifies the location where eventually related files can be downloaded.

Example 3. As an example of data acquisition, we consider the scenario in Figure 5. Collecting the results of the knowledge discovery process, P_1 decides to interact with P_2 in order to acquire the data related to the Article concept. To this end, P_1 sends to P_2 the search query shown in Figure 8.

P_2 processes the incoming search query by using a wrapper service and composes a XML document that contains the answer. In Figure 9, we show a portion of the returned XML document that describes two articles that fit the search query, together with the URIs that P_1 can use for downloading the associated files.

Copyright © 2006, Idea Group Inc. Copying or distributing in print or electronic forms without written permission of Idea Group Inc. is prohibited.

Figure 9. A portion of the XML document sent back to P₁

```
...
<Article>
  <title>
    From Surface to Intensive Matching of Semantic Web Ontologies
  </title>
  <author>S. Castano, A. Ferrara, S. Montanelli, G. Racca</author>
  <year>2004</year>
  <publisher>IEEE</publisher>
  <contentURI>http://islab.dico.unimi.it/WEBS04.pdf</contentURI>
</Article>
<Article>
  <title>
    On Combining a Semantic Engine and Flexible Network Policies
    for P2P Knowledge Sharing Networks
  </title>
  <author>
    S. Castano, A. Ferrara, S. Montanelli, E. Pagani, G. P. Rossi,
    S. Tebaldi
  </author>
  <year>2004</year>
  <publisher>IEEE</publisher>
  <contentURI>http://islab.dico.unimi.it/GLOBE04.pdf</contentURI>
</Article>
...
```

The Helios Tool Architecture for Knowledge Discovery in Open Distributed Systems

In this section, we describe the tool architecture of Helios (Helios EvoLving Interaction-based Ontology knowledge Sharing) for knowledge discovery and sharing in peer-based open distributed systems (Castano, Ferrara, Montanelli, & Zucchelli, 2003a). As described in Figure 10, each peer of the Helios network provides a semantically rich representation of the information resources to be shared by means of a peer ontology described according to the H-Model specification. Each Helios peer is equipped with the Helios toolkit and adopts the probe/search approach for interacting with the other parties of the network in order to enforce knowledge discovery and information acquisition functionalities.

In the context of Helios, many peers can be involved in the network. In a typical scenario, a peer P activates a knowledge discovery process when it needs to dynamically identify the subset of peers that contain relevant knowledge (i.e.,

Copyright © 2006, Idea Group Inc. Copying or distributing in print or electronic forms without written permission of Idea Group Inc. is prohibited.

Figure 10. The HELIOS architecture

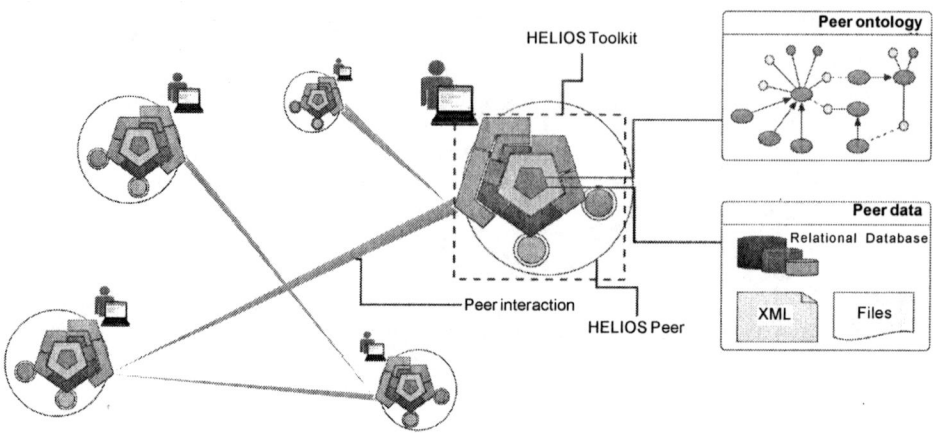

matching concepts) with respect to the target concept(s) specified in a probe query Q. The peer P exploits collected answers for defining semantic routing rules in order to restrict the subsequent propagation of other related queries only to subset of peers that have answered to Q. In other words, the peer P exploits collected answers to a probe query Q for storing information about its *semantic neighbors* just discovered. To this end, a peer ontology is organized in a two-layer architecture where the upper layer represents the *content knowledge* and the lower layer represents the *network knowledge* of a given peer, respectively. The content knowledge layer describes the knowledge a peer brings to the network as a network of H-MODEL concepts. The network knowledge layer describes the knowledge that peer P has of other peers of the network it has interacted with. When peer P receives a concept from another peer P_1 as an answer to a probe query, it stores in the network knowledge layer a description of the peer P_1. Peer descriptions are given in form of network concepts, characterized by a set of properties describing the network features of P_1. The knowledge discovery process can proceed in different ways. The peer P can send search queries asking for data about target resources related to the probe query Q. Search queries will be propagated only to semantic neighbors thus enforcing a semantic query distribution process and reducing the overall network traffic. Preliminary results on this topic can be found in (Castano et al., 2004a). Another way of exploiting semantic neighbors discovered through a probe query Q is for continuing the knowledge discovery process by sending further probe queries, devoted to describe more focused target concepts than those specified in query Q. Also these further probe queries will be propagated only to semantic

Copyright © 2006, Idea Group Inc. Copying or distributing in print or electronic forms without written permission of Idea Group Inc. is prohibited.

neighbors previously discovered, since they contain potentially relevant knowledge (Castano et al., 2003a).

In the following, we describe the HELIOS components related to ontology metadata management and dynamic knowledge discovery (see Figure 11).

Ontology Metadata Management in HELIOS

In HELIOS, each peer can define its own peer ontology by acquiring an external, pre-defined, ontology or by composing a new one using H-MODEL. HELIOS supports the acquisition of ontologies represented in a Semantic Web ontology language (i.e., RDF(S), DAML+OIL, and OWL) by means of the *ontology wrapper manager* that maps the construct of the original ontology language onto the H-MODEL constructs. The ontology wrapping manager is conceived for acquiring ontological descriptions represented using both the OWL/DAML+OIL languages for ensuring compatibility with the Semantic Web sources. In order to make the transformed ontology representation persistent, the peer ontology elements are stored in the H-MODEL metadata repository, implemented through a relational database (*H-MODEL DB*). In Figure 12, we show the ER schema of

Figure 11. The HELIOS toolkit

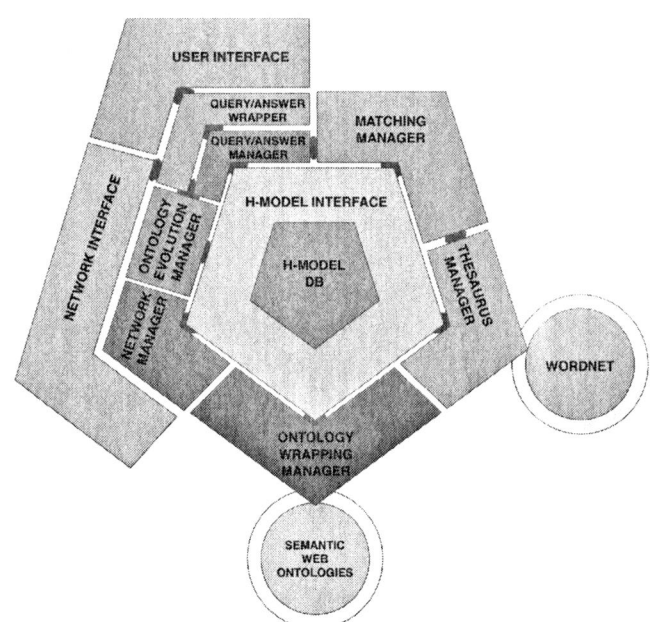

Copyright © 2006, Idea Group Inc. Copying or distributing in print or electronic forms without written permission of Idea Group Inc. is prohibited.

the metadata repository, which is conceived for storing the ontological description composed or acquired by a peer, and for storing information required for backward compatibility towards the original representation language.

The Ontology entity of the metadata schema represents information about the different ontology descriptions stored in the H-MODEL repository. This entity corresponds to the ontology element of OWL. In addition to the metadata information supported by OWL, the originalLang attribute is used for storing information about the original representation language of the stored ontology. Each ontology is identified by a URI that represents its default namespace in OWL, and can contain references to other external namespaces. Each namespace URI is stored into the XMLNS entity, and can be associated with ontologies by specifying a prefix, according to the XML/RDF syntax of OWL. An ontological element is represented by the Element entity that is associated with a namespace, which specifies that the element belongs to a stored ontology. Each element is identified by a H-MODEL unique identifier (ID), by the date of the last acquisition of information about it, by the original specification used for representing it in the original language, and by the element name. An ontology element can be a Concept, a Property, a Dataype, or an Instance. A concept can be a Content Concept or a Network Concept. Each content concept can have semantic relations with other content concepts. The semantic relation association between content concepts represents a semantic relation in the ontology, and is characterized by a type, which is one of the semantic relations supported by H-MODEL. When a semantic relation represented in OWL is stored in the repository, its original type is recorded by means of the originalType attribute. The location association is used for representing the fact that a content concept is located at zero or more peers (represented as network concepts) on the network. A property is represented in the metadata repository by the Property entity, which is characterized by information about the property definition in the original language. In particular, properties can be object or datatype properties, for backward compatibility towards OWL. The functional, transitive, and symmetric attributes store information about the original definition of the property type in OWL. The property relation association is used for representing the property hierarchy that can be specified in OWL, by setting a subPropertyOf or an inverseOf relation between two properties. In H-MODEL, each concept is associated with a set of properties by means of property constraints. In the repository, the hasProperty association between the Concept and the Property entity represents property constraints. The hasProperty association provides information about the cardinality associated with the property constraint. In particular for OWL restrictions, we consider the fact that the OWL ObjectProperty is used for linking two concepts by means of a concept property. In this case, we store the filler of the relation specified by the OWL restriction (i.e., AllValuesFrom, SomeValuesFrom, hasValue), and the value associated with the property. The

Copyright © 2006, Idea Group Inc. Copying or distributing in print or electronic forms without written permission of Idea Group Inc. is prohibited.

Figure 12. The ER schema of the H-MODEL metadata repository

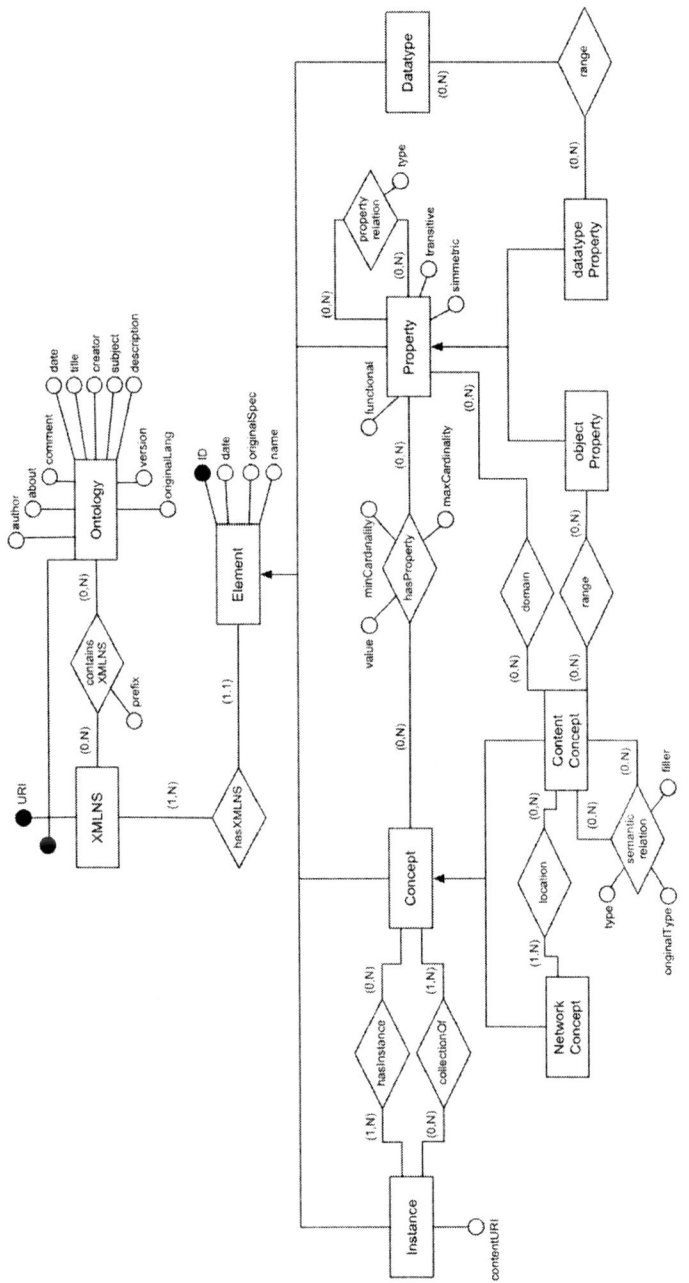

Copyright © 2006, Idea Group Inc. Copying or distributing in print or electronic forms without written permission of Idea Group Inc. is prohibited.

domain and range associations between content concepts and properties are used for storing information provided by the OWL domain and range clauses, respectively. OWL individuals are represented by means of the Instance entity that can be associated with a contentURI that refers to the XML document describing data represented by the instance. The Concept entity is associated with the Instance entity by the hasInstance and by the collectionOf associations, respectively. The hasInstance association stores information about the concept instances, while the collectionOf association stores information about those concepts that are defined as a collection of instances in the OWL representation.

Dynamic Knowledge Discovery in HELIOS

Dynamic knowledge discovery is supported in HELIOS by exploiting the toolkit components for probe query processing and ontology matching. The *query/ answer wrapper* and the *query/answer manager* are exploited for supporting the query processing and the answer composition according to the probe approach. Probe query processing is performed by exploiting the *matching manager* considering the parameters associated with the query (e.g., threshold, matching model). The matching manager performs concept matching against a peer ontology by exploiting the H-MATCH algorithm. It exploits the *thesaurus manager* for acquiring terminological relationships among the terms that constitute the linguistic features of concepts. The thesaurus manager interfaces WORDNET in order to capture the terminological relationships holding between the terms used as concept or property names in a given peer ontology. Moreover, the thesaurus manager supports the inclusion of new, user-defined, terminological relationships in the thesaurus. The *network manager* is responsible for defining a query propagation policy. To this end, a set of semantic routing rules are be defined in order to restrict the query propagation to a subset of nodes following semantic criteria. Some initial results on this topic can be found in Castano et al. (2004a). The *network interface* is responsible for managing query propagation in the network. Given a query from the query/answer manger, the network interface interacts with the network manager to exploit appropriate semantic routing rules in order to select the best recipients for each query. Finally, the *user interface* provides a graphical access to the functionalities of the HELIOS toolkit for supporting user interaction. HELIOS users can perform queries and visualize answers by exploiting the query/answer wrapper and manager, and can manage the peer ontology through the ontology evolution manager. The *ontology evolution manager* is responsible for managing the peer ontology and supporting the acquisition of new knowledge from the network. When a new concept is received in a query answer from other peers, the ontology evolution manager will support the acquisition of the new knowledge

Copyright © 2006, Idea Group Inc. Copying or distributing in print or electronic forms without written permission of Idea Group Inc. is prohibited.

according to different levels of severity. The ontology evolution manager will rely on the matching results produced by the matching manager for suggesting to the user the most appropriate location of the new concept in the peer ontology and possible semantic relations among it and existing concepts. The ontology evolution manager of HELIOS is at an initial stage of development and future research work will be devoted to this matter.

Future Trends

In this section, we discuss main directions and trends related to the use of ontologies in the context of open distributed systems, with reference to ontology-based resource discovery, ontology matching and evolution, and ontology-based routing in open distributed systems.

- **Ontology-Based Resource Discovery.** Resource discovery capabilities play a key role for open distributed systems. Recent research focuses on providing techniques for evolving from basic systems supporting only file sharing using simple filenames as metadata and exact matching techniques, to more complex systems, capable of supporting the exchange of structured contents (e.g., documents, relational data) by exploiting explicit schemas to describe knowledge, usually using RDF and thematic ontologies as metadata. In literature, research issues related to ontology-based resource discovery have been addressed in Nejdl et al. (2002) where a Peer-to-Peer infrastructure for metadata sharing in RDF format is developed. Further approaches based on pure or hybrid P2P architectures have been proposed also (Halevy et al., 2003; Kementsietsidis, Arenas, & Miller, 2003). The ontology-based resource discovery problem has also been addressed in Grid systems where resources are assigned to tasks according to given task requirements and resource policies which dynamically change (Tangmunarunkit, Decker, & Kesselman, 2003). Generally, open distributed architectures are organized according to two main approaches: the hierarchical approach supporting one or more centralized authorities with discovery capabilities (Halevy et al., 2003) and the completely distributed approach where nodes dynamically interact in order to identify members storing relevant concepts (Broekstra et al., 2003). The former approach is more efficient, but requires appropriate structures to manage knowledge discovery functionalities which can be difficult to maintain (e.g., integrated and centralized ontologies, predefined mappings between ontologies). The latter approach is more flexible and reliable, but requires query propagation

Copyright © 2006, Idea Group Inc. Copying or distributing in print or electronic forms without written permission of Idea Group Inc. is prohibited.

strategies and matching techniques in order to identify relevant members with respect to each target query. Research work in this direction is related to the development of advanced approaches where ontology-based matching techniques and semantic routing rules are used to improve the efficiency of knowledge discovery and sharing capabilities of each peer (Castano et al., 2004a).

- **Ontology Matching and Evolution.** To enable resource discovery and sharing in distributed contexts with a multitude of autonomous ontologies, appropriate matching and evolution techniques are required. In literature, research issues related to ontology matching and mapping in open systems has been addressed in Bouquet et al. (2003) and Doan et al. (2002), where intelligent techniques based on a Description Logics approach are described, which compare the knowledge contained in different concept ontologies, by looking for semantic mappings denoting similar concepts. Additional issues related to ontology matching regard the problem of inferring hidden properties of concepts and related mappings, for ontology merging purposes (Maedche, Motik, Stojanovic, Studer, & Volz, 2003a, 2003b). Research work in this area is related to the development of flexible and scalable ontology matching techniques in order to cope with ontologies which evolve quickly and are characterized by different levels of accuracy in resource description typical of dynamic scenarios like open distributed systems. Regarding the topic of ontology evolution, the OntoWeb project (Gómez-Pérez, Fernández-López, & Corcho, 2002) has created a survey of the most relevant methodologies and methods used for building, maintaining, evaluating and re engineering ontologies. In particular, it addresses some of the main ontology evolution methods and tools, such as ONIONS (Gangemi, Pisanelli, & Steve, 1999), PROMPT (Noy & Musen, 2000), FCA-Merge (Stumme & Maedche, 2001). The main problem addressed by these works is to provide methods and techniques for adding new concepts to a previously defined ontology, by contemporary preserving the ontology consistency. In the context of knowledge discovery the application of ontology evolution techniques is useful in order to integrate into a peer ontology the concepts acquired for the other peers of the network.

- **Ontology-Based Routing in Open Distributed Systems.** In open distributed systems with many independent peers and without a centralized authority, a key problem is represented by performances and effectiveness in query propagation. In particular, the query propagation algorithm should ensure that (1) peers containing relevant information with respect to a target query are always contacted; (2) non relevant peers are not bothered; (3) the delay between query dispatch and query reply is acceptable for an interactive user of the application. In literature, research issues related to query propagation in distributed systems has been addressed in Crespo and

Copyright © 2006, Idea Group Inc. Copying or distributing in print or electronic forms without written permission of Idea Group Inc. is prohibited.

Garcia-Molina (2003) and Ratnasamy, Francis, Handley, Karp, and Shenker (2001) where peers and contents are organized according to a predefined overlay structure in order to enforce efficiency in query propagation. Another approach is described in Tempich, Staab, and Wranik (2004) where a social metaphor is exploited effectively and efficiently to select the right recipients for a given target request. Generally, a semantics-based approach is required to manage query propagation, and ontology matching techniques and tools (e.g., H-MATCH) can be used to improve the effectiveness of the knowledge discovery process. In this context, research work is related to the development of an ontology-based routing protocol capable to select query recipients according to information extracted from a peer ontology.

References

Aberer, K., Cudrè-Mauroux, P., & Hauswirth, M. (2003). The chatty Web: Emergent semantics through gossiping. *Proceedings of the 12th International World Wide Web Conference, WWW 2003*, Budapest, Hungary (pp. 197-206).

Arenas, M., Kantere, V., Kementsietsidis, A., Kiringa, I., Miller, R. J., & Mylopoulos, J. (2003). The Hyperion project: From data integration to data coordination. *SIGMOD Record, Special Issue on Peer-to-Peer Data Management, 32*(3), 53-58.

Bernstein, P. A., Giunchiglia, F., Kementsietsidis, A., Mylopoulos, J., Serafini, L., & Zaihrayeu, I. (2002). Data management for peer-to-peer computing: A vision. *Proceedings of the 5th International Workshop on the Web and Databases, WebDB 2002*, in conjunction with ACM PODS/SIGMOD 2002, Madison, WI (pp. 89-94).

Bouquet, P., Magnini, B., Serafini, L., & Zanobini, S. (2003). A SAT-based algorithm for context matching. *Proceedings of the 4th International and Interdisciplinary Conference, CONTEXT 2003*, Stanford, CA (pp. 66-79).

Broekstra, J., Ehrig, M., Haase, P., van Harmelen, F., Kampman, A., Sabou, M., et al. (2003). A metadata model for semantics-based peer-to-peer systems. *Proceedings of the 1st WWW International Workshop on Semantics in Peer-to-Peer and Grid Computing, SemPGRID 2003*, Budapest, Hungary.

Copyright © 2006, Idea Group Inc. Copying or distributing in print or electronic forms without written permission of Idea Group Inc. is prohibited.

Castano, S., De Antonellis, V., & De Capitani Di Vimercati, S. (2001). Global viewing of heterogeneous data sources. *IEEE Transactions on Knowledge and Data Engineering, 13*(2), 277-297.

Castano, S., Ferrara, A., Montanelli, S., & Zucchelli, D. (2003a). HELIOS: A general framework for ontology-based knowledge sharing and evolution in P2P systems. *Proceedings of the 2nd International DEXA Workshop on Web Semantics, WebS 2003*, Prague, Czech Republic (pp. 597-603).

Castano, S., Ferrara, A., & Montanelli, S. (2003b). H-MATCH: An algorithm for dynamically matching ontologies in peer-based systems. *Proceedings of the 1st International VLDB Workshop on Semantic Web and Databases, SWDB 2003*, Berlin (pp. 231-250).

Castano, S., Ferrara, A., Montanelli, S., Pagani, E., Rossi, G. P., & Tebaldi, S. (2004a). On combining a semantic engine and flexible network policies for P2P knowledge sharing networks. *Proceedings of the 1st International DEXA Workshop on Grid and Peer-to-Peer Computing Impacts on Large Scale Heterogeneous Distributed Database Systems, GLOBE 2004*, Zaragoza, Spain (pp. 529-535).

Castano, S., Ferrara, A., Montanelli, S., & Racca, G. (2004b). From surface to intensive matching of Semantic Web ontologies. *Proceedings of the 3rd International DEXA Workshop on Web Semantics, WebS 2004*, Zaragoza, Spain (pp. 140-144).

Castano, S., Ferrara, A., Montanelli, S., & Racca, G. (2005). *Matching ontologies in open networked systems: Techniques and applications.* Technical Report. Web-Minds FIRB project. Milan, Italy: University of Milan (submitted for publication).

Connolly, D., Van Harmelen, F., Horrocks, I., McGuinness, D. L., Patel-Schneider, P. F., & Stein, L. A. (2001). *DAML+OIL (March 2001) reference description.* Retrieved from http://www.w3.org/TR/daml+oil-reference/

Cranefield, S., & Purvis, M. K. (1999). UML as an ontology modeling language. *Proceedings of the Workshop on Intelligent Information Integration, IJCAI-99*, in conjunction with the 16th International Joint Conference on Artificial Intelligence, Stockholm, Sweden.

Crespo, A., & Garcia-Molina, H. (2003). *Semantic overlay networks for P2P systems* (Technical Report 2003-75). CA: Stanford University.

Doan, A., Madhavan, J., Domingos, P., & Halevy, A. (2002). Learning to map between ontologies on the Semantic Web. *Proceedings of the 11th International World Wide Web Conference, WWW 2002*, Honolulu, HI (pp. 662-673).

Copyright © 2006, Idea Group Inc. Copying or distributing in print or electronic forms without written permission of Idea Group Inc. is prohibited.

Gangemi, A., Pisanelli, D. M., & Steve, G. (1999). An overview of the ONIONS project: Applying ontologies to the integration of medical terminologies. *Data & Knowledge Engineering, 31*(2), 183-220.

Gómez-Pérez, A., Fernández-López, M., & Corcho, O. (2002). *Technical Roadmap D.1.1.2.* OntoWeb Consortium.

Halevy, A., Ives, Z., Suciu, D., & Tatarinov, I. (2003). Schema mediation in peer data management systems. *Proceedings of the 19th International Conference on Data Engineering, ICDE 2003*, Bangalore, India.

Kementsietsidis, A., Arenas, M., & Miller, R. J. (2003). Mapping data in peer-to-peer systems: Semantics and algorithmic issues. *Proceedings of the 2003 ACM SIGMOD International Conference on Management of Data*, San Diego, CA (pp. 325-336).

Maedche, A., Motik, B., Stojanovic, L., Studer, R., & Volz, R. (2003a). Ontologies for enterprise knowledge management. *IEEE Intelligent Systems, 18*(2), 26-33.

Maedche, A., Motik, B., Stojanovic, L., Studer, R., & Volz, R. (2003b). An infrastructure for searching, reusing and evolving distributed ontologies. *Proceedings of the 12th International World Wide Web Conference, WWW 2003*, Budapest, Hungary (pp. 439-448).

Miller, G. A. (1995). WordNet: A lexical database for English. *Communications of the ACM, CACM, 38*(11), 39-41.

Nejdl, W., Wolf, B., Qu, C., Decker, S., Sintek, S., Naeve, A., et al. (2002). EDUTELLA: A P2P networking infrastructure based on RDF. *Proceedings of the 11th International World Wide Web Conference, WWW 2002*, Honolulu, HI (pp. 604-615).

Noy, N. F., & Musen, M. A. (2000). PROMPT: Algorithm and tool for automated ontology merging and alignment. *Proceedings of the 17th National Conference on Artificial Intelligence and 12th Conference on Innovative Applications of Artificial Intelligence*, Austin, TX (pp. 450-455).

Ratnasamy, S., Francis, P., Handley, M., Karp, R., & Shenker, S. (2001). A scalable content-addressable network. *Proceedings of the ACM SIGCOMM 2001 Conference on Applications, Technologies, Architectures, and Protocols for Computer Communication*, San Diego, CA (pp. 161-172).

Stumme, G., & Maedche, A. (2001). FCA-MERGE: Bottom-up merging of ontologies. *Proceedings of the 17th International Joint Conference on Artificial Intelligence, IJCAI 2001*, Seattle, WA (pp. 225-234).

Copyright © 2006, Idea Group Inc. Copying or distributing in print or electronic forms without written permission of Idea Group Inc. is prohibited.

Tangmunarunkit, H., Decker, S., & Kesselman, C. (2003). Ontology-based resource matching in the Grid — The Grid meets the Semantic Web. *Proceedings of the 1st WWW International Workshop on Semantics in Peer-to-Peer and Grid Computing, SemPGRID 2003*, Budapest, Hungary.

Tempich, C., Staab, S., & Wranik, A. (2004). REMINDIN': Semantic query routing in peer-to-peer networks based on social metaphors. *Proceedings of the Thirteenth International Conference on World Wide Web, WWW 2004*, New York (pp. 640-649).

W3C RDF Core Working Group. (2004). *RDF vocabulary description language 1.0: RDF schema*. Retrieved from http://www.w3.org/TR/rdf-schema/

W3C Web Ontology Working Group. (2004). *OWL Web Ontology Language guide*. Retrieved from http://www.w3.org/TR/owl-guide/

W3C XML Schema Working Group. (2004). *XML schema part 2: Datatypes*. Retrieved from http://www.w3.org/TR/xmlschema-2/

Endnotes

[1]   These considerations are based on our research work in developing the ARTEMIS information integration tool environment (Castano, De Antonellis, & De Capitani Di Vimercati, 2001).

[2]   Terminological relationship weights have been taken from the ARTEMIS schema matching algorithm, since such weights have been extensively tested and experimented on real schema matching cases.

[3]   Property and semantic relation weights have been chosen on the basis of experimentation of the algorithm on real ontology matching cases.

[4]   Each peer enforces a semantic query propagation strategy in order to select the recipients of each query. Further issues related to this topic are subsequently discussed in the context of the HELIOS system.

Copyright © 2006, Idea Group Inc. Copying or distributing in print or electronic forms without written permission of Idea Group Inc. is prohibited.

Chapter IX

Metadata- and Ontology-Based Semantic Web Mining

Marie Aude Aufaure, Supélec, France

Bénédicte Le Grand, Laboratoire d'Informatique de Paris 6, France

Michel Soto, Laboratoire d'Informatique de Paris 6, France

Nacera Bennacer, Supélec, France

Abstract

The increasing volume of data available on the Web makes information retrieval a tedious and difficult task. The vision of the Semantic Web introduces the next generation of the Web by establishing a layer of machine-understandable data, e.g., for software agents, sophisticated search engines and Web services. The success of the Semantic Web crucially depends on the easy creation, integration and use of semantic data. This chapter is a state-of-the-art review of techniques which could make the Web more "semantic". Beyond this state-of-the-art, we describe open research areas and we present major current research programs in this domain.

Copyright © 2006, Idea Group Inc. Copying or distributing in print or electronic forms without written permission of Idea Group Inc. is prohibited.

Introduction

This section presents the context and the challenges of semantic information retrieval. We also introduce the goals of Semantic Web (Berners-Lee et al., 2001) and data mining. Available data have become more and more complex; spatiotemporal parameters contribute to this complexity, as well as data's lack of structure, multidimensionality, large volume, and dynamic evolution. Moreover, data formats and models are numerous, which makes their interoperability challenging. Biological databanks illustrate this situation. In the domain of tourism, queries can entail computations (e.g., in order to find the best path to a destination) including constraints which are not necessarily precisely formulated. Answers may be provided through the use of Web services, and should be customized according to a user profile. Several Web mining techniques have been proposed to enhance these different types of information retrieval, among which methods deriving from data analysis and from conceptual analysis. All these methods aim at making the Web more understandable but they differ in the way they deal with the complexity of data.

The increasing interest in Web information retrieval led to the Semantic Web initiative from the World Wide Web Consortium. The Semantic Web is not a new Web, but an extension of the existing one to make it more understandable to machines. The main goal is thus to express semantic information about data formally, so that this information may be processed and used by computers. Semantic information may appear as semantic annotations or metadata. Several formats have been designed to meet this goal, among which the Resource Description Framework (W3C, 1999) from the W3C and Topic Maps (ISO, 1999) from the International Standardization Organization. Both formats aim at describing resources and establish relationships among them. RDF can be enriched with an RDF schema which expresses class hierarchies and typing constraints, e.g., to specify that a given relation type can connect only specific classes. The semantic tagging provided by RDF and Topic Maps may be extended by references to external knowledge coming from controlled vocabularies, taxonomies, and ontologies. An ontology (Gruber, 1993) is an abstract model which represents a common and shared understanding of a domain. Ontologies generally consist of a list of interrelated terms and inference rules and can be exchanged between users and applications. They may be defined in a more or less formal way, from natural language to description logics. The Web Ontology Language (OWL) belongs to the latter category. OWL is built upon RDF and RDFS and extends them to express class properties.

Metadata and ontologies are complementary and constitute the Semantic Web's building blocks. They avoid meaning ambiguities and provide more precise

Copyright © 2006, Idea Group Inc. Copying or distributing in print or electronic forms without written permission of Idea Group Inc. is prohibited.

answers. In addition to a better accuracy of query results, another goal of the Semantic Web is to describe the semantic relationships between these answers.

The promises of the Semantic Web are numerous, but so are its challenges, starting with scalability. Semantic Web data are likely to increase significantly and associated techniques will have to evolve. The new tagging and ontology formats require new representation and navigation paradigms. The multiplicity of ontologies raises the issue of their integration; this area has been widely explored and solutions have been proposed, even though some problems still remain. The highly dynamic nature of the Semantic Web makes the evolution and maintenance of semantic tagging and ontologies difficult. The ultimate challenge is the automation of semantics extraction. This subject is developed in a whole section of this chapter. We study how traditional Web approaches might be used for a partial automation of knowledge extraction. Pages content and usage analysis are complementary to expand knowledge databases. However, this automation requires an evaluation of the extracted information.

This chapter is organized as follows: first, we introduce the notions of semantic metadata in general and ontologies in particular. Then we raise the issue of Semantic Web mining (Berendt et al., 2002) and data integration, before studying how and to what extent the knowledge extraction process can be automated. We finally suggest some research directions for the future before concluding by presenting the limits of the Semantic Web's extension.

Metadata and Ontologies

This section presents metadata representation formats, in particular RDF and Topic Maps, and their application to complex data. We also describe the concept of ontology and one associated standard, the Web Ontology Language (OWL). We study the added value of ontologies in comparison with simple metadata, in terms of expressivity and inference.

Let us first define metadata and annotations: metadata are data about data. An annotation is an explicative or critical note attached to a document, text or image. Web pages annotations become metadata when they are stored into a database or a server. We distinguish information attached to a resource from information stored and handled independently.

The Semantic Web can be divided into various layers of metadata, each level providing different degrees of expressivity, as shown in Figure 1 (Berners-Lee, 1998). In the following of this section, we describe Semantic Web formalisms, starting from the bottom of the stack.

Copyright © 2006, Idea Group Inc. Copying or distributing in print or electronic forms without written permission of Idea Group Inc. is prohibited.

Figure 1. Semantic Web stack (Berners-Lee, 1998)

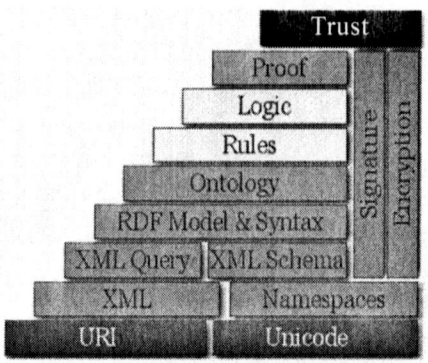

XML, Namespaces, and Controlled Vocabularies

XML is a first level of semantics which allows users to structure data with regard to their content rather than their presentation (Yergeau et al., 2004). XML tags may represent the meaning of data whereas HTML tags indicate the way data should be displayed.

Namespaces allow the unambiguous use of several vocabularies within a single document, by indicating explicitly which set a term belongs to. A *controlled vocabulary* is a set of terms defined by a community without giving any sense or organization among these terms. As an example, a book index is a controlled vocabulary. A very popular controlled vocabulary is the *Dublin Core*.

Dublin Core (www.dublincore.org) is a set of very simple elements used to describe various resources in terms of content (Title, Description, Subject, Source, Coverage, Type, Relationship), of intellectual property (Creator, Contributor, Editor, Rights) and of version (Date, Format, Identifier, Language). Dublin Core is composed of 15 elements which semantics have been established by an international consortium. This norm presents all the descriptive information found in traditional archive research systems, while preserving hierarchical relationships that exist between the different description levels. It facilitates the navigation into the hierarchical information structure.

Moreover, Dublin Core defines the categories of information that may be attached to a resource (Web page, document, or image) in order to enhance information retrieval. Dublin Core is used by a large community due to the following advantages:

Copyright © 2006, Idea Group Inc. Copying or distributing in print or electronic forms without written permission of Idea Group Inc. is prohibited.

- The set of elements is very simple, which makes this norm very easy to use for an efficient information retrieval;

- Its semantics is also easily understandable: Dublin Core helps beginner users find their way within data by providing a common set of well defined and understood elements;

- Dublin Core is widely used; as an example, in 1999, it was translated into 20 languages;

- This norm is extensible; Dublin Core elements may be enriched with domain-specific information for particular communities.

RDF and Topic Maps

XML, controlled vocabularies and namespaces provide a first level of metadata. However, more semantics can be added with the *Resource Description Framework* (RDF) or *Topic Maps* standards. RDF was developed by the World Wide Web Consortium (W3C, 1999) whereas Topic Maps were defined by the International Organization for Standardization (ISO, 1999). Topic Maps do not appear on the Semantic Web stack shown on Figure 1, because there is not a W3C recommendation. On this figure, Topic Maps would be at the same level as RDF. The Topic Map paradigm was adapted to the Web by the TopicMaps.Org Consortium (TopicMaps.Org, 2001). Both RDF and Topic Maps aim at representing knowledge about information resources by annotating them. These paradigms are presented in the following subsections.

RDF

The *Resource Description Framework* (RDF) (W3C, 1999) syntax was designed to represent information about resources in the World Wide Web. Examples of such metadata are the *author, creation* and *modification dates* of a Web page. RDF provides a common framework for expressing semantic information about data so that it can be exchanged between applications without loss of meaning. RDF identifies things with Web identifiers (called *Uniform Resource Identifiers*, or *URIs*), and describes *resources* in terms of *properties* and property *values*.

Figure 2 shows the graphical RDF description of a Web page. This semantic annotation indicates that this page belongs to John Smith and that it was created on January 1, 1999 and modified on August 1, 2004. This corresponds to three RDF *statements*, giving information respectively on the *author, creation*, and

Copyright © 2006, Idea Group Inc. Copying or distributing in print or electronic forms without written permission of Idea Group Inc. is prohibited.

Figure 2. Example RDF graph

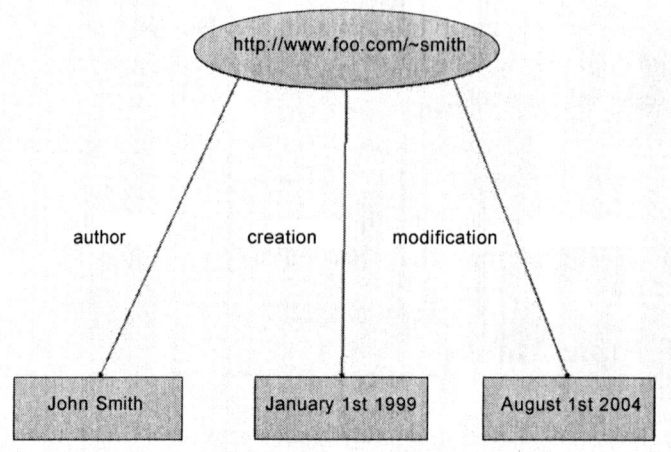

modification dates of this page. Each statement consists of a (*Resource, Property, Value*) triplet. In our example:

- *http://www.foo.com/~smith* is a **resource**,
- The element *<author>* is a **property**,
- The string « *John Smith* » is a **value**.

A statement may also be described in terms of (*Subject, Predicate, Object*):

- The resource *http://www.foo.com/~smith* is the **subject**,
- The property *<author>* is the **predicate**,
- The value « *John Smith* » is the **object**.

As shown in Figure 2, statements about resources can be represented as a *graph* of nodes and arcs corresponding to the resources, their properties and their values. RDF provides an XML syntax (called serialization syntax) for these graphs. The following code is the XML translation of the graph in Figure 2:

Copyright © 2006, Idea Group Inc. Copying or distributing in print or electronic forms without written permission of Idea Group Inc. is prohibited.

```
<?xml version="1.0"?>
<RDF>
<Description about="http://www.foo.com/~smith">
        <author>John Smith</author>
        <created>January 1, 1999</created>
        <modified>August 1, 2004</modified>
</Description>
</RDF>
```

Topic Maps

Topic Maps (ISO, 1999) are an ISO standard which describes knowledge and links it to existing information resources. RDF and Topic Maps thus have similar goals.

Although Topic Maps allow organizing and representing very complex structures, the basic concepts of this model (*topics*, *occurrences*, and *associations*) are simple. A *topic* is a syntactic construct which corresponds to the expression of a real-world concept in a computer system. Figure 3 represents a very small Topic Map which contains four topics: *EGC 2005*, *Paris*, *Ile-de-France* and *France*. These topics are instances of other topics: *EGC 2005* is a *conference*, *Paris* is a *city*, *Ile-de-France* is a *region*, and *France* is a *country*. A topic type is a topic itself, which means that *conference*, *city*, *region*, and *country* are also topics.

A topic may be linked to several information resources (e.g., Web pages) which are considered to be somehow related to this topic. These resources are called *occurrences* of a topic. Occurrences provide means of linking real resources to abstract concepts, which helps organize data and understand their context.

An *association* adds semantics to data by expressing a relationship between several topics, such as *EGC 2005 takes place in Paris, Paris is located in Ile-de-France*, and so on. Every topic involved in an association plays a specific role in this association, for example, *Ile de France* plays the role of *container* and *Paris* plays the role of *containee*.

It is interesting to notice that topics and information resources belong to two different layers. Users may navigate at an abstract level (the topic level) instead of navigating directly within data.

RDF and Topic Maps both add semantics to existing data without modifying them. They are two compatible formalisms: Moore (2001) stated that RDF could be used to model Topic Maps and vice versa. There are slight differences, for

Copyright © 2006, Idea Group Inc. Copying or distributing in print or electronic forms without written permission of Idea Group Inc. is prohibited.

Figure 3. Example topic map

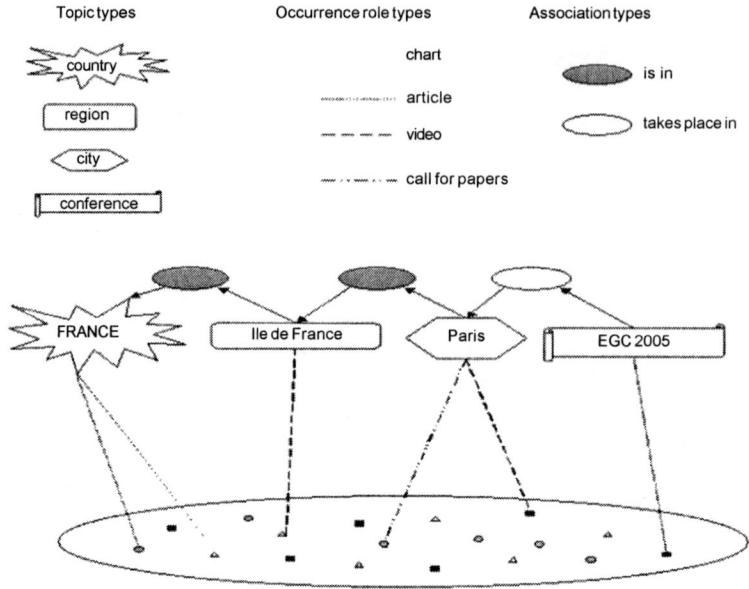

example, the notion of *scope* (context) exists in Topic Maps and not in RDF. RDF is more synthetic and better adapted to queries whereas Topic Maps are better for navigation purposes.

So far, we have described the lower layers of the Semantic Web stack; in the next section, we will describe more expressive formalisms: ontologies. We will also describe two other formalisms, which are not specific to the Web: taxonomies and thesauri.

Taxonomies, Thesauri, and Ontologies

Taxonomies and Thesauri

Taxonomies and thesauri do not appear on the Semantic Web stack as they were not specifically designed for the Web; they, however, belong to the Semantic Web picture. In this section, we define these notions and we indicate their level in the stack.

A *taxonomy* is a hierarchically-organized controlled vocabulary. The world has many taxonomies, because human beings naturally classify things. Taxonomies

Copyright © 2006, Idea Group Inc. Copying or distributing in print or electronic forms without written permission of Idea Group Inc. is prohibited.

are semantically weak. According to Daconta et al. (2003), taxonomies are commonly used when navigating without a precise research goal in mind.

A *thesaurus* is a "controlled vocabulary arranged in a known order and structured so that equivalence, homographic, hierarchical, and associative relationships among terms are displayed clearly and identified by standardized relationship indicators" (ANSI/NISO Z39.19-1993 [R1998], p. 1). The purpose of a thesaurus is to facilitate documents retrieval. The *WordNet* thesaurus (Miller, 1995) organizes English nouns, verbs, adverbs, and adjectives into a set of synonyms and defines relationships between synonyms.

Both taxonomies and thesauri provide a vocabulary of terms and simple relationships between these terms. Therefore, taxonomies and thesauri are above XML, namespaces and controlled vocabulary in the Semantic Web stack. However, the relationships they express are not as rich as the ones provided by RDF or Topic Maps and consequently by ontologies.

Ontologies

Definitions

As we saw earlier, Berners-Lee (1998) proposed a layered architecture for the Semantic Web languages, among which were XML, XML schema, RDF, and RDF Schema (RDFS). RDFS defines classes and properties (binary relation), range and domain constraints on properties, subclass and subproperty as subsumption relations. However, RDFS is insufficient in terms of expressivity; this is also true for Topic Maps. On the other hand, ontologies allow a better specification of constraints on classes. They also make reasoning possible, as new knowledge may be inferred, e.g., by transitivity. *Ontologies* aim at formalizing domain knowledge in a generic way and provide a common agreed understanding of a domain, which may be used and shared by applications and groups.

In computer science, the word *ontology*, borrowed from philosophy, represents a set of precisely defined terms (vocabulary) about a specific domain and accepted by this domain's community. An ontology thus enables people to agree upon the meaning of terms used in a precise domain, knowing that several terms may represent the same concept (synonyms) and several concepts may be described by the same term (ambiguity). Ontologies consist in a hierarchical description of important concepts of a domain, and in a description of each concept's properties. Ontologies (Gómez-Pérez et al., 2003) are at the heart of information retrieval from nomadic objects, from the Internet and from heterogeneous data sources.

Ontologies generally consist of a taxonomy — or vocabulary — and of inference rules such as transitivity and symmetry. They may be used in conjunction with

Copyright © 2006, Idea Group Inc. Copying or distributing in print or electronic forms without written permission of Idea Group Inc. is prohibited.

RDF or Topic Maps, e.g., to allow consistency checking or to infer new information.

According to Gruber (1993), "*an ontology is an explicit specification of a conceptualization.*"

Jeff Heflin, editor of the OWL Use Cases and Requirements (Heflin, 2004), considers that "*an ontology defines the terms used to describe and represent an area of knowledge. [...] Ontologies include computer-usable definitions of basic concepts in the domain and the relationships among them. [...] Ontologies are usually expressed in a logic-based language, so that detailed, accurate, consistent, sound, and meaningful distinctions can be made among the classes, properties, and relations.*"

Berners-Lee et al. (2001) say that "*Artificial-intelligence and Web researchers have co-opted the term for their own jargon, and for them an ontology is a document or file that formally defines the relations among terms. The most typical kind of ontology for the Web has a taxonomy and a set of inference rules.*"

Ontologies may be classified as follows:

Guarino (1998) classifies ontologies according to their level of dependence with regard to a specific task or point of view. He distinguishes four categories: high-level, domain, task, and application ontologies.

Lassila and McGuinness (2001) categorize ontologies according to their expressiveness and to the richness of represented information. Depending on the domain and the application, an ontology may be more or less rich, from a simple vocabulary to real knowledge bases; it may be a glossary where each term is associated to its meaning in natural language. It may also be a thesaurus in which terms are connected through semantic links (synonyms in *WordNet*) or even genuine knowledge bases comprising notions of concepts, properties, hierarchical links, and properties constraints.

After defining the concept of ontology, we now present ontology languages.

Ontology Languages

The key role that ontologies are likely to play in the future of the Web has led to the extension of Web markup languages. In the context of the Semantic Web, an ontology language should:

- Be compatible with existing Web standards,
- Define terms precisely and formally with adequate expressive power,
- Be easy to understand and use,

Copyright © 2006, Idea Group Inc. Copying or distributing in print or electronic forms without written permission of Idea Group Inc. is prohibited.

- Provide automated reasoning support,

- Provide richer service descriptions which could be interpreted by intelligent agents,

- Be sharable across applications.

Ontology languages can be more or less formal. The advantage of formal languages is the reasoning mechanisms which appear in every phase of conception (satisfiability, subsumption, etc.), use (query, instantiation) and maintenance of an ontology (consistency checking after an evolution). The complexity of underlying algorithms depends on the power and the semantic richness of the used logics.

When querying an ontology, a user does generally not have the global knowledge of the ontology schema. The language should thus allow him to query both the ontology schema and its instances in a consistent manner. The use of description logics (DL), a subset of first-order logic, unifies the description and the manipulation of data. In DL, the knowledge base consists of a TBox (Termino-logical-Box) and of a ABox (Assertional-Box). The TBox defines concepts and relationships between concepts, whereas the ABox consists of assertions describing a situation (Nakabasami, 2002).

At the description level, concepts and roles are defined; at the manipulation level, the query is seen as a concept and reasoning mechanisms may be applied. For instance, the description of a query may be compared to an inconsistent description. If they are equivalent, this means that the user made a mistake in the formulation of his query (remind that he does not know the ontology schema). The query may also be compared (by subsumption) to the hierarchy of concepts (the ontology). One limit of description logics is that queries can only return existing objects, instead of creating new objects, as database query languages such as SQL can do.

In the next section, we focus on a specific ontology language: the Web Ontology Language (OWL).

OWL

To go beyond the "plain text" searching approach it is necessary to specify the semantics of the Web resources content in a way that can be interpreted by intelligent agents. The W3C has designed the Web Ontology Language: OWL (W3C, 2004) (Dean & Schreiber, 2003), a semantic markup language for Web resources, as a revision of the DAML+OIL (Horrocks, 2002). It is built on W3C standards XML, RDF/RDFS (Brickley & Guha, 2003; Lassila & Swick, 1999)

Copyright © 2006, Idea Group Inc. Copying or distributing in print or electronic forms without written permission of Idea Group Inc. is prohibited.

and extends these languages with richer modeling primitives. Moreover, OWL is based on description logics (Baader et al., 2003; Horrocks & Patel-Schneider, 2003; Horrocks et al., 2003); OWL may then use formal foundations of description logic, mainly known reasoning algorithms and implemented systems (Volker & Möller, 2001; Horrocks, 1998).

OWL allows:

- The formalization of a domain by defining classes and properties of those classes,
- The definition of individuals and the assertion of properties about them, and
- The reasoning about these classes and individuals.

We saw in the previous section that RDF and Topic Maps lacked expressive power; OWL, layered on top of RDFS, extends RDFS's capabilities. It adds various constructors for building complex class expressions, cardinality restrictions on properties, characteristics of properties, and mapping between classes and individuals (W3C, 2004) (Dean & Schreiber, 2003). An ontology in OWL is a set of axioms describing classes, properties, and facts about individuals.

The following basic example of OWL illustrates these concepts:

In this example, *Man* and *Woman* are defined as subclasses of the *Person* class; *hasParent* is a property that links two persons. *hasFather* is a subproperty of *hasParent* and its range is constrained to the *Man* Class. *hasChild* is the inverse property of *hasParent*.

```
<owl:Class rdf:ID="Person"/>

<owl:Class rdf:ID="Man">
    <rdfs:subClassOf rdf: resource="#Person"/>
</owl:Class>

<owl:Class rdf:ID="Woman">
    <rdfs:subClassOf rdf: resource="#Person"/>
    <rdfs:disjointWith rdf:resource="#Man" />
</owl:Class>
```

Copyright © 2006, Idea Group Inc. Copying or distributing in print or electronic forms without written permission of Idea Group Inc. is prohibited.

```
<owl:ObjectProperty rdf:ID="hasParent">
    <rdfs:domain rdf:resource="#Person"/>
    <rdfs:range rdf:resource="#Person"/>
</owl:ObjectProperty>

<owl:ObjectProperty rdf:ID="hasFather">
    <rdfs:subPropertyOf rdf:resource="#hasParent"/>
    <rdfs:range rdf:resource="#Man"/>
</owl:ObjectProperty>

<owl:ObjectProperty rdf:ID="hasChild">
    <owl:inverseOf rdf:resource="#hasParent">
</owl:ObjectProperty>
```

Although OWL is more expressive than RDFS or Topic Maps, it still has limitations; in particular, it lacks a more powerful language to better describe properties, in order to provide more inference capabilities. An extension to OWL with Horn-style rules has been proposed by Horrocks and Patel Schneider (2004), called ORL: OWL Rules Language. ORL itself may be further extended if more expressive power is needed.

Semantic Web Information Retrieval

Semantic Web mining aims at integrating the areas of Semantic Web and Web mining (Berendt et al., 2002). The purpose is twofold:

- Improve Web mining efficiency by using semantic structures such as ontologies, metadata, thesauri, and
- Use Web mining techniques and learn ontologies from Web resources as automatically as possible and thus help building the Semantic Web.

We present the benefits of metadata and ontologies for a more relevant information retrieval, as shown on Figure 4 (Decker et al., 2000). The use of controlled vocabularies avoids meaning conflicts, whereas ontologies allow

Copyright © 2006, Idea Group Inc. Copying or distributing in print or electronic forms without written permission of Idea Group Inc. is prohibited.

semantic data integration. Results can be customized through the use of semantic annotations.

Figure 4 shows the various components of a semantic information retrieval from Web pages. Automated *agents* use various Semantic Web mechanisms in order to provide relevant information to *end users* or *communities of users*. To achieve this goal, Web pages must be *annotated*, using the terms defined in an ontology (*Ontology Construction Tool*). Once the pages are semantically annotated, agents use existing *metadata* and *inference engines* to answer queries. If a query is formulated with a different ontology, a semantic integration is performed with the *Ontology Articulation Toolkit*.

Information Retrieval in the Semantic Web

In this section, we show how semantic metadata enhance information retrieval. References to ontologies avoid ambiguities and therefore allow advanced

Figure 4. Use of metadata on the Semantic Web (information food chain [Decker et al., 2000])

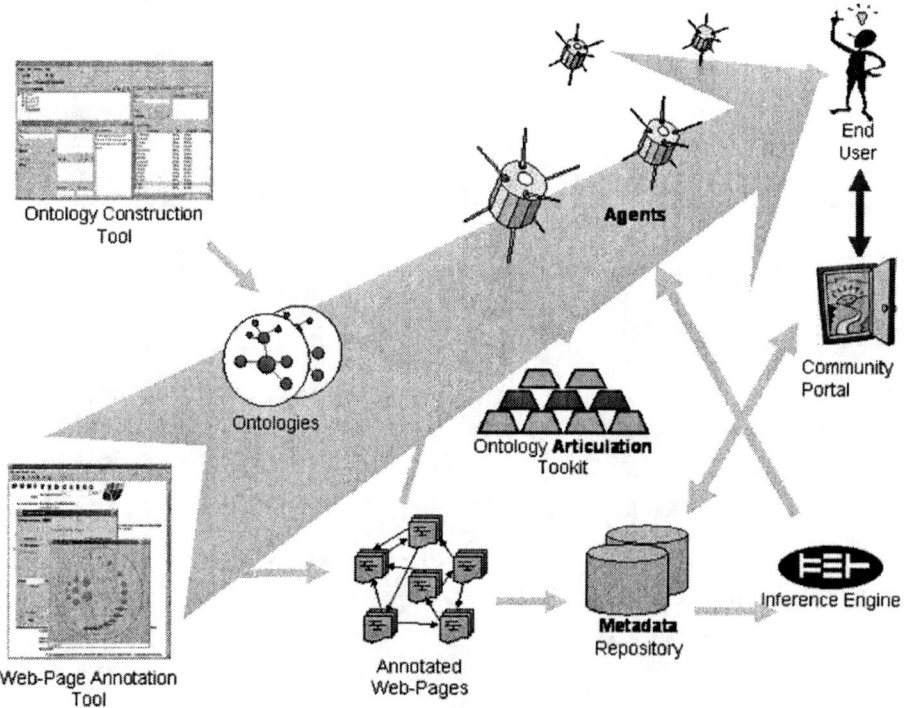

Copyright © 2006, Idea Group Inc. Copying or distributing in print or electronic forms without written permission of Idea Group Inc. is prohibited.

queries and provide more relevant answers to *precise* information needs. We define a search as *precise* if the information need can be formally specified with a query language. However, it is not always possible to formulate a precise query, for example if what is looked for is an overview of a set of Web pages. Typically, this is the case when one follows HTTP hyperlinks during a Web navigation. In order to meet the goals of these *fuzzy* searches, the semantic relationships defined by RDF graphs or Topic Maps are very helpful, as they connect related concepts. Thus, Semantic Web techniques are complementary and they benefit both precise and fuzzy searches.

An implementation of information retrieval prototypes based on RDF and Topic Maps was achieved in the *OmniPaper* project (Paepen et al., 2002), in the area of electronic news publishing. In both cases, the user submits a natural-language query to a large set of digital newspapers. Searches are based on linked keywords which form a navigation layer. User evaluation showed that the semantic relations between articles were considered very useful and important for relevant content retrieval.

Semantic Web methodologies and tools have also been implemented in an IST/CRAFT European Program called *Hi-Touch*, in the domain of tourism (Euzénat et al., 2003). In the *Hi-Touch* platform, Semantic Web Technologies are used to store and organize information about customers' expectations and tourism products. This knowledge can be processed by machines as well as by humans in order to find the best matching between supply and demand. The knowledge base system combines RDF, Topic Maps, and ontologies.

Figure 5 illustrates a semantic query performed with Mondeca's *Intelligent Topic Manager* (http://www.mondeca.com), in the context of the *Hi-Touch* project. Users can express their queries with keywords, but they can also specify the type of result they expect, or provide more details about its relationships with other concepts. Figure 5 also shows the graphical environment, centered on the query result, which allows users to see its context.

Semantic Integration of Data

The Web is facing the problem of accessing a dramatically increasing volume of information generated independently by individual groups, working in various domains of activity with their own semantics. The integration of these various semantics is necessary in the context of the Semantic Web because it allows the capitalization of existing semantic repositories such as ontologies, taxonomies, and thesauri. This capitalization is essential for reducing cost and time on the Semantic Web pathway.

Copyright © 2006, Idea Group Inc. Copying or distributing in print or electronic forms without written permission of Idea Group Inc. is prohibited.

Figure 5. Formulation of a semantic query and graphical representation (Hi-Touch Project)

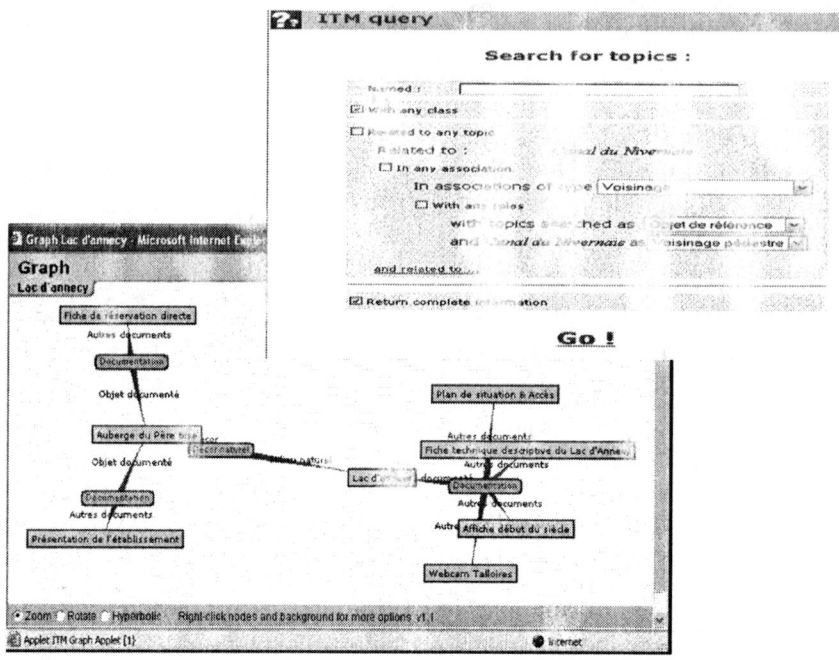

A semantic integration allows to share data that exhibits a high degree of semantic heterogeneity, notably when related or overlapping data encompass different levels of abstraction, terminologies or representations. Data available in current information systems is heterogeneous both in its content and its representation formalism. Two common data integration methods are mediators and data warehouses.

The warehouse approach provides a global view, by centralizing relevant data. Access to data is fast and easy and thus data warehouses are useful when complex queries and analyses are needed. However, the limits of this approach are the required storage capability and the maintenance of the warehouse content. With this approach updates may be performed using different techniques:

- Periodical full reconstruction: this is the most commonly used and simplest technique, but this is also the most time-consuming method.

Copyright © 2006, Idea Group Inc. Copying or distributing in print or electronic forms without written permission of Idea Group Inc. is prohibited.

- Periodical update: incremental approach for updating the data warehouse with the difficulty of detecting the changes within the multiple sources of data.

- Immediate update: another incremental approach which aims at keeping the data as consistent as possible. This technique may consume a lot of communication resources; thus it can only be used for small data warehouses built on data sources with a low update rate.

On the other hand, the mediator approach keeps the initial distribution of data. The mediator can be seen as an interface between users and data sources during a query. The data mediator architecture provides a transparent access to heterogeneous and distributed data sources and eliminates the problem of data update (Ullman, 1997). Initial queries are expressed by users with the global schema provided by the mediator and reformulated in sub queries on the data sources. Answers are then collected and merged according to the global schema (Halevy, 2001).

There are currently two approaches for building a global schema for a mediator: *global as view (GAV)* and *local as view (LAV)*. With the first approach, a global schema is built using the terms and the semantics of data sources. As a consequence, query reformulation is simple but the addition of a new data source modifies the global schema. Thus, the global as view approach does not scale very well. With the second approach, the global schema is built independently from the data sources. Each data source is defined as a view on the global schema using the terms and the semantic of the global schema. Thus, adding or removing a new data source is easy but query reformulation is much more complex. Nevertheless, the local as view approach is currently preferred for its scaling capabilities. Consequently, a lot of work is done on query reformulation where ontologies play a central role as they help to express queries. A third approach named *GLAV* aims at combining advantages of GAV and LAV by associating views over the global schema to views over the data sources (Cali, 2003). Both GAV and LAV approaches consider data sources as sets of relations from data bases. This appears to be inadequate in the context of Web data integration because of the necessary navigation through hyperlinks to obtain the data. Combining the expressivity of GAV and LAV allows to formulate query execution plans which both query and navigate the Web data sources (Friedman et al., 1999).

Copyright © 2006, Idea Group Inc. Copying or distributing in print or electronic forms without written permission of Idea Group Inc. is prohibited.

Ontologies Integration and Evolution

The success of the Semantic Web depends on the expansion of ontologies. While many people and organizations develop and use knowledge and information systems, it seems obvious that they will not use a common ontology. As ontologies will proliferate, they will also diverge; many personalized and small-scale conceptualizations will appear. Accessing the information available on the Semantic Web will be possible only if these multiple ontologies are reconciled.

Ontology integration and evolution should take advantage of the work already done in the database field for schema integration and evolution (Rahm & Bernstein, 2001; Parent & Spaccapietra, 1998). Automatic schema matching led to a lot of contributions in schema translation and integration, knowledge representation, machine learning, and information retrieval. Schema integration and ontology integration are quite similar problems. Database schemas can be well-structured (relational databases) or semistructured (XML schemas). Integrity constraints and cardinality are of great importance in these structures. However, ontologies are semantically richer than database schemas; they may also integrate rules and be defined formally (using description logics). Database schema integration takes instances into account while instances are less important in the case of ontologies (we do not always have instances for a given ontology).

Schema integration is studied since the beginning of the 1980s. The goal is to obtain a global view of a set of schemas developed independently. The problem is that structures and terminologies are different because these schemas have been designed by different persons. The approach consists in finding relationships between different schemas (matching), and then in unifying the set of correspondences into an integrated and consistent schema.

Mechanisms for ontologies integration aim at providing a common semantic layer in order to allow applications to exchange information in semantically sound manners. Ontologies integration has been the focus of a variety of works originating from diverse communities entailing a large number of fields from machine learning and formal theories to heuristics, database schema and linguistics. Relevant terms encountered in these works include *merging, alignment, integration, mapping*, and *matching*. Ontology *merging* aims at creating a new ontology from several ontologies. The objective is to build a consistent ontology containing all the information from the different sources. Ontology *alignment* makes several ontologies consistent through a mutual agreement. Ontology *integration* creates a new ontology containing only parts of the source ontologies. Ontology *mapping* defines equivalence relations between similar concepts or relations from different ontologies. Ontology *matching* (Doan et al., 2003) aims at finding the semantic mappings between two given ontologies.

Copyright © 2006, Idea Group Inc. Copying or distributing in print or electronic forms without written permission of Idea Group Inc. is prohibited.

(Hammed et al., 2004) review several architectures for multiple-ontology systems at a large scale. The first architecture is "bottom-up" and consists in mappings between pairs of ontologies. In this case, the reconciliation is done only when necessary and not for all ontologies. The advantages of such an approach are its simplicity (because of the absence of a common ontology) and its flexibility (the mappings are performed only if necessary and can be done by the designers of the individual ontologies). The main drawback comes from the number of mappings to do when many ontologies are taken into account. Another drawback is that there is no attempt to find common conceptualizations.

The second approach maps the ontologies towards a common ontology. In this case, mapping an ontology O1 to another ontology O2 consists firstly in mapping O1 to the common ontology and secondly in mapping from the common ontology to O2. The advantage is that it reduces the number of mappings and the drawback is the development cost of the common ontology which has to be sufficiently expressive to allow mappings from the individual ontologies. An alternative approach consists in building clusters of common ontologies and in defining mappings between these clusters. In this case, individual ontologies map with one common ontology and mappings between the common ontologies are also defined. This approach reduces the number of mappings and finds common conceptualizations, which seems more realistic in the context of the Semantic Web.

Several tools have been developed to provide support for the construction of semantic mappings. Underlying approaches are usually based on heuristics that identify structural and naming similarities. They can be categorized according to the type of inputs required for the analysis: descriptions of concepts in *OB-SERVER* (Mena & al., 2000), concept hierarchies in *iPrompt* and *AnchorPrompt* (Noy et al., 2003) and instances of classes in *GLUE* (Doan et al., 2003) and *FCA-Merge* (Stumme & Maedche, 2001). The automated support provided by these tools significantly reduces the effort required by the user. Approaches designed for mapping discovery are based upon machine learning techniques and compute similarity measures to extract mappings. In this section, we present the *FCA-Merge* method (Stumme et al., 2001) for ontology merging, the *GLUE* system (Doan et al., 2003) based on a machine learning approach and the *iPrompt* method.

FCA-Merge (Stumme & Maedche, 2001) is based on formal concept analysis and lattice generation and exploration. The input of the method is a set of documents, representative of a particular domain, from which concepts and the ontologies to merge are extracted. This method is based on the strong assumption that the documents cover all concepts from both ontologies. The concept lattice is then generated and pruned. Then, the construction of the merged ontology is semi-automatic.

Copyright © 2006, Idea Group Inc. Copying or distributing in print or electronic forms without written permission of Idea Group Inc. is prohibited.

GLUE (Doan et al., 2003) employs machine learning techniques to find mappings between two ontologies; for each concept from one ontology, *GLUE* finds the most similar concept in the other ontology using probabilistic definitions of several similarity measures. The similarity measure between two concepts is based on conditional probabilities. A similarity matrix is then generated and GLUE uses some common knowledge and domain constraints to extract the mappings between two ontologies. That knowledge includes domain-independent knowledge such as "two nodes match if nodes in their neighborhood also match" as well as domain-dependant knowledge such as "if node Y is a descendant of node X, and Y matches *professor*, then it is unlikely that X matches *assistant professor*". *GLUE* uses a multilearning strategy and exploits the different types of information a learner can obtain from the training instances and the taxonomic structure of ontologies.

The *iPrompt* method (Noy et al., 2003) is dedicated to ontology merging; it is defined as a plug-in on Protégé2000 (Noy et al., 2001). The semi-automatic algorithm is the following:

- Make initial suggestions for merging (executed manually by the user),
- Select an operation (done by the user according to a particular focus),
- Perform automatic updates,
- Find conflicts,
- Update the initial list of suggestions.

Other approaches focus on the specification and formalization of inter-schema correspondences. Calvanese et al. (2001) propose a formal framework for Ontology Integration Systems. Ontologies in their framework are expressed as Description Logic (DL) knowledge bases, and mappings between ontologies are expressed through suitable mechanisms based on queries. Two approaches are proposed to realize this query/view-based mapping: global-centric and local-centric. In the global-centric approach, the mapping is specified by associating to each relation in the global schema one relational query over source relations; on the other hand, the local-centric approach relies on reformulation of the query in terms of the queries to the local sources.

Ontology evolution (Noy et al., 2004) is rather similar to ontology merging; the difference relies in finding differences rather than similarities between ontologies. Ontology evolution and versioning should also benefit from the work done in the database community. Ontologies change over time. Noy et al. (2004) describe changes that can occur in an ontology: changes in the domain (comparable with database schema evolution), changes in conceptualization (application

Copyright © 2006, Idea Group Inc. Copying or distributing in print or electronic forms without written permission of Idea Group Inc. is prohibited.

or usage points of view), and changes in the explicit specification (transformation from a knowledge representation language to another). The compatibility between different versions is defined as follows: instance-data preservation, ontology preservation (a query result obtained with the new version is a superset of those obtained with the old version), consequence preservation (in the case of an ontology treated as a set of axioms, the inferred facts from the old version can also be inferred with the new version), and consistency preservation (the new version of the ontology does not introduce logical inconsistencies). An open research issue in this field is the development of algorithms for automatically finding differences between versions.

In this section we explained how the Semantic Web will enhance information retrieval and data mining. However, we have seen that the success of the Semantic Web required the integration of data and ontologies. Another (obvious) prerequisite is the existence of semantic metadata. The next section presents current techniques and open research areas in the domain of automatic extraction of semantic metadata.

Automatic Semantics Extraction

Information retrieval provides answers to precise queries, whereas data mining brings an additional view for the understanding and the interpretation of data, which can be materialized with metadata. This section is more prospective and tackles current work in the field of the extraction of concepts, relationships between concepts, and metadata. We show how ontologies may enhance knowledge extraction through data mining methods. This will allow a partial automation of semantic tagging and will ease the update and maintenance of metadata and ontologies. Evaluation methods will have to be defined in order to check the validity of extracted knowledge.

Tools and Methods for Manual Ontology Construction

Most existing ontologies have been built manually. The first methodologies we can find in the literature (Ushold & King, 1995; Grüninger & Fox, 1995) have been defined taking into account enterprise ontologies development.

Based on the experience of the *Tove* project, Grüninger and Fox's methodology is inspired by the development of knowledge-based systems using first order logic. They first identify the main scenarios and they elaborate a set of informal competency questions that the ontology should be able to answer. The set of

Copyright © 2006, Idea Group Inc. Copying or distributing in print or electronic forms without written permission of Idea Group Inc. is prohibited.

questions and answers are used to extract the main concepts and their relation-ships and properties which are formalized using first-order logic. Finally, we must define the conditions under which the solutions of the questions are complete.

This methodology provides a basis for ontology construction and validation. Nevertheless, some support activities such as integration and acquisition are missing, as well as management functions (e.g., planification, quality control).

Methontology (Gómez-Pérez et al., 2003) builds ontologies from scratch; this methodology also enables ontology re-engineering (Gómez-Pérez & Rojas, 1999). Ontological re-engineering consists in retrieving a conceptual model from an ontology, and transforming it in a more suitable one. *Methontology* enables the construction of ontologies at the "knowledge level". This methodology consists in identifying the ontology development process with the following main activities: evaluation, configuration, management, conceptualization, integration, and implementation. A lifecycle is based on evolving prototypes. The methodology specifies the steps to perform each activity, the techniques used, the products to be output and how they are to be evaluated. This methodology is partially supported by *WebODE* and many ontologies have been developed in different fields.

The *DOGMA* modeling approach (Jarrar & Meersman, 2002) comes from the database field. Starting from the statement that integrity constraints may vary from one application to another and that the schema is more constant, they propose to split the ontology in two parts. The first one holds the data structure and is application-independent, and the second one is a set of commitments dedicated to one application.

On-To-Knowledge is a process-oriented methodology for introducing and maintaining ontology-based knowledge management systems (Staab et al., 2001); it is supported by the *OntoEdit* Tool. *On-To-Knowledge* has a set of techniques, methods, and principles for each of its processes (feasibility study, ontology kickoff, refinement, evaluation, and maintenance) and indicates the relationships between the processes. This methodology takes usage scenarios into account and is consequently highly application-dependant.

Many tools and methodologies exist for the construction of ontologies. Their differences are the expressiveness of the knowledge model, the existence of an inference and query engine, the type of storage, the formalism generated and its compatibility with other formalisms, the degree of automation, consistency checking, etc.

These tools may be divided into two groups:

- Tools for which the knowledge model is directly formalized in an ontology language:

Copyright © 2006, Idea Group Inc. Copying or distributing in print or electronic forms without written permission of Idea Group Inc. is prohibited.

> ➢ *Ontolingua Server* (Ontolingua et KIF),

> ➢ *OntoSaurus* (Loom),

> ➢ *OILed* (OIL then DAML+OIL then OWL) DL, consistency checking and classification using inference engines such as *Fact* and *Racer*.

- Tools for which the knowledge model is independent from the ontology language:

  > ➢ *Protégé-2000*,
  > ➢ *WebODE*,
  > ➢ *OntoEdit*,
  > ➢ *KAON*.

The most frequently cited tools for ontology management are *OntoEdit*, *Protégé-2000*, and *WebODE*. They are appreciated for their n-tiers architecture, their underlying database support, their support of multilingual ontologies and for their methodologies of ontology construction.

In order to reduce the effort to build ontologies, several approaches for the partial automation of the knowledge acquisition process have been proposed. They use natural language analysis and machine learning techniques.

Concepts and Relationships Extraction

Ontology learning (Maedche, 2002) can be seen as a plug-in in the ontology development process. It is important to define which phases may be automated efficiently. Appropriate data for this automation should also be defined. Existing ontologies should be reused using fusion and alignment methods. A priori knowledge may also be used. One solution is to provide a set of algorithms to solve a problem and combine results. An important issue about ontologies is their adaptation to different domains, as well as their extension and evolution.

When data is modeled with schemas, the work achieved during the modeling phase can be used for ontology learning. If a database schema exists, existing structures may be combined into more complex ones, and they may be integrated through semantic mappings. If data is based on Web schemas, such as DTDs or XML schemas, ontologies may be derived from these structures. If data is defined with instances, ontology learning may be done with conceptual clustering and ABox mining (Nakabasami, 2002). With semistructured data, the goal is to find the implicit structure.

Copyright © 2006, Idea Group Inc. Copying or distributing in print or electronic forms without written permission of Idea Group Inc. is prohibited.

The most common type of data used for ontology learning is natural language data, as can be found in Web pages. In recent years, research aimed at paving the way and different methods have been proposed in the literature to address the problem of (semi) automatically deriving a concept hierarchy from text. Much work in a number of disciplines (computational linguistics, information retrieval, machine learning, databases, software engineering) has actually investigated and proposed techniques for solving part of the overall problem.

The notion of ontology learning is introduced as an approach that may facilitate the construction of ontologies by ontology engineers. It comprises complementary disciplines that feed on different types of unstructured and semistructured data in order to support a semiautomatic, cooperative ontology engineering process characterized by a coordinated interaction with human modelers.

Resource processing consists in generating a set of pre-processed data as input for the set of unsupervised clustering methods for automatic taxonomy construction. The texts are preprocessed, enriched by background knowledge using stopword, stemming, and pruning techniques. Strategies for disambiguation by context are applied.

Clustering methods organize objects into groups whose members are similar in some way. These methods operate on vector-based semantic representations which describe the meaning of a word of interest in terms of counts of its co-occurrence with context words appearing within some delineation around the target word. The use of a similarity/distance measure in order to compute the similarity/distance between vectors of terms in order to decide if they are semantically similar and thus should be clustered or not.

In general, counting frequencies of terms in a given set of linguistically preprocessed documents of a corpus is a simple technique that allows extracting relevant lexical entries that may indicate domain concepts. The underlying assumption is that a frequent term in a set of domain-specific texts indicates the occurrence of a relevant concept. The relevance of terms is measured according to the information retrieval measure tfidf (term frequency inverted document frequency).

More elaborated approaches are based on the assumption that terms are similar because they share similar linguistic contexts and thus give rise to various methods which group terms based on their linguistic context and syntactic dependencies.

We now present related work in the field of ontology learning.

Faure and Nedellec (1998) present an approach called ASIUM, based on an iterative agglomerative clustering of nouns appearing in similar contexts. The user has to validate the clusters built at each iteration. ASIUM method is based on conceptual clustering; the number of relevant clusters produced is a function of the percentage of the corpus used.

Copyright © 2006, Idea Group Inc. Copying or distributing in print or electronic forms without written permission of Idea Group Inc. is prohibited.

In Cimiano et al. (2004) the linguistic context of a term is defined by the syntactic dependencies that it establishes as the head of a subject, of an object or of a PP-complement with a verb. A term is then represented by its context using a vector, the entries of which count the frequency of syntactically dominating verbs.

Pereira et al. (1993) present a divisive clustering approach to build a hierarchy of nouns. They make use of verb-object relations to represent the context of a noun. The results are evaluated by considering the entropy of the produced clusters and also in the context of a linguistic decision task.

Caraballo (1999) uses an agglomerative technique to derive an unlabeled hierarchy of nouns through conjunctions of nouns and appositive constructs. The approach is evaluated by presenting the hypernyms and the hyponym candidates to users for validation.

Bisson et al. (2000) present a framework and its corresponding workbench (*Mo'K*) that supports the development of conceptual clustering methods to assist users in an ontology building task. It provides facilities for evaluation, comparison, characterization of different representations, as well as pruning parameters and distance measures of different clustering methods.

Most approaches have focused only on discovering taxonomic relations, although nontaxonomic relations between concepts constitute a major building block in common ontology definitions. In Maedche et al. (2000) a new approach is described to retrieve nontaxonomic conceptual relations from linguistically processed texts using a generalized association rule algorithm. This approach detects relations between concepts and determines the appropriate level of abstraction for those relations. The underlying idea is that frequent couplings of concepts in sentences can be regarded as relevant relations between concepts. Two measures evaluate the statistical data derived by the algorithm: *Support* measures the quota of a specific coupling within the total number of couplings. *Confidence* denotes the part of all couplings supporting both domain and range concepts within the number of couplings that support the same domain concept. The retrieved measures are propagated to super concepts using the background knowledge provided by the taxonomy. This strategy is used to emphasize the couplings in higher levels of the taxonomy. The retrieved suggestions are presented to the user. Manual work is still needed to select and name the relations.

Verbs play a critical role in human languages. They constrain and interrelate the entities mentioned in sentences. The goal in Wiemer-Hastings et al. (1998) is to find out how to acquire the meanings of verbs from context.

In this section, we focused on the automation of semantics extraction. The success of such initiatives is crucial to the success of the Semantic Web, as the volume of data does not allow a completely manual annotation. This subject remains an open research area.

Copyright © 2006, Idea Group Inc. Copying or distributing in print or electronic forms without written permission of Idea Group Inc. is prohibited.

In the next section, we present other research areas which we consider as strategic for the Semantic Web.

Future Trends

Web Content and Web Usage Mining Combination

One interesting research topic is the exploitation of users profiles and behavior models in the data mining process, in order to provide personalized answers. Web mining (Kosala & Blockeel, 2000) is a data mining process applied to the Web. Vast quantities of information are available on the Web and Web mining has to cope with its lack of structure. Web mining can extract patterns from data trough content mining, structure mining and usage mining. *Content mining* is a form of text mining applied to Web pages. This process allows to discover relationships related to a particular domain, co-occurrences of terms in a text, and so forth. Knowledge is extracted from a Web page. *Structure mining* is used to examine data related to the structure of a Web site. This process operates on Web pages' hyperlinks. Structure mining can be considered as a specialization of Web content mining. *Web usage mining* is applied to usage information such as logs files. A log file contains information related to the queries executed by users to a particular Web site. Web usage mining can be used to modify the Web site structure or give some recommendations to the visitor. Personalization can also be enhanced by usage analysis.

Web mining can be useful to add semantic annotations (ontologies) to Web documents and to populate these ontological structures. As stated below, Web content and Web usage mining should be combined to extract ontologies and to adapt them to the usage.

Ontology creation and evolution require the extraction of knowledge from heterogeneous sources. In the case of the Semantic Web, the knowledge extraction is done from the content of a set of Web pages dedicated to a particular domain. Web pages are semistructured information. Web usage mining extracts navigation patterns from Web log files and can also extract information about the Web site structure and user profiles. Among Web usage mining applications, we can point out personalization, modification and improvement of Web site, detailed description of a Web site usage. The combination of Web content and usage mining could allow to build ontologies according to Web pages content and refine them with behavior patterns extracted from log files.

Web usage mining provides more relevant information to users and it is therefore a very powerful tool for information retrieval. Another way to provide more

Copyright © 2006, Idea Group Inc. Copying or distributing in print or electronic forms without written permission of Idea Group Inc. is prohibited.

accurate results is to involve users in the mining process, which is the goal of visual data mining, described in the next section.

Visualization

Topic Maps, RDF graphs, and ontologies are very powerful but they may be complex. Intuitive visual user interfaces may significantly reduce the cognitive load of users when working with these complex structures. Visualization is a promising technique for both enhancing users' perception of structure in large information spaces and providing navigation facilities. According to Gershon and Eick (1995), it also enables people to use a natural tool of observation and processing (their eyes as well as their brain) to extract knowledge more efficiently and find insights.

The goal of semantic graphs visualization is to help users locate relevant information quickly and explore the structure easily. Thus, there are two kinds of requirements for semantic graphs visualization: representation and navigation. A good representation helps users identify interesting spots whereas an efficient navigation is essential to access information rapidly. We both need to understand the structure of metadata and to locate relevant information easily.

A study of representation and navigation metaphors for Semantic Web visualization has been studied by Le Grand and Soto (to appear). Figure 6 shows two example metaphors for the Semantic Web visualization: a 3D cone-tree and a virtual city. In both cases, the semantic relationships between concepts appear on the display, graphically, or textually.

Many open research issues remain in the domain of Semantic Web visualization; in particular, evaluation criteria must be defined in order to compare the various existing approaches. Moreover, scalability must be addressed, as most current visualization tools can only represent a limited volume of data.

Semantic Web services are also an open research area and are presented in the next section.

Semantic Web Services

Web services belong to the broader domain of *service-oriented computing* (Papazoglou, 2003) where the application development paradigm relies on a loosely coupling of services. A service is defined by an abstract interface independently of any platform technology. Services are then published in directories where they can be retrieved and used alone or composed with other services. Web services (W3C, 2004) are an important research domain as they

Copyright © 2006, Idea Group Inc. Copying or distributing in print or electronic forms without written permission of Idea Group Inc. is prohibited.

Figure 6. Example visualization metaphors for the Semantic Web

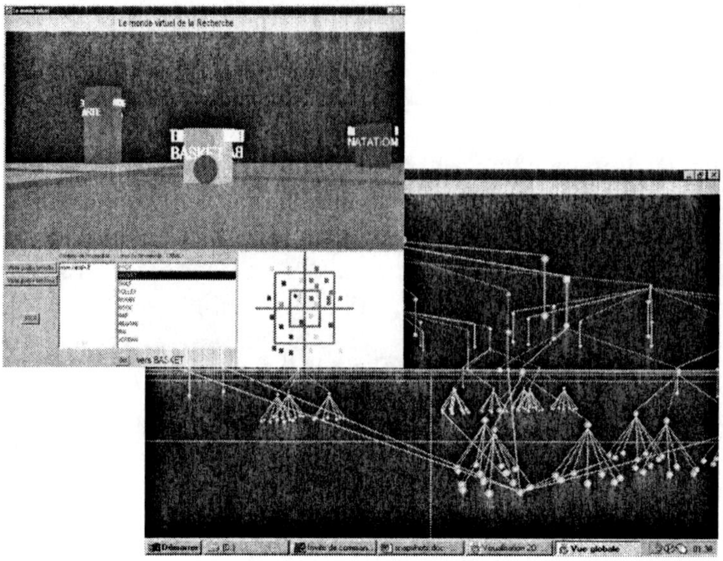

are designed to make the Web more dynamic. Web services extend the browsable Web with computational resources named services. Browsable Web connects people to documents, whereas Web services connect applications to other applications (Mendelsohn, 2002). One goal of Semantic Web services (Fensel et al., 2002; Mc Ilraith et al., 2001) is to make Web services interact in an intelligent manner. Two important issues are Web services discovery and composition, as it is important to find and combine the services in order to do a specific task.

The Semantic Web can play an important role in the efficiency of Web services, especially in order to find the most relevant Web services for a problem or to build ad hoc programs from existing ones.

Web services and the Semantic Web both aim at automating a part of the process of information retrieval by making data usable by computers and not only by human beings. In order to achieve this goal, Web services semantics must be described formally and Semantic Web standards can be very helpful. Semantics are involved in various phases: the description of services, the discovery and the selection of relevant Web services for a specific task, and the composition of several Web services in order to create a complex Web service. The automatic discovery and composition of Web services is addressed in the SATINE European project.

Copyright © 2006, Idea Group Inc. Copying or distributing in print or electronic forms without written permission of Idea Group Inc. is prohibited.

Towards A Meta-Integration Scheme

We have addressed the semantic integration of data in section 3.2. But as the Semantic Web grows, we now have to deal with the integration of metadata. We have presented ontology merging and ontology mapping techniques in this chapter. In this section, we propose a meta-integration scheme, which we call *metaglobal semantic model*.

Semantic integration may valuably be examined in terms of *interoperability* and *composability*. *Interoperability* may be defined as the interaction capacity between distinct entities, from which a system emerges. Interoperability in the context of the Semantic Web, will allow, for example, to make several semantic repositories work together to satisfy a user request. On the other side, *composability* may be defined as the capacity to reuse existing third party components to build any kind of system. Composability will allow building new semantic repositories from existing ones in order to cope with specific groups of users. Automatic adaptation of different components will be necessary and automatic reasoning capabilities are needed for this purpose. This requires a deep understanding of the nature and the structure of semantics repositories. Currently, there is neither a "global" vision nor a formal specification of semantic repositories. The definitions of taxonomies, thesauri, and ontologies, mentioned in the above sections, are still mostly in natural language and, as a paradox, there is not always a computer usable definition of these strategic concepts. This may be the main reason why semantic integration is so difficult to achieve. An effort from the Semantic Web community is needed to provide the Semantic Web community with a meta-global semantic model of the data.

Metamodeling for the Semantic Web: A Global Semantic Model

A global metamodel should be provided above data to overcome the semantic repositories' complexity and to make global semantic emerge. It is important to understand that the goal here is not only to integrate existing semantic objects such as ontologies, thesauri or dictionaries but to create global semantic framework consistency for the Semantic Web. Ontologies, thesauri, or dictionaries must considered as a first level of data semantics; we propose to add a more generic and abstract conceptual level allowing to express data semantics but also to locate these data in the context the Semantic Web

This global semantic framework is necessary to:

Copyright © 2006, Idea Group Inc. Copying or distributing in print or electronic forms without written permission of Idea Group Inc. is prohibited.

- Exhibit global coherence of the data of any kind,
- Get insight on the data,
- Navigate at a higher level of abstraction,
- Provide users with an overview of data space and help them find relevant information rapidly, and
- Improve communication and cooperation between different communities and actors.

Requirements for a Metamodel for Semantic Data Integration

A metamodel is a model for models, i.e., a domain-specific description for designing any kind of semantic model. A metamodel should specify the components of a semantic repository and the rules for the interactions between these components as well as their environment, for example the others existing or future semantic repositories. This metamodel should encompass the design of any kind of ontology, taxonomy or thesaurus. The design of such a metamodel is driven by the need to understand the functioning of semantic repositories over time in order to take into account their necessary maintenance and their deployments. With this respect, the metamodeling of semantic repository requires to specify the properties of their structure (for example, elementary components, i.e., object, class, modeling primitives, relations between components, description logic, etc.). Thanks to these specifications, the use of a metamodel allows the semantic integration of data on the one hand, and the transformation into formal models (mathematical, symbolic, logical, etc.) for interoperability and composability purpose, on the other hand. Integration and transformation of the data is made easier by the use of a modeling language.

Technical Implementation of the Global Semantic Level

The global semantic level could be implemented with a variety of formalisms but the Unified Modeling Language (UML) has already been successfully used in the context of interoperability and composability.

The Unified Modeling Language is an industry standard language with underlying semantics for expressing object models. It has been standardized and developed under the auspices of the Object Management Group (OMG), which is a consortium of more than 1.000 leading companies producing and maintaining computer industry specifications for interoperable applications. The UML formalism provides a syntactic and semantic language to specify models in a rigorous, complete and dynamic manner. The customization of UML (UML

Copyright © 2006, Idea Group Inc. Copying or distributing in print or electronic forms without written permission of Idea Group Inc. is prohibited.

profile) for the Semantic Web may be of value for semantic data specification and integration.

It is worth pointing out that the current problem of semantic data integration is not specific to the Semantic Web. For example, in post genomic biology, semantic integration is also a key issue and solutions based on metamodeling and UML are also under study in the life sciences community.

Conclusion

In this chapter, we presented a state of the art of techniques which could make the Web more "semantic". We described the various types of existing semantic metadata, in particular XML, controlled vocabularies, taxonomies, thesauri, RDF, Topic Maps and ontologies; we presented the strengths and limits of these formalisms.

We showed that ontology was undoubtedly a key concept on Semantic Web pathway. Nevertheless, this concept is still far from being machine-understandable. The future Semantic Web development will depend on the progress of ontologies engineering.

A lot of work is currently in progress within the Semantic Web community to make ontology engineering an operational and efficient concept. The main problems to be solved are ontologies integration and automatic semantics extraction. Ontologies integration is needed because there are already numerous existing ontologies in many domains. Moreover, the use of a common ontology is neither possible nor desirable. As creating an ontology is a very time-consuming task, existing ontologies must be capitalized in the Semantic Web; several integration methods were presented in this chapter. Since an ontology may also be considered as a model for a domain knowledge, the Semantic Web community should consider existing work on metamodeling from the OMG (Object Modelling Group) as a possible way to build a global semantic metamodel to achieve ontology reconciliation.

In the large-scale context of the Semantic Web, automatic semantic integration is necessary to quicken the creation and the updating processes of ontologies. We presented current initiatives aiming at automating the knowledge extraction process. This remains an open research area, in particular the extraction of relationships between concepts. The evaluation of ontology learning is a hard task because of its unsupervised character. In Cimiano et al. (2004) and Maedche and Staab (2002) the hierarchy obtained by applying clustering techniques is evaluated using handcrafted reference ontology. The two ontolo-

Copyright © 2006, Idea Group Inc. Copying or distributing in print or electronic forms without written permission of Idea Group Inc. is prohibited.

gies are compared at a lexical and at a semantic level using lexical overlap/recall measures and taxonomic overlap measure.

The success of the Semantic Web depends on the deployment of ontologies. The goal of ontology learning is to support and to facilitate the ontology construction by integrating different disciplines in particular natural language processing and machine learning techniques. The complete automation of ontology extraction from text is not possible regarding the actual state of research and an interaction with human modeler remains primordial.

We finally presented several research directions which we consider as strategic for the future of the Semantic Web. One goal of the Semantic Web is to provide answers which meet end users' expectations. The definition of profiles and behavior models through the combination of Web content and Web usage mining could provide very interesting results.

More and more data-mining techniques involve end users, in order to take advantage of their cognitive abilities; this is the case in visual data mining, in which the knowledge extraction process is (at least partially) achieved through visualizations.

Another interesting application domain for the Semantic Web is the area of Web services, which have become very popular, especially for mobile devices. The natural evolution of current services is the addition of semantics, in order to benefit from all Semantic Web's features.

The interest and the need of the Semantic Web have already been proven, the next step is to make the current Web more semantic, with all the techniques we presented here.

References

Baader, F., Horrocks, I., & Sattler, U. (2003). *Description logics as ontology languages for the Semantic Web. Lecture Notes in Artificial Intelligence*. Springer.

Berendt, B., Hotho, A., & Stumme, G. (2002, June 9-12). Towards Semantic Web mining. *Proceedings of First International Semantic Web Conference (ISWC)*, Sardinia, Italy (pp. 264-278).

Berners-Lee, T., Hendler, J., & Lassila, O. (2001). The Semantic Web. *Scientific American, 284*(5), 34-43.

Bisson, G., Nedellec, C., & Canamero, L. (2000). Designing clustering methods for ontology building: The Mo'K workbench. *Proceedings of the ECAI Ontology Learning Workshop*.

Copyright © 2006, Idea Group Inc. Copying or distributing in print or electronic forms without written permission of Idea Group Inc. is prohibited.

Brickley, D., & Guha, R. V. (2003). *RDF vocabulary description language 1.0: RDF schema*. World Wide Web Consortium. Retrieved from http://www.w3.org/TR/rdf-schema/

Cali, A. (2003). Reasoning in data integration system: Why LAV and GAV are siblings. *Proceedings of the 14th International Symposium on Methodologies for Intelligent Systems (ISMIS 2003)*.

Calvanese, D., De Giacomo, G., & Lenzerini, M. (2001). A framework for ontology integration. *Proceedings of the 1st Internationally Semantic Web Working Symposium (SWWS)* (pp. 303-317).

Caraballo, S. A. (1999). Automatic construction of a hypernym-labeled noun hierarchy from text. *Proceedings of the 37th Annual Meeting of the ACL*.

Cimiano, P., Hotho, A., & Staab, S. (2004, August). Comparing conceptual, partitional and agglomerative clustering for learning taxonomies from text. *Proceeding of ECAI-2004*, Valencia.

Daconta, M., Obrst, L., & Smith, K. (2003). *The Semantic Web: A guide to the future of XML*. Wiley.

Dean, M., & Schreiber, G. (2003). *OWL Web Ontology Language: Reference*. World Wide Web Consortium. Retrieved from http://www.w3.org/TR/2003/CR-owl-ref-20030818/

Decker, S., Jannink, J., Melnik, S., Mitra, P., Staab, S., Studer, R., & Wiederhold, G. (n.d.). An information food chain for advanced applications on the WWW. *ECDL 2000*, 490-493.

Doan, A., Madhavan, J., Dhamankar, R., Domingos, P., & Halevy, A. (2003). Learning to match ontologies on the Semantic Web. *VLDB Journal, 12*(4), 303-319.

Euzénat, J., Remize, M., & Ochanine, H. (2003). *Projet Hi-Touch*. Le Web sémantique au secours du tourisme. Archimag.

Faure, D., & Nedellec, C. (1998). A corpus-based conceptual clustering method for verb frames and ontology. In P. Verlardi (Ed.), *Proceedings of the LREC Workshop on Adapting Lexical and Corpus Resources to Sublanguages and Applications*.

Fensel, D., Bussler, C., & Maedche, A. (2002). Semantic Web enabled Web services. *International Semantic Web Conference*, Italy (pp. 1-2).

Friedman, M., Levy, A., & Millstein, T. (1999). Navigational plans for data integration. *Proceedings of of AAAI'99* (pp. 67-73).

Gershon, N., & Eick, S. G. (1995). Visualisation's new tack: Making sense of information. *IEEE Spectrum*, 38-56.

Copyright © 2006, Idea Group Inc. Copying or distributing in print or electronic forms without written permission of Idea Group Inc. is prohibited.

Gómez-Pérez, A., Fernández-López, M., & Corcho O. (2003). *Ontological engineering.* Springer.

Gómez-Pérez, A., & Rojas, M. D. (1999). Ontological reengineering and reuse. In D. Fensel, & R. Studer (Eds.), *The 11th European Workshop on Knowledge Acquisition, Modeling and Management (EKAW'99), Lecture Notes in Artificial Intelligence LNAI 1621*, Germany (pp. 139-156). Springer-Verlag.

Guarino, N. (1998). Formal ontology in information systems. In N. Guarino (Ed.), *First International Conference on Formal Ontology in Information Systems*, Italy (pp. 3-15).

Gruber, T. (1993, August). Toward principles for the design of ontologies used for knowledge sharing. In N. Guarino, & R. Poli (Eds.), *International Journal of Human-Computer Studies, Special issue on Formal Ontology in Conceptual Analysis and Knowledge Representation.*

Grüninger, M., & Fox, M. S. (1995). Methodology for the design and evaluation of ontologies. In D. Skuce (Ed.), *IJCAI'95 Workshop on Basic Ontological Issues in Knowledge Sharing*, Canada.

Halevy, A. Y. (2001). Answering queries using views: A survey. *The VLDB Journal, 10*(4), 270-294.

Hameed, A., Preece, A., & Sleeman, D. (2004). Ontology reconciliation. In S. Staab, & R. Studer (Eds.), *Handbook on ontologies* (pp. 231-250).

Harsleev, V., & Möller, R. (2001). Racer system description. *Proceedings of the International Joint Conference on Automated Reasoning (IJCAR 2001), Lecture Notes in Artificial Intelligence 2083* (pp. 701-705). Springer.

Heflin, J. (2004). *OWL Web Ontology Language use cases and requirements.* W3C Recommendation. Retrieved from http://www.w3.org.

Horrocks I. (1998). Using an expressive description logic: FaCT or fiction? *Proceedings of the 6th International Conference on Principles of Knowledge Representation and Reasoning (KR' 98)* (pp. 636-647).

Horrocks, I. (2002). DAML+OIL: A reasonable Web ontology language. *Proceedings of EDBT 2002, Lecture Notes in Computer Science 2287* (pp. 2-13). Springer.

Horrocks, I., & Patel-Schneider, P. F. (2003). Reducing OWL entailment to description logic satisfiability. *Proceedings of International Semantic Web Conference (ISWC 2003), Lecture Notes in Computer Science number 2870* (pp. 17-29). Springer.

Horrocks, I., & Patel-Schneider, P. F. (2004). A proposal for an owl rules language. *Proceedings of the 13th International World Wide Web Conference (WWW 2004).* ACM.

Copyright © 2006, Idea Group Inc. Copying or distributing in print or electronic forms without written permission of Idea Group Inc. is prohibited.

Horrocks, I., Patel-Schneider, P. F., & van Harmelen, F. (2003). From SHIQ and RDF to OWL: The making of a Web ontology language. *Journal of Web Semantics.*

International Electrotechnical Commission (IEC). (1999). International Organization for standardization (ISO). Topic Maps. International Standard ISO/IEC 13250.

Jarrar, M., & Meersman, R. (2002). Formal ontology engineering in the DOGMA approach. In Meersman, Tari, et al. (Eds.), *Proceedings of the Confederated International Conferences: On the Move to Meaningful Internet Systems (Coopis, DOA and ODBASE 2002). Lecture Notes in Computer Science 2519* (pp. 1238-1254). Springer.

Kosala, R., & Blockeel, H. (2000). Web mining research: A survey. *SIGKDD Explorations — Newsletter of the ACM Special Interest Group on Knowledge Discovery and Data Mining, 2*(1), 1-15.

Lassila, O., & McGuiness, D. (2001). *The role of frame-based representation on the Semantic Web.* Technical Report KSL-01-02. Stanford, CA.

Lassila, O., & Swick, R. (1999). *Resource description framework (RDF) model and syntax specification.* Retrieved February 22, 1999, from http://www.w3.org/TR/REC-rdf-syntax/

Le Grand, B., & Soto, M. (2005). Topic maps visualization. In V. Geroimenko, & C. Chen (Eds.), *Visualizing the Semantic Web* (2nd ed.). Springer.

McIlraith, S., Son, T. C., & Zeng, H. (2001). Semantic Web services. *IEEE Intelligent Systems, Special Issue on the Semantic Web, 16*(2), 46-53.

Maedche, A. (2002). *Ontology learning for the Semantic Web.* Kluwer Academic Publishers.

Maedche, A., & Staab, S. (2000). Discovering conceptual relations from text. In W. Horn (Ed.), *Proceedings of the 14th European Conference on Artificial Intelligence,* Berlin (pp. 21-25). IOS Press.

Maedche, A., & Staab, S. (2002). Measuring similarity between ontologies. *Proceedings of EKAW'02.* Springer.

Mendelsohn, N. (2002). *Web services and the World Wide Web.* Retrieved from http://www.w3.org/2003/Talks/techplen-ws/w3cplenaryhowmany webs.htm

Mena, E., Illarramendi, A., Kashyap, V., & Sheth, A. (2000). An approach for query processing in global information systems based on interoperation across preexisting ontologies. *Distributed and Parallel Databases — An International Journal, 8*(2).

Miller, G. A. (1995). WordNet: A lexical database for English. *Communications of the ACM, 11,* 39-41.

Copyright © 2006, Idea Group Inc. Copying or distributing in print or electronic forms without written permission of Idea Group Inc. is prohibited.

Moore, G. (2001). RDF and topic maps: An exercise in convergence. *XML Europe 2001*, Germany.

Nakabasami, C. (2002). An inductive approach to assertional mining of Web ontology revision. *Proceedings of the International Workshop on Rule Markup Languages for Business Rules on the Semantic Web (RuleML-2002),* Italy.

Noy, N. F., & Klein, M. (2004). Ontology evolution: Not the same as schema evolution. *Knowledge and Information Systems, 6*(4), 428-440.

Noy, N. F., & Musen, M. A. (2003). The PROMPT Suite: Interactive tools for ontology merging and mapping. *International Journal of Human-Computer Studies.*

Noy, N. F., Sintek, M., Decker, S., Crubezy, M., Fergerson, R. W., & Musen, M. A. (2001). Creating Semantic Web contents with Protege-2000. *IEEE Intelligent Systems, 16*(2), 60-71.

Paepen, B., et al. (2002). OmniPaper: Bringing electronic news publishing to a next level using XML and artificial intelligence. *Proceedings* (pp. 287-296).

Papazoglou, M. P. (2003). Service-oriented computing: Concepts, characteristics and directions. *Proceeding of 4th International Conference on Web Information Systems Engineering (WISE 2003).*

Parent, C., & Spaccapietra, S. (1998). Issues and approaches of database integration, *CACM, 41*(5), 166-178.

Pereira, F., Tishby, N., & Lee, L. (1993). Distributional clustering of English words. *Proceedings of the 31st Annual Meeting of the ACL.*

Rahm, E., & Bernstein, P. A. (2001). A survey of approaches to automatic schema matching. *The VLDB Journal, 10*, 334-350.

SATINE. (n.d.). *Semantic-based interoperability infrastructure for integrating Web service platforms to peer-to-peer networks.* IST Project. Retrieved from http://www.srdc.metu.edu.tr/Webpage/projects/satine/

Staab, S., Studer, R., & Sure, Y. (2001). Knowledge processes and ontologies. *IEEE Intelligent Systems, 16*(1), 26-34.

Stumme, G., & Maedche, A. (2001). FCA-MERGE: Bottom-up merging of ontologies. In B. Nebel (Ed.), *Proceedings of the 17th International Conference on Artificial Intelligence (IJCAI '01)* (pp. 225-230).

TopicMaps.Org XTM Authoring Group. (2001). *XTM: XML Topic Maps (XTM) 1.0. TopicMaps.Org specification.*

Ullman, J. D. (1997). Information integration using logical views. In F. N. Afrati, & Kolaitis (Eds.), *Proceedings of the 6th International Conference on*

Copyright © 2006, Idea Group Inc. Copying or distributing in print or electronic forms without written permission of Idea Group Inc. is prohibited.

Database Theory (ICDT'97), Lecture Notes in Computer Science 1186 (pp. 19-40).

Uschold, M., & King, M. (1995). Towards a methodology for building ontologies. In D. Skuce (Ed.), *IJCAI'95 Workshop on Basic Ontological Issues in Knowledge Sharing* (6.1-6.10).

W3C (W3C Working Group Note). (2004, February 11). *Web Services Architecture.* Retrieved from http://www.w3.org/TR/ws-arch/

W3C (World Wide Web Consortium). (1999). *Resource Description Framework (RDF) model and syntax specification.* W3C.

W3C (World Wide Web Consortium). (2004). In D. L. McGuinness, & F. van Harmelen, *OWL Web Ontology Language — Overview.* W3C Recommendation.

Wiemer-Hastings, P., Graesser, A., & Wiemer-Hastings, K. (1998). Inferring the meaning of verbs from context. *Proceedings of the 20th Annual Conference of the Cognitive Science Society* (pp. 1142-1147). Mahwah, NJ: Lawrence Erlbaum Associates.

Yergeau, F., Bray, T., Paoli, J., Sperberg-McQueen, S., & Maler, E. (2004). *Extensible Markup Language (XML) 1.0* (3rd ed.). W3C Recommendation.

Copyright © 2006, Idea Group Inc. Copying or distributing in print or electronic forms without written permission of Idea Group Inc. is prohibited.

Section IV

Applications and Policies

Copyright © 2006, Idea Group Inc. Copying or distributing in print or electronic forms without written permission of Idea Group Inc. is prohibited.

Chapter X

Translating the Web Semantics of Georeferences

Stephan Winter, The University of Melbourne, Australia

Martin Tomko, The University of Melbourne, Australia

Abstract

This chapter presents a review of the ways of georeferencing in Web resources, as opposed to the georeferencing of other information communities, specifically in route directions for wayfinders. The different information needs of the two information communities, reflected by their different semantics of georeferences, are identified. In a case study, we investigate the possibilities of translating the semantics of georeferences in Web resources to landmarks in route directions. We show that interpreting georeferences in Web resources enhances the perceivable properties of described features. Finally, we identify open questions for future research.

Copyright © 2006, Idea Group Inc. Copying or distributing in print or electronic forms without written permission of Idea Group Inc. is prohibited.

Introduction

The Web consists of a large amount of predominantly weakly structured and organized resources, with only a few resources having an explicit and structured description of the content. We can think of the Web as an informal network of diverse heterogeneous data sources, including simple files as well as modern object-relational and semantic databases. Many, if not most of these resources provide some form of reference to geographic space. Georeferences link the features of physical or social reality described in the content of the resources to particular locations in geographic space.

The descriptions of features together with their georeferences can be seen as a map inherent in the Web. This map has some properties, particularly heterogeneity. The types of features described, the ways the features are described, and the ways the georeferences are made are diverse, and the links between features and georeferences are implicit and diverse as well. Without a specified semantics of features and their reference to geographic space, the Web-inherent map cannot be translated automatically into an explicit map of general or specific purpose. This is true of the opposite as well. In general, search engines have problems with geographic searches when looking simply for keywords and not considering semantics of natural language structures.

In this chapter we will investigate georeferences in Web resources for a very specific purpose: exploiting the wealth of inherent geographic knowledge of Web resources for route directions. Choosing a specific purpose for (re-)constructing the inherent map in the Web allows for the identification of the fundamental challenges for research by a single case-based study. The case-based approach limits the complexity of the reconstruction at least by selecting an appropriate destination domain — the source domain, the Web, remains a heterogeneous domain. Choosing wayfinding as the destination domain does not limit the generality of our findings; other destination domains have to address the same challenges.

In wayfinding, people generate travel routes from their mental maps, and communicate these routes by relating movement and orientation actions to landmarks at selected points along the route. In comparison, wayfinding services generate travel routes on metric travel networks. The metric travel networks cannot communicate these routes by referring to landmarks due to a lack of landmark knowledge. There is neither a clear understanding of what constitutes a landmark, nor is there a ready-made directory of landmarks available. In this situation, the map inherent in the Web is a rich pool of geospatial features, which potentially can be used by wayfinding services for searching for landmarks.

Our hypothesis is: Referencing to geographic space is fundamentally different for Web content providers and wayfinders; nevertheless, links can be established

Copyright © 2006, Idea Group Inc. Copying or distributing in print or electronic forms without written permission of Idea Group Inc. is prohibited.

between what is represented in Web resources and what is looked for by wayfinding services. In this regard, the following question stands out: Can we generate orientation and wayfinding information out of ordinary Web resources? Or, more specifically:

- How can we identify features in Web resources spatially related to a location or route?
- How can we assess the (spatial) relevance of these features for orientation?
- How can we refer to selected features, or relate selected features to the wayfinder?

In order to approach the above research questions, the next section brings some definitions and a scenario to introduce the topic in more detail. We then investigate georeferences in the destination domain, wayfinding, and compare them with the current and emerging ways Web resources refer to geographic space. We use a case study derived from a scenario to identify issues of translating the semantics of georeferences, and to direct to solutions. For a street segment in Melbourne, Australia, we collect all available Web resources, identify and categorize their ways of georeferencing, and derive knowledge from these georeferences that is relevant for wayfinders along this segment. The procedure allows for the identification of research challenges for semantic translation of georeferences. Finally we will summarize and discuss our findings.

Background

Georeferences

We use the term *georeferencing* in a broad sense, extending here a recent definition by Hill (2004): Georeferencing is relating information (e.g., documents, datasets, maps, images, biographical information, artifacts, specimens, directions) to geographic locations through place names (i.e., toponyms), place descriptions (e.g., "the green building"), place relations (e.g., "the building opposite to the church"), place codes (e.g., postal codes), or through geocode (e.g., geographic coordinates).

For all these kinds of georeferences we can find examples in Web resources. Web resources can refer to geographic space either in the content or in tags.

Copyright © 2006, Idea Group Inc. Copying or distributing in print or electronic forms without written permission of Idea Group Inc. is prohibited.

Georeferences in the content are made in natural language. They address the human reader, and require a proper understanding of some semantic and world knowledge to be shared between content provider and reader. In principle, all the listed types of georeferences can occur in the content in some context. For example, a common type of georeferencing on commercial Web sites can be found under the label "contact us", frequently referencing to a telephone number, a post address, etc.; one prototype can be found at http://www.geom.unimelb.edu.au/contact/. Details of georeferences in Web content are discussed below. Georeferences hidden in tags address automatic processing tools, and follow some externally defined and shared formal tag type definitions. For automated processing one would select only types of georeferences with a formal semantics, e.g., geocodes. Examples are discussed below.

In contrast, georeferences made in route directions for wayfinders exist only in form of natural language, and hence, in all the variety and complexity of natural language (Klein, 1979; Weissensteiner & Winter, 2004). In route directions, the information is a direction to move or to turn, and this information is related to geographic locations through features along the route. Hence, georeferences are used to evoke some wayfinding behavior at specific locations along the route. The direction "at the church turn right" uses a place description by referring to a feature by a categorical term. It could use the place name — the feature's name — as well ("at the Trinity Church turn right") if the route direction is given to a reader who is familiar with the place. Less common are place codes like postal addresses ("at 11 Main Street turn right"), but one can find references to street names ("at corner Collins Street/Spring Street turn right"). Uncommon are geocodes. Route directions are studied in more detail.

Semantics and Ontologies of Georeferences

A wayfinder's semantic of georeferences is based on the capabilities of human perception and experience of space. To perform a wayfinding behavior at the place intended by the sender of the route direction, the reader (a wayfinder) has to recognize the place referred to from his or her perspective in an unambiguous manner. In contrast, the context of Web resource content suggests other purposes of georeferencing. The context is not specified, but it is typically not wayfinding (although there are some Web resources giving route directions). More often, Web resources use georeferences to identify individuals or institutions, to help contacting them by mail, or to help locating them by street address. For instance, "XY Ltd, 1 Collins Street, Melbourne" identifies a specific company XY Ltd, namely the one that is located at 1 Collins Street, Melbourne. It is also sufficient to address a letter to the company, or to look up a city map

Copyright © 2006, Idea Group Inc. Copying or distributing in print or electronic forms without written permission of Idea Group Inc. is prohibited.

for that company. Perceptual aspects of local features are irrelevant for that purpose.

So far, we deal with different information communities. An information community is a group of people sharing a semantics (Bishr, 1998). The information communities identified here are:

1. People seeking orientation or wayfinding information. The way this group refers to places is driven by their motor, visual, or other senses' experience. It is related to how people learn and memorize space, i.e., landmark and route experiences, which is categorically different from postal address knowledge.

2. People seeking location information from Web resources, for many purposes. The most frequent georeferences in this community (but not the only ones) are postal addresses; however, the intended meaning of a postal address or any other georeference in a Web resource can be quite diverse according to their specific purpose.

From a formal perspective, each of the two information communities has its own ontology of georeferencing. An ontology in this sense is a specification of a conceptualization (Gruber, 1993). Conceptualizations represent ways in which an information community understands the world. A wayfinder for example uses concepts like *church*, *intersection*, and so forth, and a Web content provider uses concepts like *house number*, *street name*, or *post code*. A specification is some abstract description of those concepts, and includes at least a vocabulary of terms and some specification of their meaning (Bittner, Donnelly, & Winter, 2004).

With all the fuzziness in the definition of the information community of Web users, this chapter explores the possibilities and challenges to use the georeferences in Web resources for helping wayfinders by enriching route directions with perceivable and cognitively identifiable georeferences. In this respect, our final goal is a translation of terms using available ontologies, or by creating new ones.

Wayfinding

To illustrate the problem of georeferencing in wayfinding contexts imagine the following scenario, which will be referred to throughout this chapter. Hillary, a tourist in Melbourne, Australia, found a recommendation for the Indulgence Afternoon Tea at the Hotel Windsor in her travel guide, together with the address: 111 Spring Street. This information does not help her to find the place.

Copyright © 2006, Idea Group Inc. Copying or distributing in print or electronic forms without written permission of Idea Group Inc. is prohibited.

So she asks at her hotel reception for the route:

"To the Hotel Windsor?"

Note that Hillary does refer to the institution, Hotel Windsor, not to its address. Even locals might not know which building 111 Spring Street is, but they have an experience of the Hotel Windsor.

"Ok, when you leave the Hyatt, turn right and walk down to the end of the street. At that intersection you can see to your left the Parliament, and opposite to the Parliament is the Hotel Windsor."

The direction giver refers to landmarks ("Hyatt", "Parliament") and to the structure of the street network ("end of street", "intersection"), remembering experiences he assumes to be shared by Hillary on location ("walk down", "you can see"). The reference to the Parliament in this context is an interesting one, since Hillary will have no idea what the Parliament in Melbourne looks like. The direction giver seems to be convinced that she will recognize it instantly, which means that the Parliament building in Melbourne must have a prototypical appearance (Lakoff, 1987; Rosch, 1978). Hillary is now sure to find her way.

Now imagine that Hillary would have asked her mobile device for directions, instead of the receptionist. The user interface of her wayfinding service insists on a destination address; she has to go back to her travel guide entry to be reminded that this is 111 Spring Street. Then, the service starts: "From 123 Collins Street, walk ...". Trying to make sense of this she recalls that her actual location, the Hyatt, has the address 123 Collins Street. It turns out that this dialog demands some cognitive effort from Hillary.

Finally imagine Hillary, a tourist in some near future, declaring to her personal service *WebGuide*: "Guide me to the Hotel Windsor!" Can WebGuide, with its real-time access to all Web resources, provide better service than the current mobile device? Services like WebGuide will need to relate street address information, which is given by route planning services, to spatial features of physical or social reality that can be experienced by travelers. Furthermore, it needs to select from the pool of found features the ones that are of relevance for wayfinding and for the route.

Copyright © 2006, Idea Group Inc. Copying or distributing in print or electronic forms without written permission of Idea Group Inc. is prohibited.

Georeferencing of Wayfinders

In this section we review the relevant literature to identify the semantics of georeferences by wayfinders. These semantics will be contrasted later with the semantics of georeferences in Web resources. *Wayfinding* as a basic human activity is investigated in spatial cognition and related disciplines (Freksa, Brauer, & Habel, 2000; Freksa, Brauer, Habel, & Wender, 2003; Freksa, Habel, & Wender, 1998; Golledge, 1999; Golledge & Stimson, 1997; Jarvella & Klein, 1982; Kaplan & Kaplan, 1982). We are particularly interested in how people refer to geographic space in route directions, which is directly coupled with how people experience and memorize space.

Landmarks

People learn, memorize, and communicate their environment by experiences (Golledge, Rivizzigno, & Spector, 1976; Golledge & Stimson, 1997; Siegel & White, 1975; Weissensteiner & Winter, 2004), an understanding that goes hand in hand with the embodied mind (Johnson, 1987) and the theory of affordance (Gibson, 1979). Wayfinding experience is acquired by motor, visual, and other senses, which relate the activity to perceived environmental features that can become landmarks. Landmarks have a particular role in learning, memorizing and communicating routes (Cornell, Heth, & Broda, 1989; Habel, 1988; Michon & Denis, 2001). As Denis et al. (1999) have shown, people prefer to determine the place for reorientation during wayfinding (decision points) by landmarks; they rarely use distances for that purpose. This observation conforms to the understanding that human landmark, route and survey knowledge (Siegel & White, 1975) is primarily of a topological nature. In general, landmarks can be classified into landmarks at decision points, along route segments (route marks), and distant, off-route landmarks (Lovelace, Hegarty, & Montello, 1999; Presson & Montello, 1988). References to landmarks are made at preferred places along the route, at least at decision points (Denis et al., 1999; Habel, 1988; Klippel, 2003; Michon & Denis, 2001).

Already Lynch speaks of landmarks in his classic categorization of the structuring elements of a city (1960): he distinguishes landmarks, places, paths, barriers and regions. However, his concept of a landmark is a narrow one; from a cognitive point of view one can argue that the latter four structuring elements can form landmarks as well. But even his distinction is based on human experience of space.

The route direction Hillary got at the hotel reception referred to landmarks ("Hyatt", "intersection", "Parliament"). The Hyatt is a landmark at the start

Copyright © 2006, Idea Group Inc. Copying or distributing in print or electronic forms without written permission of Idea Group Inc. is prohibited.

point of the route, which is a distinguished decision point. The intersection with sight of the Parliament forms a structural landmark (Lynch would call it a place), and the Parliament is first a distant landmark (when standing at the street intersection), but later a landmark at another distinguished decision point, the destination.

Identification of Landmarks

People refer to space in wayfinding situations preferably by landmarks. Current wayfinding services lack that ability, but next-generation services will be able to communicate routes by landmarks as well. Since the notion of a landmark is subjective, bound to shared or sharable experience, services cannot identify landmarks. They can only implement generic methods to identify *salient features* in spatial data sets, as best matches.

A formal measure of salience is based on three qualities of landmarks: visual, semantic and structural ones (Sorrows & Hirtle, 1999). For each of these qualities some parameters can be defined as observables. Observing these parameters for all features, the features that have most distinct parameter values from others are called salient features, and will be considered for inclusion in route directions (Elias & Brenner, 2004; Nothegger, Winter, & Raubal, 2004; Raubal & Winter, 2002). Landmarks can be chosen dependent on route properties (Winter, 2003), and dependent on the context of the recipient (Winter, Raubal, & Nothegger, 2004).

Hillary's wayfinding service can only observe parameters of qualities of features. A service can for example identify that Hillary's current location, measured in geographic coordinates, is in the Hyatt. This conclusion can be done by reverse geocoding (deriving a postal address from geographic coordinates), and looking up business directories. The business directory reveals the business type: *hotel* forms a semantic quality of the building, which distinguishes it from other buildings close-by, and hence, contributes to its salience.

Functions of Landmarks in Route Directions

Landmarks are to be included in route directions on specific locations within the route direction. The basic actions of a wayfinder are to change direction ("turn"), and to change location ("move"). Route directions can follow many grammars; Frank (2003), for example, distinguishes between "turn and move n segments", "turn and move distance", "turn and move until", and several others. If the grammar utilizes landmarks at decision points ("at x turn and move"), these landmarks function as the anchors of an action.

Copyright © 2006, Idea Group Inc. Copying or distributing in print or electronic forms without written permission of Idea Group Inc. is prohibited.

Investigating the verbs or actions in route directions is a valuable task in itself. Results are action ontologies that define the degrees of freedom of a traveler in a particular mode of traveling (Kuhn, 2001; Timpf, 2002). By that way action ontologies co-determine the form and frequency of landmarks required in the route directions.

Action ontologies provide the mean to link and assign human activities to static objects. In the case of wayfinding, the most common use is to serialize the features used as wayfinding references along the path in a chronological manner and to provide specific directions to the wayfinder. The terms used change not only depending on the direction given, but also on the nature of the feature the direction is anchored to: a landmark, street segment, or start or end of the path.

In our scenario, Hillary is a pedestrian tourist. The local expert at the hotel reception uses action verbs like "leave", "turn right", and "walk down". Leaving is an action that refers to image schemata of a container and a path from inside the container to outside (Fauconnier & Turner, 2002; Johnson, 1987). A pedestrian in a building has clear categories of inside and outside, and will typically change between them through exits or doors. The next action is bound to the act of leaving: "when you leave the Hyatt, turn right". We expect that outside of the hotel lobby is a street, and Hillary, when leaving through the hotel exit door, will find herself on a sidewalk. In that situation, she has to make a decision how to continue traveling towards the destination. As a pedestrian, she has choices such as turning left or right, crossing the street, or entering a tram.

Landmarks and the Web

From studying Web resources it becomes clear that they do not provide observations of visual, semantic, or structural quality parameters of features of the environment in a first instance. They do not because their intention is different from providing route directions. Web resources intend to identify features (e.g., by a unique postal address), to find features (e.g., by route directions), or to establish trust in institutions (e.g., by naming an expensive location). General wayfinding, and particularly measures of salience are not on this list. Nevertheless, georeferences in Web resources can be used to make conclusions on visual, semantic or structural qualities of features along routes.

Copyright © 2006, Idea Group Inc. Copying or distributing in print or electronic forms without written permission of Idea Group Inc. is prohibited.

Georeferencing on the Web

In the early stages of the development of the Internet, little attention was paid to the spatial location of both the network nodes and the content. The Internet is closely coupled with its geography, and with the geography of the features described by the contents of the Web. This relation continues to deepen, as the Internet moves to its ubiquitous age. Mobile services increased the demand for context-aware applications, leading to a boom of location-based and location-aware services, and consecutively, to the spread of localised content. Today, it is estimated that 20-35% of all the searches performed on the Web seek geographically related results (Young, 2004). That means the reality proved the need for grounding the content in real world context.

Early georeferencing on the Web, still the prevalent method of georeferencing, mirrors the approaches of more traditional media. It is restricted to text (e.g., addresses, postcodes, or telephone numbers) and images (e.g., photographs, sketches, maps). These methods are less suitable for automated processing and interpretation. Textual descriptions or images are forms of narratives, and their semantics is inferred in active reading processes. While automatic natural language understanding is an active field of research in artificial intelligence, in-depth understanding is still the challenge of the discipline (McCarthy, 1990). Address patterns are often integrated with parts of text containing natural language statements, often referring to relations between features of the environment (e.g., "close to"). Furthermore, these address patterns are often imprecise, inconsistent, or incomplete, as, e.g., in the string "in Collins Street". For instance, our traveler Hillary would encounter problems when searching by keyword for "Hotels" and "Melbourne" on current search engines. Results point to places as distinct as Melbourne, Australia, and Melbourne, Florida. Similarly she would experience problems with a search string "111 Spring Street" due to its incompleteness with respect to a full postal address, which, on the other hand, might as well not exist in the corresponding Web resource.

A step towards more formal forms of georeferences is represented by the various national address standards, providing some degree of unification of georeferencing, be it solely for mailing purposes. Only the Semantic Web (Berners-Lee, Hendler, & Lassila, 2001) brings the structure necessary to process Web resources automatically, by introducing formal language structures for annotating the content. These formal languages enable the creation of consistent models of all aspects of interest, or ontologies. These can be used to annotate the content and will enable machine aided reasoning and linking between various independent ontologies. In this way, the Web will change into a giant, intelligent knowledge base. However, we cannot expect that a sudden massive conversion of current Web resources to match Semantic Web requirements will happen. Sophisticated algorithms parsing textual information on the

Copyright © 2006, Idea Group Inc. Copying or distributing in print or electronic forms without written permission of Idea Group Inc. is prohibited.

Web are therefore required. Parsers seeking address structures and other georeferences have to be developed, as well as framework ontologies mapping and interconnecting these patterns. This section provides further details on existing georeferencing technologies on the Web and efforts to extract the location information from Web resources.

Informal Georeferencing

The location of a host of a Web resource does not provide reliable clues about the location of the features described by the content of a Web resource. Therefore, we have to derive the location of the features described by the content by other means. Insufficient spread of semantically annotated Web resources forces Web users to extract the georeferences from the text content. This task is not only affected by Web resource layout issues, but also by the problems of parsing natural language content affected by language and cultural differences. The only patterns with more structured content related to georeferencing are postal addresses.

Parsing and understanding is complicated by the initial uncertainty of the reader at which level of detail the reference to geographical space is made. There is a difference between the level of detail provided by a general tourist guide describing Australia, and the same location described by regional, local, or community Web resources. However, a model for place name-based information retrieval was proposed in Jones, Alani, and Tudhope (2001). Similarly, a useful notion of localness was introduced by Ma, Matsumoto, and Tanaka (2003), describing the extent to which the site provides regional information, the level of detail of the resource (localness degree), and the ubiquity of the resource.

The successful extraction of location information from Web resources enables us to create a candidate set of potential features in the selected environment, and the localness analysis can contribute to filter only the most relevant features and assess their salience and relevance to the specified location. Still, there is no certainty that the identified resources address an existent, permanent and salient feature that is useful for a wayfinder. Therefore, we propose to assess additionally the action ontologies associated with the georeferencing information.

Natural Language Statements

Natural language statements in Web resources are as flexible and various in georeferencing as people are in speech acts. Particularly private Web resources

Copyright © 2006, Idea Group Inc. Copying or distributing in print or electronic forms without written permission of Idea Group Inc. is prohibited.

show narrative forms of georeferencing. These georeferences are frequently given in the context of wayfinding information, and hence, are a valuable source of data for intelligent analysis.

Such natural language statements communicate locations, frequently by subjective personal experience with the environment. They also communicate spatial configurations: different elements of the environment can be related through natural language statements describing their spatial relationships. These statements provide the wayfinder with information often accessible with less cognitive effort than when relying on formal georeferences. The reason probably lies in the aptitude of the narrator to communicate the elements relevant to the context of the assumed recipient. These personal sites are often designed for a specific information community, and are of specific value for this community.

Humans often refer to features in the environment by providing their spatial context, in particular with regard to nearby landmarks. In natural language descriptions, one will find frequent usage of terms providing a description of a spatial relation between described features. Fuzzy expression as "close to", "nearby", "further down" are used as often as more exact terms as "next to", "opposite", "within a distance", "after". The context-dependent interpretation of fuzzy topological relations is beyond the focus of this chapter, but for a start see Worboys (2001).

The spatial relations between georeferenced features are also a valuable source of information to be interpreted by automated wayfinding services. Not only they enable to reconstruct a more adequate cognitive image of the environment and present it to the wayfinder, but they also provide this information in context, and select the reference points with the highest salience in the specific situation.

Semantics of Postal Addresses

Compared to natural language statements, the more formal semantics of postal addresses bring some structure to the Web content. Postal addresses are open to standardization at national levels; see for example the Australian and New Zealand standard on a Geocoded National Address File, GNAF (ICSM, 2003). A postal address characterizes a piece of land that can be represented geometrically by a polygon. Due to the different land administration legislatives in the world and over time, this polygon consists not necessarily of a single cadastral parcel or a single land register property.

International standardization is more complicated and is only at early stages. Currently, efforts concentrate on postal address interchange formats in XML, with address visualization templates adapted to local specifics, assuring the interoperability of dedicated postal database systems. Among the most widely

Copyright © 2006, Idea Group Inc. Copying or distributing in print or electronic forms without written permission of Idea Group Inc. is prohibited.

known attempts to standardized, global address content grammars in XML are the standards of the Universal Postal Union (UPU, 2002) and of OASIS (Organization for the Advancement of Structured Information Standards) (Kumar et al., 2002). A future Geospatial Semantic Web (discussed in a separate section) has to provide consistent ontologies for postal addresses in a global manner.

Postal addresses can be geocoded by selecting a representative point for the polygon. Geocodes are currently becoming part of address databases, such as GNAF. Note that geocodes have a semantics of their own: some are centroids of polygons, some street front center points, some building entrance points, and some simply arbitrary points inside of the polygons. Where no geocoded address files exist, geocoding can be calculated from street network datasets. Street segments contain typically two attributes representing the house number intervals for the two sides of the segment. These attributes can be linearly interpolated to calculate a geocode of a postal address. Note that in this case the geocode represents a point on the street network, which is not necessarily inside or on the boundary of the polygon. Typically this point is used for route planning, which requires dynamic segmentation of the street network.

Address files provide exactly one geocode per postal address, and no finer distinction is made. Particularly in rural areas the position defined by the geocode and the position of a building on that ground might differ significantly, in Australia's outback, for example, for some kilometers.

Taking the perspective of a wayfinder, limitations of addresses are manifold. Wayfinders are interested in features along a route. These features might have no postal address at all (e.g., monuments, or public land/crown land), or have an address that is not along the route (e.g., buildings at street intersections with an address of the crossroad, or rear sides of buildings), and vice versa, features with a specific street address might be located in backyards or malls, and therefore be invisible from the street itself.

Formal Georeferencing on the Web

Relating Content to Host Location

The location of the host IP address — or the entity-based geographical context (McCurley, 2001) — is at best in indirect relation with the content served, and the localization accuracy of IP addresses is unreliable. Depicting the spatial reference of a resource by assuming that it is mostly relevant to geographically close users provides only a low level of accuracy and reliability (Buyukkokten,

Copyright © 2006, Idea Group Inc. Copying or distributing in print or electronic forms without written permission of Idea Group Inc. is prohibited.

Cho, Garcia-Molina, Gravano, & Shivakumar, 1999). The analysis of the location of users' IP addresses can help to locate a larger region of interest, but is not sufficient for wayfinding applications.

A different attempt, advocated by GIS specialists and building upon interoperability initiatives of the Open Geospatial Consortium (OGC) and the DigitalEarth initiative, was supposed to lead to the GeoWeb (Leclerc, Reddy, Iverson, & Eriksen, 2002). The attempt consisted of a new Internet top-level domain .geo, with special URLs containing the encoded georeferenced tile covering the queried area. In the proposed system, the URL http://4e7s.14e3s.geo would denote a Web resource containing information about a 1 by 1 degree area with the longitude 144 degrees east and 37 degrees south. This approach was meant to ease geographical queries on interoperable distributed OGC compliant data sources. Ideally, geographic 3D encoded content would be distributed over this Internet subnetwork. The project was abandoned after rejection of the .geo top level domain name.

Geospatial Interoperability Initiatives

The need for automated and interoperable processing of location information enabled to coordinate efforts among special professional interest groups, led by OGC and the International Standardization Organization (ISO) in cooperation. OGC's location specifications are strongly focused on expert users and thus lack the support for general content providers' use. The developed standards focus on sharing specialized spatial data (raster and vector), metadata, and service interoperability. This focus influenced the design approach. The inherent complexity of spatial information is reflected by the standards, but makes them difficult to understand and implement. The complex structure of spatial data descriptions, namely the Geography Markup Language (GML) (Cox, Daisey, Lake, Portele, & Whiteside, 2003) also virtually prevents wider adoption by the general public for annotating public Web resources. The geographic information community shares a more consistent view on geographic data semantics and understands details that might be irrelevant for general users. This is further underlined by GML being an Extensible Markup Language (XML) grammar (W3C, 2004a), with the structure formalized uniquely in XML schema (XSD) (W3C, 2004c). The fact that the semantics of the GML grammar are only formalized through an XML schema limits the possibilities of direct inclusion of GML statements in, for instance, RDF or OWL annotation of Web resources. This restricts the usage of GML mostly to specialized OGC compliant GIS applications.

Therefore, the resulting technology is not directly usable for semantically enhanced storage and publication of general Web resources, and a need for

Copyright © 2006, Idea Group Inc. Copying or distributing in print or electronic forms without written permission of Idea Group Inc. is prohibited.

additional ontologies for spatial content categorization remains. The interoperability of these general usage ontologies should, however, have the possibility to be interoperable with future possible implementations of GML using the Resource Description Format (RDF) (W3C, 1999) or the Web Ontology Language (OWL) (W3C, 2004b), to enhance the reusability of the content. In general, pure XML based approaches are suitable for fast deployment of interoperable, but strongly specialized services, while RDF encoded content is service agnostic and widely reusable, but often not designed with any particular service in mind. Consider the example in Figure 1: it shows a point geometry referencing the GML namespace, integrated with a pure XML description of the Hotel Windsor, where our traveler Hillary wants to get. However, the elements used to annotate the content are only described in the windsor.xsd schema. The semantics of such a description are well interpretable by humans or specifically engineered systems, but there is no automated mean to transport the semantics to a different application. For instance, the element represented by the tag <stars> has no semantics objectively declared, and bears some meaning only to the designer of this specific system. In a global hotel database, the same attribute of a hotel may be represented by

Figure 1. Example of an XML description of Hotel Windsor with a GML part

```xml
<?xml version="1.0"?>
<MelbourneHotels
xmlns="http://www.geom.unimelb.edu.au/tomko"
xmlns:gml="http://www.opengis.net/gml"
xmlns:xlink="http://www.w3.org/1999/xlink"
xmlns:xsi="http://www.w3.org/2000/10/XMLSchema-instance"
xsi:schemaLocatio="http://www.geom.unimelb.edu.au/tomko/
MelbourneHotels.xsd">
        <Hotel>
            <gml:name>The Windsor Hotel</gml:name>
            <gml:description>
                The Windsor Hotel A Melbourne Hotel -
                5 star Luxury Accommodation Melbourne
            </gml:description>
            <address>103-115 Spring Street</address>
            <gml:location>
                    <gml:Point
                    srsName="http://www.opengis.net/gml/srs/
                    epsg.xml#4326">
                            <gml:coordinates>
                                    37.8206S,144.9674E
                            </gml:coordinates>
                    </gml:Point>
            </gml:location>
        </Hotel>
</MelbourneHotels>
```

Copyright © 2006, Idea Group Inc. Copying or distributing in print or electronic forms without written permission of Idea Group Inc. is prohibited.

a element <category>. Deeper semantics have to be derived as a result of first-order predicate logic. Better support for richer semantical reasoning is possible in RDF.

Geographic Annotation

Geocoding the content of Web resources is also possible through a wide spectrum of different tagging conventions, and is applied for example by photography enthusiasts annotating their photographs with geo-tags, some spatial location search engines, worldwide postcodes initiatives, and others. Some of them use a simple syntax and loose structure based on HTML, others are more sophisticated and support XML (and consecutively RDF), are strongly structured and partially approach the vision of the Semantic Web.

HTML tags are the simplest way to insert machine readable content in Web resources. Despite being machine readable, they do not allow inclusion of structures and specific ontologies, or even simpler, unstructured vocabularies. As any XML based markup language, HTML allows the creation of custom made tags by content providers. As HTML parsers are made not to be vulnerable to these additions, it may be an appealing way to enhance the content of Web resources. On the other hand, the lack of structure and standardized tags limits their usability by general search engines. Usually inserted in the <meta> tags of the header of the Web resource or around the annotated element, only two major sets of tags gained more widespread use, and both approaches are not maintained anymore. These were represented by geotags.com and GeoURL.com, both with associated search engines. Figure 2 shows the Hotel Windsor example in GeoURL. As long as a standardized set of markup tags is not adopted as a W3C specification and further implemented by major search engines, widespread use of geo-annotation through HTML tags will not be successful.

Geospatial Semantic Web

The Geospatial Semantic Web (Egenhofer, 2002) exceeds simple geo-tagging, and will avoid the problems of HTML/XML tags in many ways: no single

Figure 2. GeoURL tag example

```
<meta name="ICBM" content="144.9674, -37.8206">
<meta name="DC.title" content="The Windsor Hotel">
```

Copyright © 2006, Idea Group Inc. Copying or distributing in print or electronic forms without written permission of Idea Group Inc. is prohibited.

geographic annotation vocabulary needs to be standardized anymore. Knowledge sharing across different application domains is possible through interoperable, semantically enriched markup in RDF-S or OWL. Formalized ontologies enable use of reasoners for querying asserted ontologies and enable the linking of independent knowledge bases. While many research groups are dealing with formalizing geographical ontologies (Fonseca, Egenhofer, Agouris, & Camara, 2002; Fonseca, Egenhofer, Davis, & Borges, 2000; Grenon & Smith, 2004; Smith & Mark, 2001), only a few relate their research to Semantic Web applications.

The majority of resources of the current Web are primarily focused on providing information related to what, when, where and how. The *where*, crucial in our context, is one of several categories of content, and frequently not the most important one. With the Geospatial Semantic Web, georeferencing becomes the focus of interest. Ontologies of places are linked together through ontologies of relations and provide the georeference to resource specific content profiting of consistent action ontologies.

Georeferencing Ontologies for the Geospatial Semantic Web

Ontologies formalizing place descriptions were the first to be developed within the Geospatial Semantic Web area. One of the first attempts to encode location information in RDF is represented by the simple RDFGeo vocabulary of the W3C RDF Interest Group (W3C, 2003) enabling annotation of point geographies with latitude, longitude, and altitude in RDF. This vocabulary even enables tagging of resources that have not been totally ported to RDF. It does this by enabling the insertion of a subset of tags in XHTML. However, the limitation to point data makes RDFGeo hardly usable for more specialized applications. Figure 3 shows the Hotel Windsor example in RDFGeo annotation. Consecutively, Figure 4 shows the structure of our own OWL ontology of tourist accommodation, including a use of the RDFGeo for the description of point locations. The full code of this ontology can be found in the Appendix. This ontology has more advanced semantics included, using the class properties of the OWL language. The example of Hotel Windsor is included (note, that despite it being possible to merge in one file the OWL ontology definition and its instances, this is in general not recommended).

The proposal for a similar vocabulary encoded in RDF Site Summary (RSS) (Beged-Dov et al., 2001) can be found in the work of Singh (2004), focusing on the ease of annotating blog resources with coordinate attributes for community mapping applications. Current efforts led by the W3C Semantic Web Advanced Development for Europe project focus on the support for fuzzy geographical regions and resulting fuzzy relations between geographical objects (i.e., interpretation of terms such as "near"). It further focuses on ontologies for enabling the

Copyright © 2006, Idea Group Inc. Copying or distributing in print or electronic forms without written permission of Idea Group Inc. is prohibited.

Figure 3. Example of RDFGeo *annotation of Hotel Windsor*

```
<rdf:RDF
    xmlns:dc="http://purl.org/dc/elements/1.1/"
    xmlns:foaf="http://xmlns.com/foaf/0.1/"
    xmlns:geo="http://www.w3.org/2003/01/geo/wgs84_pos#"
    xmlns:rdf="http://www.w3.org/1999/02/22-rdf-syntax-ns#">
    <rdf:Description rdf:about="http://www.thewindsor.com.au/">
        <dc:title>The Windsor Hotel A Melbourne Hotel -
                5 star Luxury Accommodation Melbourne</dc:title>
        <foaf:topic rdf:parseType="Resource">
            <geo:lat>-37.8206</geo:lat>
            <geo:long>144.9674</geo:long>
        </foaf:topic>
    </rdf:Description>
</rdf:RDF>
```

Figure 4. Structure of the accommodation.owl ontology

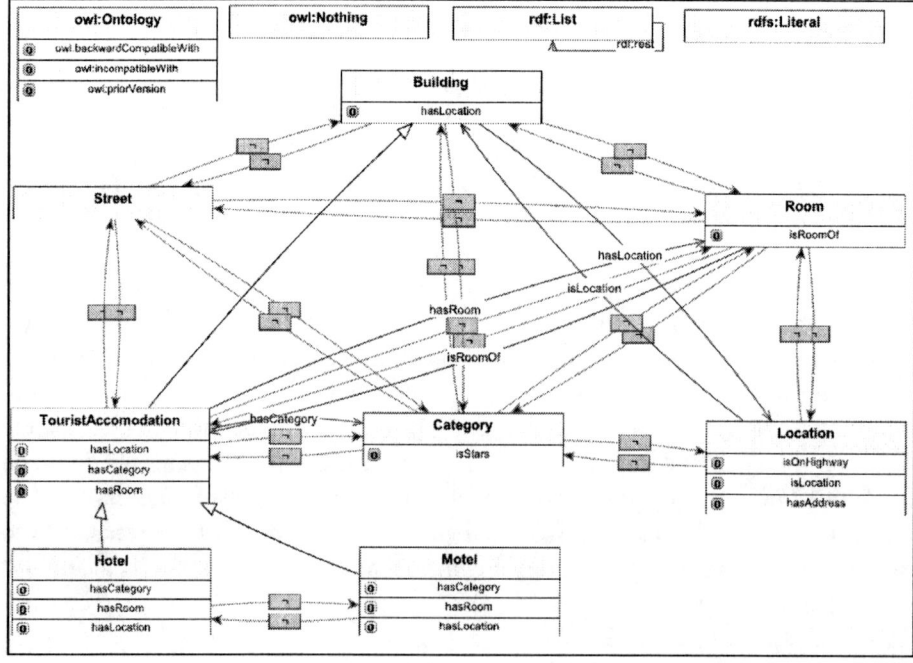

interoperability of different postal addressing standards (McCathieNevile, 2004). Other projects deal with the conversion of the OGC GML specification in RDF and OWL (Defne, Islam, & Piasecki, 2004). Goad (2004) has a critical view on GML translations to RDF and advocates simplicity over the complexity of the

Copyright © 2006, Idea Group Inc. Copying or distributing in print or electronic forms without written permission of Idea Group Inc. is prohibited.

original OGC specification, which leads him to the proposal of RDFGeom, an alternative markup grammar, but with maintained support for a subset of the GML capabilities.

Spatial Operations and Spatial Relation Ontologies

Basic needs of geographical searches on location-based data retrieval are summarized and solutions are proposed in the project of the Alexandria Digital Library (Goodchild, 2004), with the focus on retrieving map files from a database. Location semantics are closely coupled with spatial operations. These present the mean to query the Web for content based on query strings using references to relations between several features. The system should also describe the relations between features of the environment using spatial relation references in a way close to natural language statements, in order to provide interaction simple for general users. Implementation issues, such as reference system conversions, or problems with different units should be hidden to users (Neumann, 2003). Possibilities to query spatial data are still limited, as only a few ontologies for topology, distance and orientation exists (Hiramatsu & Reitsma, 2004), and as spatial operators implemented in RQL are still not standardized (Corcoles & Gonzalez, 2003). The key is to enable the Geospatial Semantic Web to use fuzzy terms designing spatial relations in a manner consistent with human understanding of these statements. Imagine Hillary's communication with WebGuide: the system uses terms such as opposite, close to and nearby consistently, but also coherently with Hillary's understanding of the term.

Wayfinding Georeferences from Web Resources

In a case study we present an inventory and analysis of currently used georeferences in Web resources, and we demonstrate the possibilities of their use. The case study concerns a segment of Hillary's route from the Hyatt to the Hotel Windsor, namely 1 to 5 Collins Street, Melbourne, Australia. Figure 5 shows Hillary's route, including this short segment.

Imagine the actions that Hillary's next-generation WebGuide system would need to perform in order to provide her with user-friendly route directions. Hillary has a destination, and is looking for route directions. A route service can calculate a route on a street network, and this route has to be described to Hillary in a manner a human would communicate it. The route as delivered by the route

Copyright © 2006, Idea Group Inc. Copying or distributing in print or electronic forms without written permission of Idea Group Inc. is prohibited.

Figure 5. Hillary's route, including 1 to 5 Collins Street (inset), as provided by Whereis, 2005

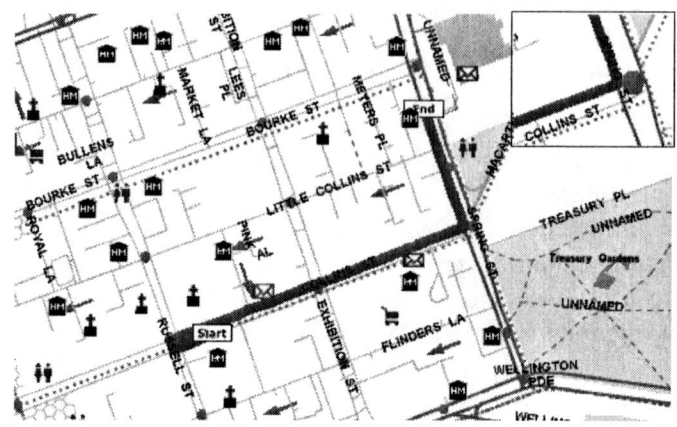

service consists at this stage of street segments, i.e., address intervals. WebGuide will use the addresses (or another formal method of georeferencing) as links to Web resources. From the content of the collected Web resources along the route, WebGuide can reconstruct the layout of the environment, identify visual, semantic, and structural qualities of local features to determine their salience, and finally present a natural language description of the route, matched to the preferences of Hillary.

Experiment

A complete directory of Web resources referring to features, institutions, or events along a specific street segment can be so far collected only manually. Keyword-based search delivers large numbers of results, among them many false hits. For example, a Google search made on the 06.10.2004 with the query string "2 Collins Street" "Melbourne" and limited search in Australian Web resources returned 93.400 links. A check of the links reveals that many of the found links refer in fact to other house numbers (non-perfect matches). Others do not refer to a postal address at all but describe some phenomena in Collins Street. Again others refer to Little Collins Street. And a fourth group are directory pages, georeferencing not one but many features (e.g., a page "Bars in Melbourne"). However, many of the found Web resources refer to the correct address: 2 Collins Street is one of the high-rise office buildings in this central business district area, with multiple Web resources describing the many busi-nesses located here. The limitation to Australian Web resources is an IP

Copyright © 2006, Idea Group Inc. Copying or distributing in print or electronic forms without written permission of Idea Group Inc. is prohibited.

address-based spatial filter, which excludes Web resources hosted elsewhere, but also excludes many false hits to any Collins Street in a Melbourne outside of Australia. Still, the query misses Web resources that contain the valid short form "2 Collins St", the street name in wrong spelling, or, although being relevant, contain no street name at all.

For the case study, we collected Web references for the addresses 1 to 5 Collins Street, and filtered out the first 50 true references for each address. We excluded any other type of georeference from consideration, and also any Web resource that is not directly accessible via Google (e.g., searchable databases like the Yellow Pages). In other work we assess the relevance of these resources for wayfinders (Tomko, 2004).

Extraction of Georeferences from Web Resources

Although georeferences are frequently found, they are not structured for general reuse. Web content providers include georeferences in form of postal addresses, maps, or place descriptions addressing a human reader. Additionally, their subjective categorizations of content describing the *where* let georeferences appear behind a variety of links from top-level Web resources: they may appear under "Contact us", "How to find us", "Next branch", and so on. For example, Hillary's place of interest, the Hotel Windsor, offers a button "Locations and attractions" at the top-level page (http://www.thewindsor.com.au/), as the only link to promise some georeferences for the place of the Hotel. When following this link, it turns out to describe some touristic hotspots of Melbourne, with a link somewhere in between to "Click here to view a map of our location", which finally provides the information searched for — for a human reader. The postal address can be found under "Contact us" — "Overview".

Further, one may consider the technical insertion of address text within the page. HTML tables, divisions, paragraphs and frames all split the content so that it may not always be straightforward to retrieve the caption and its content. An examination of postal address patterns in the Web resources for 1 to 5 Collins Street showed 29 different captions and header types when referencing to location. This does not include all the descriptive references in natural language.

Action Ontologies and Georeferencing in the Web

To assess the variety of purposes, addressed by Web content providers through inserted georeferences, we have analyzed the text in the immediate proximity of the reference (headers, sentences, captions) and categorized them in a taxonomy. This taxonomy was used to assert the action ontology of the resource

Copyright © 2006, Idea Group Inc. Copying or distributing in print or electronic forms without written permission of Idea Group Inc. is prohibited.

Table 1. Web content providers' action ontology

Action	Verb	Category
Sending	to send	Mailing
Contacting	to contact somebody	
Being in / at	to occupy a position	Wayfinding
Finding	to find something	
Parking	to park	Car driving
Selling	to sell	Identification
Winning	to win	

providers, represented by task verbs related to the captions. These terms help to understand the purpose of the georeferences targeted by the Web providers. Further, it enables to isolate those terms, and hence, resources that are most likely usable for wayfinding. Out of a set of the 29 different captions and header types found in the set of Web references, we identified the limited amount of action verbs used (Table 1).

The first four terms constitute a substantial part of the samples examined. Resources describing actions referring to parking, selling and winning present a marginal part of the examined georeferenced resources, and we can assume that resources dedicated to these actions are rare, or the importance of location information related to these activities is low. Selling, for example is a common action term on specialized Web resources, specifically real estate databases. These databases are in so far exceptional as they refer to properties to sell or to be sold, but not to the location of the site owner, the real estate agency. In general, the restricted diversity of actions relating to georeferences makes the identification of features suitable for wayfinding difficult.

The primary purpose of postal addresses (mailing to persons or institutions) is still a common action term associated with georeferencing in Web resources. However, a missing term (associated to the action term "is" and the category of "existence/identification") is even more common. It is obvious that Web authors do not distinguish between the different needs to wayfinders and mail addressing (even if one may argue that there is no difference, as the postman needs to find his way to the location). If we narrow the focus on the two categories of major relevance to wayfinding, namely to terms associated with identification and navigation, we can isolate a group of terms that constitute members of the content providers' ontology and may be asserted in the ontology of wayfinders as well (Table 2).

The verb "be in/be at" is the most common action reference related to georeferences in the examined set of Web resources. According to WordNet (Fellbaum, 1998), "to be" can have the following main meanings: to have the quality, be identical to, occupy a position, exist, be equal, constitute, follow (work

Copyright © 2006, Idea Group Inc. Copying or distributing in print or electronic forms without written permission of Idea Group Inc. is prohibited.

Table 2. Terms used for georeferencing with wayfinding related usage

Terms	Type	Category
Name and address of organization, branch, or service	Identification by feature usage	Existence, identification
Locations of buildings or institutions	Directory listings of entities of a certain type	
Buildings or institutions at locations	Directory listings of entities in a certain area	
"Where to find", "is located at"	Directions for wayfinding	Wayfinding

in a specific place, with a specific subject, or in a specific function), represent. As we can see, most of them are deeply rooted in references to location. In general usage, being is frequently associated with positional adverbs like "in" or "at", and introduces by that way a georeference describing the location of the subject.

Frequency of action terms belonging to the categories of existence or wayfinding is an indication of probability that this Web resource actually describes the feature georeferenced in the content, particularly its spatial properties. Observations like frequency of relevant action terms can be used to rank Web resources for wayfinding purposes.

Reconstruction of Perceivable Feature Properties

Translation between postal addresses or other descriptive georeferences and experiential characteristics of space needs explicit (or externally known) semantics and world knowledge.

The semantics of experiential characteristics of space is informal. For example, there is no accepted formal definition of a landmark. First steps towards a formalization are proposals for measures of salience, which apply methods from spatial analysis and computer vision to assess salience. Assessing the salience of a feature from its georeferenced Web resource follows different rules and is tangential to the previous research.

To reconstruct a perceivable image of the reality for a wayfinder from Web resources the first step is to extract the ones that are potentially relevant, by their georeferences. In the second step their (spatial) relations to each other have to be identified. For example, the large number of businesses registered at 2 Collins Street suggests an aggregated perceivable feature of a larger office building. Such spatial properties can translate into natural language descriptions. Afterwards, a selection of identified perceivable features can be made according to their relevance for a given context of a wayfinder.

Copyright © 2006, Idea Group Inc. Copying or distributing in print or electronic forms without written permission of Idea Group Inc. is prohibited.

We have partially introduced the different ontologies related to concepts of georeferencing, showed how action ontologies can be exploited to isolate those that may be relevant for wayfinding, and we pointed to the spatial relations and proximity statements that are found in natural language descriptions in Web resources. With all these tools at hand, one can derive from the examined Web resources perceivable facts about 1 to 5 Collins Street such as:

- 1 Collins Street has a name: Rialto Tower.
- 1 Collins Street is at least 16 floors high.
- 1 Collins Street is an office building.
- 1 Collins Street is a landmark of Collins Street and has won an architecture prize.
- Opposite of 1 Collins Street is 4-6 Collins Street.
- 4-6 Collins Street has a name: ANZAC house.
- 4-6 Collins Street is within 150 meters of the Parliament train station.
- 2 Collins Street is adjacent to 4-6 Collins Street.
- 2 Collins Street has a name: Alcaston House.
- 2 Collins Street has at least 6 floors.
- 2 Collins Street has also a Spring Street entrance.

> It is either a corner building, or it reaches across the block.
> Spring Street is *close* to Collins Street.

Such derived statements can be used by an intelligent service to enrich route directions with landmark references. By coupling HTML or XML parsers, datamining engines (especially those able to interpret natural language statements) and formalized ontologies (OWL ontologies of geocodes, postal addresses and of action ontologies with relation to space), these references can be automatically extracted and an ad-hoc knowledge base built and exploited by a service. Agent systems can incrementally improve these knowledge bases to match less usual patterns in georeferencing, or to adapt the system to various natural languages.

The wealth of information provided through heterogeneous Web resources provides a rich resource of descriptive information that enables the reconstruction of perceivable characteristics of the environment. Resources were particular rich in our case study since we selected an area in a business district. Other urban neighborhoods may address a lesser amount of information on the Web,

Copyright © 2006, Idea Group Inc. Copying or distributing in print or electronic forms without written permission of Idea Group Inc. is prohibited.

however, the knowledge base for processing the resources will be the same.

One of the problems faced with in our investigation is the low quality of Web resources. For instance, several Web georeferences refer to 3 Collins Street and 5 Collins Street, however, corresponding entrances of buildings cannot be found in the external world. Outdated Web resources may cause other conflicts.

Future Research

The experiment in our case study was done completely manually. It envisions some future capabilities of a Geospatial Semantic Web, but it identifies also open problems for further research. Given the two information communities of Web content providers and wayfinders, we anticipate their formal ontologies of georeferences, and we anticipate a knowledge base for mapping between these ontologies. We identify the following areas for further research:

- **Ontologies of postal address systems.** Ontologies of georeferencing are the promise of the Geospatial Semantic Web (Egenhofer, 2002). They will be used for semantically enriched annotation. Local search capabilities of the prominent search engines (e.g., Google, Yahoo, Sensis) require already some formalized understanding of postal address systems in heterogeneous resources. Hence, this research is under way.

- **Knowledge base for relevance assessment.** Given the richness and diversity of postal address georeferencing, what are the rules for ranking Web hits according to their relevance for wayfinders? Some preliminary rules follow from our case study with human wayfinding directions: preference has to be given to ground floor addresses, to some categories of businesses (both for their visibility to the wayfinder), and to some global building characteristics (function, heritage). These rules need a formalization for automatic reasoning in the framework of above described ontologies. Other rules will categorize the meaning of an address in a Web resource, particularly in cases where several addresses appear. For instance, a Web resource can contain the address of a real estate object, and the address of the real estate agent. The functions of these addresses are quite different.

- **Intelligent geocoding.** Wayfinders perceive the physical features along a route from outside. Postal addresses have some relation to the wayfinders' perception: postal addresses are unique identifiers for (postal) delivery, which is the threshold point between inside (the private space) and outside

Copyright © 2006, Idea Group Inc. Copying or distributing in print or electronic forms without written permission of Idea Group Inc. is prohibited.

(the public space). It needs a formal semantics of geocodes, and among the possible meanings of geocodes one that refers to that process of approaching a feature of a specific postal address from outside.

- **Support for spatial relations.** We need implementations of fuzzy spatial relations on postal addresses or spatial features in general, such as "near", "next", "opposite", and so on. These implementations are required for two purposes: for understanding descriptive georeferences given in Web resources, and for generating natural language georeferences from addresses.

Conclusion

We have presented a literature review and a case study of ways of georeferencing in Web resources on the one hand, and in route directions for wayfinders on the other hand. The main difference in the semantics of the two information communities is an identifier of features, institutions or persons in the Web, compared to an experiential view on the environment by wayfinders. With the goal of using Web resources for enriching route directions by landmarks, we focused on the translation of the semantics of georeferences in Web resources to meaningful georeferences in the context of route directions. For that purpose we made an experiment for a short street segment, and studied the available Web resources. We showed that an interpretation of georeferences in the content of Web resources enhances the perceivable properties of features, and hence, can be used by a tool to select appropriate features as georeferences in route directions. Open questions for research are identified.

Remember Hillary enquiring her next-generation service *WebGuide*: "Guide me to the Hotel Windsor!" WebGuide analyses the query and identifies (a) an action ("guide"), (b) the georeference included in the query ("Hotel Windsor"), and (c) the spatial relation of the georeference to Hillary ("to"), such that it identifies the Hotel Windsor as a destination of a route. From context, i.e., from Hillary's enquiry posed in a hotel lobby in Melbourne, Australia, and Hillary being a tourist with no special appointment for today in her calendar, WebGuide concludes that Hillary is looking for a Hotel Windsor in short distance, ideally in Melbourne, Australia. After locating the destination it computes the route on Melbourne's street network. Then WebGuide performs a search on the Web for all resources along the route, and reconstructs the environment Hillary will be guided through. Using natural language statements, WebGuide provides Hillary with a brief, custom made set of directions, including landmarks and other descriptions of the environment. Hillary can be sure that she will have no trouble in finding the place

Copyright © 2006, Idea Group Inc. Copying or distributing in print or electronic forms without written permission of Idea Group Inc. is prohibited.

for her Indulgence Tea, and she will even have the opportunity to admire some of the most important buildings of Melbourne on her way: "At Rialto Tower, an historic landmark building of prize-winning architecture, turn left into Spring Street".

Acknowledgments

This work has been supported by the Cooperative Research Centre for Spatial Information, whose activities are funded by the Australian Commonwealth's Cooperative Research Centres Programme. Additional funding of the first author from an internal grant of the University of Melbourne is acknowledged.

References

Beged-Dov, G., Brickley, D., Dornfest, R., Davis, I., Dodds, L., Eisenzopf, J., et al. (2001). *RDF Site Summary (RSS) 1.0.*

Berners-Lee, T., Hendler, J., & Lassila, O. (2001). The Semantic Web. *Scientific American, 284*(5), 34-43.

Bishr, Y. (1998). Overcoming the semantic and other barriers to GIS interoperability. *International Journal of Geographical Information Science, 12*(4), 299-314.

Bittner, T., Donnelly, M., & Winter, S. (2004). Ontology and Semantic Interoperability. In D. Prosperi, & S. Zlatanova (Eds.), *Large scale 3D data integration: Challenges and opportunities.* London: CRCPress.

Buyukkokten, O., Cho, J., Garcia-Molina, H., Gravano, L., & Shivakumar, N. (1999). *Exploiting geographical location information of Web pages.* Paper presented at the Proceedings of Workshop on Web Databases (WebDB'99).

Corcoles, J. E., & Gonzalez, P. (2003). *Querying spatial resources. An approach to the Semantic Geospatial Web.* Paper presented at the CAiSE'03 Workshop: Web Services, e-Business, and the Semantic Web (WES): Foundations, Models, Architecture, Engineering and Applications.

Cornell, E. H., Heth, C. D., & Broda, L. S. (1989). Children's wayfinding: Response to instructions to use environmental landmarks. *Developmental Psychology, 25*(5), 755-764.

Copyright © 2006, Idea Group Inc. Copying or distributing in print or electronic forms without written permission of Idea Group Inc. is prohibited.

Cox, S., Daisey, P., Lake, R., Portele, C., & Whiteside, A. (2003). *OpenGIS® Geography Markup Language (GML) implementation specification.* OpenGIS® Implementation Specification No. OpenGIS® Project Document OGC 02-023r4). Wayland: Open GIS Consortium, Inc.

Defne, Z., Islam, A. S., & Piasecki, M. (2004). *Ontology for Geography Markup Language (GML3.0) of Open GIS Consortium (OGC).* Retrieved from http://loki.cae.drexel.edu/~wbs/ontology/ogc-gml.htm

Denis, M., Pazzaglia, F., Cornoldi, C., & Bertolo, L. (1999). Spatial discourse and navigation: An analysis of route directions in the city of Venice. *Applied Cognitive Psychology, 13,* 145-174.

Egenhofer, M. (2002). *Toward the Semantic Geospatial Web.* Paper presented at the 10th ACM International Symposium on Advances in Geographic Information Systems, McLean, VA.

Elias, B., & Brenner, C. (2004). Automatic generation and application of landmarks in navigation data sets. In P. Fisher (Ed.), *Developments in spatial data handling* (pp. 469-480). Berlin: Springer.

Fauconnier, G., & Turner, M. (2002). *The way we think: Conceptual blending and the mind's hidden complexities.* New York: Basic Books.

Fellbaum, C. (Ed.). (1998). *WordNet: An electronic lexical database.* Cambridge, MA: The MIT Press.

Fonseca, F. T., Egenhofer, M. J., Agouris, P., & Camara, G. (2002). Using ontologies for integrated geographic information systems. *Transactions in GIS, 6*(3), 231-257.

Fonseca, F. T., Egenhofer, M. J., Davis, C. A., & Borges, K. A. V. (2000). Ontologies and knowledge sharing in urban GIS. *Computer, Environment and Urban Systems, 24*(3), 232-251.

Frank, A. U. (2003). Pragmatic information content: How to measure the information in a route description. In M. Duckham, M. F. Goodchild, & M. Worboys (Eds.), *Foundations in geographic information science* (pp. 47-68). London: Taylor & Francis.

Freksa, C., Brauer, W., & Habel, C. (Eds.). (2000). *Spatial cognition II* (Vol. 1849). Berlin: Springer.

Freksa, C., Brauer, W., Habel, C., & Wender, K. F. (Eds.). (2003). *Spatial cognition III* (Vol. 2685). Berlin: Springer.

Freksa, C., Habel, C., & Wender, K. F. (Eds.). (1998). *Spatial cognition* (Vol. 1404). Berlin: Springer.

Gibson, J. J. (1979). *The ecological approach to visual perception.* Boston: Houghton Mifflin.

Copyright © 2006, Idea Group Inc. Copying or distributing in print or electronic forms without written permission of Idea Group Inc. is prohibited.

Goad, C. (2004, September 2004). *RDF versus GML*. Retrieved December 6, 2004, from http://www.mapbureau.com/gml/

Golledge, R. G. (Ed.). (1999). *Wayfinding behavior: Cognitive mapping and other spatial processes*. Baltimore: The Johns Hopkins University Press.

Golledge, R. G., Rivizzigno, V. L., & Spector, A. (1976). Learning about a city: Analysis by multidimensional scaling. In R. G. Golledge, & G. Rushton (Eds.), *Spatial choice and spatial behavior* (pp. 95-116). Columbus: Ohio State University Press.

Golledge, R. G., & Stimson, R. J. (1997). *Spatial behavior: A geographic perspective*. New York: The Guildford Press.

Goodchild, M. (2004). The Alexandria Digital Library Project. *D-Lib Magazine, 10*.

Grenon, P., & Smith, B. (2004). SNAP and SPAN: Towards dynamic spatial ontology. *Spatial Cognition and Computation, 4*(1), 69-104.

Gruber, T. R. (1993). *Toward principles for the design of ontologies used for knowledge sharing*. Technical Report KSL 93-04. Stanford University, Knowledge Systems Laboratory.

Habel, C. (1988). Prozedurale Aspekte der Wegplanung und Wegbeschreibung. In H. Schnelle & G. Rickheit (Eds.), *Sprache in Mensch und Computer* (pp. 107-133). Opladen: Westdeutscher Verlag.

Hill, L. L. (2004). Georeferencing in digital libraries. *D-Lib Magazine, 10*(5).

Hiramatsu, K., & Reitsma, F. (2004). *GeoReferencing the Semantic Web: Ontology based markup of geographically referenced information*. Paper presented at the Joint EuroSDR/EuroGeographics Workshop on Ontologies and Schema Translation Services, Paris.

ICSM. (2003). Rural and Urban Addressing Standard AS/NZS 4819:2003. Retrieved July 12, 2004, from http://www.icsm.gov.au/icsm/street/index.html

Jarvella, R. J., & Klein, W. (Eds.). (1982). *Speech, place, and action*. Chichester, UK: John Wiley & Sons.

Johnson, M. (1987). *The body in the mind: The bodily basis of meaning, imagination, and reason*. Chicago: The University of Chicago Press.

Jones, C. B., Alani, H., & Tudhope, D. (2001). *Geographical information retrieval with ontologies of place*. Paper presented at the COSIT 2001.

Kaplan, S., & Kaplan, R. (1982). *Cognition and environment: Functioning in an uncertain world*. New York: Praeger.

Klein, W. (1979). Wegauskünfte. *Zeitschrift für Literaturwissenschaft und Linguistik, 33*, 9-57.

Copyright © 2006, Idea Group Inc. Copying or distributing in print or electronic forms without written permission of Idea Group Inc. is prohibited.

Klippel, A. (2003). Wayfinding choremes. In W. Kuhn, M. F. Worboys, & S. Timpf (Eds.), *Spatial information theory* (Vol. 2825, pp. 320-334). Berlin: Springer.

Kuhn, W. (2001). Ontologies in support of activities in geographical space. *International Journal of Geographical Information Science, 15*(7), 613-632.

Kumar, R., Webber, D. R., Bennett, J., Lubenow, J., Nyholm, N., Goncalves, M., et al. (2002). *xNAL, Extensible Name and Address Language (NAML).* (Standard): OASIS.

Lakoff, G. (1987). *Women, fire, and dangerous things: What categories reveal about the mind.* Chicago: The University of Chicago Press.

Leclerc, Y. G., Reddy, M., Iverson, L., & Eriksen, M. (2002). Discovering, modeling and visualizing global grids over the Internet. In M. Goodchild, & A. J. Kimerling (Eds.), *Discrete global grids.* Santa Barbara, CA: National Center for Geographic Information & Analysis.

Lovelace, K. L., Hegarty, M., & Montello, D. R. (1999). Elements of good route directions in familiar and unfamiliar environments. In C. Freksa, & D. M. Mark (Eds.), *Spatial information theory* (Vol. 1661, pp. 65-82). Berlin: Springer.

Lynch, K. (1960). *The image of the city.* Cambridge, MA: MIT Press.

Ma, Q., Matsumoto, C., & Tanaka, K. (2003). A localness-filter for searched Web pages. In X. Zhou, Y. Zhang, & M. E. Orlowska (Eds.), *APWeb 2003* (Vol. 2642, pp. 525-536). Berlin: Springer-Verlag.

McCarthy, J. (1990). An example for natural language understanding and the AI problems it raises. In J. McCarthy (Ed.), *Formalizing common sense* (pp. 70-76). Norwood, NJ: Ablex.

McCathieNevile, C. (2004). *SWAD-Europe extra deliverable 3.20: Report on developer workshop 9 — Geospatial information on the Semantic Web.* Workshop Report No. 9.

McCurley, K. S. (2001). *Geospatial mapping and navigation of the Web.* Paper presented at the the 10th International World Wide Web Conference WWW10, Hong Kong.

Michon, P.-E., & Denis, M. (2001). When and why are visual landmarks used in giving directions? In D. R. Montello (Ed.), *Spatial information theory* (Vol. 2205, pp. 292-305). Berlin: Springer.

Neumann, M. (2003). *Spatially navigating the Semantic Web for user adapted presentations of cultural heritage information in mobile environments.* Paper presented at the SWDB'03. The First International Workshop on Semantic Web and Databases. Co-located with VLDB, Berlin.

Copyright © 2006, Idea Group Inc. Copying or distributing in print or electronic forms without written permission of Idea Group Inc. is prohibited.

Nothegger, C., Winter, S., & Raubal, M. (2004). Selection of salient features for route directions. *Spatial Cognition and Computation, 4*(2), 113-136.

Presson, C. C., & Montello, D. R. (1988). Points of reference in spatial cognition: Stalking the elusive landmark. *British Journal of Developmental Psychology, 6*, 378-381.

Raubal, M., & Winter, S. (2002). Enriching wayfinding instructions with local landmarks. In M. J. Egenhofer, & D. M. Mark (Eds.), *Geographic information science* (Vol. 2478, pp. 243-259). Berlin: Springer.

Rosch, E. (1978). Principles of categorization. In E. Rosch, & B. B. Lloyd (Eds.), *Cognition and categorization* (pp. 27-48). Hillsdale, NJ: Lawrence Erlbaum Associates.

Siegel, A. W., & White, S. H. (1975). The development of spatial representations of large-scale environments. In H. W. Reese (Ed.), *Advances in child development and nehavior* (Vol. 10, pp. 9-55). New York: Academic Press.

Singh, R. (2004). *GeoBlogging: Collaborative, peer-to-peer geographic information sharing.* Paper presented at the URISA Public Participation in GIS 3rd Annual Conference.

Smith, B., & Mark, D. M. (2001). Geographic categories: An ontological investigation. *International Journal of Geographical Information Science, 15*(7), 591-612.

Sorrows, M. E., & Hirtle, S. C. (1999). The nature of landmarks for real and electronic spaces. In C. Freksa, & D. M. Mark (Eds.), *Spatial information theory* (Vol. 1661, pp. 37-50). Berlin: Springer.

Timpf, S. (2002). Ontologies of wayfinding: a traveler's perspective. *Networks and Spatial Economics, 2*(1), 9-33.

Tomko, M. (2004). *Case study: Assessing spatial distribution of Web resources for navigation services.* Paper presented at the 4th International Workshop on Web and Wireless Geographical Information Systems, Goyang, Korea.

UPU. (2002). *International postal address components and templates.* No. S42-1. Universal Postal Union.

W3C. (1999). *Resource Description Framework (RDF).* Retrieved December 15, 2004, from http://www.w3.org/RDF/

W3C. (2003). *RDFIG Geo vocab workspace.* Retrieved December 15, 2004, from http://www.w3.org/2003/01/geo/

W3C. (2004a). *Extensible Markup Language (XML) 1.1.* Retrieved December 15, 2004, from http://www.w3.org/TR/2004/REC-xml11-20040204/

Copyright © 2006, Idea Group Inc. Copying or distributing in print or electronic forms without written permission of Idea Group Inc. is prohibited.

W3C. (2004b). *OWL Web Ontology Language reference*. Retrieved December 15, 2004, from http://www.w3.org/2004/OWL/

W3C. (2004c). *XML schema part 1: Structures*. Retrieved December 15, 2004, from http://www.w3.org/TR/xmlschema-1/

Weissensteiner, E., & Winter, S. (2004). Landmarks in the communication of route instructions. In M. Egenhofer, C. Freksa, & H. J. Miller (Eds.), *Geographic information science* (Vol. 3234, pp. 313-326). Berlin: Springer.

Winter, S. (2003). Route adaptive selection of salient features. In W. Kuhn, M. F. Worboys, & S. Timpf (Eds.), *Spatial information theory* (Vol. 2825, pp. 320-334). Berlin: Springer.

Winter, S., Raubal, M., & Nothegger, C. (2004). Focalizing measures of salience for wayfinding. In L. Meng, A. Zipf, & T. Reichenbacher (Eds.), *Map-based mobile services: Theories, methods and implementations* (pp. 127-142). Berlin: Springer Geosciences.

Worboys, M. F. (2001). Nearness relations in environmental space. *International Journal of Geographical Information Science, 15*(7), 633-652.

Young, R. (2004). *Google local search and the impact on natural optimization*. Retrieved from http://www.WebPronews.com

Appendix

```
<?xml version="1.0"?>
<rdf:RDF
xmlns:protege="http://protege.stanford.edu/plugins/owl/protege#"
xmlns:rdf="http://www.w3.org/1999/02/22-rdf-syntax-ns#"
xmlns:xsd="http://www.w3.org/2001/XMLSchema#"
xmlns:rdfs="http://www.w3.org/2000/01/rdf-schema#"
xmlns:owl="http://www.w3.org/2002/07/owl#"
xmlns:geo="http://www.w3.org/2003/01/geo/wgs84_pos#"
xmlns:daml="http://www.daml.org/2001/03/daml+oil#"
xmlns:dc="http://purl.org/dc/elements/1.1/"
xml:base="http://www.geom.unimelb.edu.au/tomko/accomodation#">
 <owl:Ontology rdf:about="Accomodation ontology">
  <owl:imports rdf:resource="http://www.w3.org/2003/01/geo/wgs84_pos"/>
 </owl:Ontology>
```

Copyright © 2006, Idea Group Inc. Copying or distributing in print or electronic forms without written permission of Idea Group Inc. is prohibited.

```
<owl:Class rdf:ID="TouristAccomodation">
 <owl:disjointWith>
   <owl:Class rdf:ID="Category"/>
 </owl:disjointWith>
 <owl:disjointWith>
   <owl:Class rdf:ID="Room"/>
 </owl:disjointWith>
 <rdfs:subClassOf rdf:resource="http://www.w3.org/2002/07/owl#Thing"/>
 <rdfs:subClassOf>
   <owl:Class rdf:ID="Building"/>
 </rdfs:subClassOf>
</owl:Class>
<owl:Class rdf:ID="Motel">
 <rdfs:subClassOf>
   <owl:Restriction>
    <owl:hasValue rdf:datatype="http://www.w3.org/2001/XMLSchema#boolean"
     >true</owl:hasValue>
     <owl:onProperty>
      <owl:FunctionalProperty rdf:ID="isOnHighway"/>
     </owl:onProperty>
   </owl:Restriction>
 </rdfs:subClassOf>
 <rdfs:subClassOf rdf:resource="#TouristAccomodation"/>
 <owl:disjointWith>
   <owl:Class rdf:ID="Hotel"/>
 </owl:disjointWith>
</owl:Class>
<owl:Class rdf:about="#Hotel">
 <rdfs:subClassOf rdf:resource="#TouristAccomodation"/>
 <owl:disjointWith rdf:resource="#Motel"/>
 <rdfs:subClassOf>
   <owl:Restriction>
     <owl:onProperty>
      <owl:FunctionalProperty rdf:about="#isOnHighway"/>
     </owl:onProperty>
    <owl:hasValue rdf:datatype="http://www.w3.org/2001/XMLSchema#boolean"
```

Copyright © 2006, Idea Group Inc. Copying or distributing in print or electronic forms without written permission of Idea Group Inc. is prohibited.

```
      >false</owl:hasValue>
     </owl:Restriction>
    </rdfs:subClassOf>
  </owl:Class>
  <owl:Class rdf:ID="Location">
   <owl:disjointWith>
     <owl:Class rdf:about="#Category"/>
   </owl:disjointWith>
   <rdfs:subClassOf
rdf:resource="http://www.w3.org/2003/01/geo/wgs84_pos#Point"/>
   <owl:disjointWith>
     <owl:Class rdf:about="#Room"/>
   </owl:disjointWith>
  </owl:Class>
  <owl:Class rdf:about="#Building">
   <owl:disjointWith>
     <owl:Class rdf:about="#Room"/>
   </owl:disjointWith>
   <owl:disjointWith>
     <owl:Class rdf:about="#Category"/>
   </owl:disjointWith>
  </owl:Class>
  <owl:Class rdf:ID="Street"/>
  <owl:Class rdf:about="#Room">
   <owl:disjointWith rdf:resource="#Building"/>
   <owl:disjointWith rdf:resource="#Location"/>
   <owl:disjointWith>
     <owl:Class rdf:about="#Category"/>
   </owl:disjointWith>
   <owl:disjointWith rdf:resource="#TouristAccomodation"/>
  </owl:Class>
  <owl:Class rdf:about="#Category">
   <owl:disjointWith rdf:resource="#Building"/>
   <owl:disjointWith rdf:resource="#Room"/>
   <owl:disjointWith rdf:resource="#TouristAccomodation"/>
   <owl:disjointWith rdf:resource="#Location"/>
```

Copyright © 2006, Idea Group Inc. Copying or distributing in print or electronic forms without written permission of Idea Group Inc. is prohibited.

```
</owl:Class>
<owl:ObjectProperty rdf:ID="hasCategory">
  <rdfs:range rdf:resource="#Category"/>
  <rdfs:domain rdf:resource="#TouristAccomodation"/>
  <rdf:type rdf:resource="http://www.w3.org/2002/07/owl#FunctionalProperty"/>
</owl:ObjectProperty>
<owl:FunctionalProperty rdf:ID="isRoomOf">
  <owl:inverseOf>
    <owl:InverseFunctionalProperty rdf:ID="hasRoom"/>
  </owl:inverseOf>
  <rdf:type rdf:resource="http://www.w3.org/2002/07/owl#ObjectProperty"/>
  <rdfs:domain rdf:resource="#Room"/>
  <rdfs:range rdf:resource="#TouristAccomodation"/>
</owl:FunctionalProperty>
<owl:FunctionalProperty rdf:ID="isStars">
  <rdf:type rdf:resource="http://www.w3.org/2002/07/owl#DatatypeProperty"/>
  <rdfs:range rdf:resource="http://www.w3.org/2001/XMLSchema#int"/>
  <rdfs:domain rdf:resource="#Category"/>
</owl:FunctionalProperty>
<owl:FunctionalProperty rdf:ID="hasLocation">
  <owl:inverseOf>
    <owl:FunctionalProperty rdf:ID="isLocation"/>
  </owl:inverseOf>
  <rdf:type rdf:resource="http://www.w3.org/2002/07/owl#ObjectProperty"/>
  <rdfs:range rdf:resource="#Location"/>
  <rdf:type rdf:resource="http://www.w3.org/2002/07/owl#InverseFunctionalProperty"/>
  <rdfs:domain rdf:resource="#Building"/>
</owl:FunctionalProperty>
<owl:FunctionalProperty rdf:about="#isOnHighway">
  <rdfs:domain>
    <owl:Class>
     <owl:unionOf rdf:parseType="Collection">
       <owl:Class rdf:about="#Location"/>
       <owl:Class rdf:about="#Motel"/>
       <owl:Class rdf:about="#Hotel"/>
     </owl:unionOf>
```

Copyright © 2006, Idea Group Inc. Copying or distributing in print or electronic forms without written permission of Idea Group Inc. is prohibited.

```
    </owl:Class>
   </rdfs:domain>
   <rdfs:range rdf:resource="http://www.w3.org/2001/XMLSchema#boolean"/>
   <rdf:type rdf:resource="http://www.w3.org/2002/07/owl#DatatypeProperty"/>
  </owl:FunctionalProperty>
  <owl:FunctionalProperty rdf:ID="hasAddress">
   <rdfs:range rdf:resource="http://www.w3.org/2001/XMLSchema#string"/>
   <rdf:type rdf:resource="http://www.w3.org/2002/07/owl#DatatypeProperty"/>
   <rdfs:domain rdf:resource="#Location"/>
  </owl:FunctionalProperty>
  <owl:FunctionalProperty rdf:about="#isLocation">
   <rdfs:range rdf:resource="#Building"/>
   <rdf:type rdf:resource="http://www.w3.org/2002/07/owl#InverseFunctionalProperty"/>
   <owl:inverseOf rdf:resource="#hasLocation"/>
   <rdfs:domain rdf:resource="#Location"/>
   <rdf:type rdf:resource="http://www.w3.org/2002/07/owl#ObjectProperty"/>
  </owl:FunctionalProperty>
  <owl:InverseFunctionalProperty rdf:about="#hasRoom">
   <owl:inverseOf rdf:resource="#isRoomOf"/>
   <rdfs:range rdf:resource="#Room"/>
   <rdf:type rdf:resource="http://www.w3.org/2002/07/owl#ObjectProperty"/>
   <rdfs:domain rdf:resource="#TouristAccomodation"/>
  </owl:InverseFunctionalProperty>
  <Category rdf:ID="Five">
   <isStars rdf:datatype="http://www.w3.org/2001/XMLSchema#int"
   >5</isStars>
  </Category>
  <Category rdf:ID="Four">
   <isStars rdf:datatype="http://www.w3.org/2001/XMLSchema#int"
   >4</isStars>
  </Category>
  <Hotel rdf:ID="Windsor">
   <hasLocation>
    <Location rdf:ID="Location1">
     <isLocation rdf:resource="#Windsor"/>
     <geo:lat rdf:datatype="http://www.w3.org/2001/XMLSchema#string"
```

Copyright © 2006, Idea Group Inc. Copying or distributing in print or electronic forms without written permission of Idea Group Inc. is prohibited.

```
    >-37.8206</geo:lat>
   <geo:long rdf:datatype="http://www.w3.org/2001/XMLSchema#string"
    >144.9674</geo:long>
   <hasAddress rdf:datatype="http://www.w3.org/2001/XMLSchema#string"
    >Spring street</hasAddress>
   <isOnHighway rdf:datatype="http://www.w3.org/2001/XMLSchema#boolean"
    >false</isOnHighway>
   </Location>
  </hasLocation>
 <rdfs:comment rdf:datatype="http://www.w3.org/2001/XMLSchema#string"
 >Hotel Windsor Melbourne, 5 star luxury accomodation</rdfs:comment>
  <hasCategory rdf:resource="#Five"/>
 </Hotel>
 <Street rdf:ID="Spring"/>
</rdf:RDF>
```

Copyright © 2006, Idea Group Inc. Copying or distributing in print or electronic forms without written permission of Idea Group Inc. is prohibited.

Chapter XI

Ontological Engineering in Pervasive Computing Environments

Athanasios Tsounis, University of Athens, Greece

Christos Anagnostopoulos, University of Athens, Greece

Stathes Hadjiethymiades, University of Athens, Greece

Izambo Karali, University of Athens, Greece

Abstract

Pervasive computing is a broad and compelling research topic in computer science that focuses on the applications of technology to assist users in everyday life situations. It seeks to provide proactive and self-tuning environments and devices to seamlessly augment a person's knowledge and decision making ability, while requiring as little direct user interaction as possible. Its vision is the creation of an environment saturated with seamlessly integrated devices with computing and communication capabilities. The realisation of this vision requires that a very large number of devices and software components interoperate seamlessly. As these devices and the associated software will pervade everyday life, an increasing number of software and hardware providers will deploy functionality in

Copyright © 2006, Idea Group Inc. Copying or distributing in print or electronic forms without written permission of Idea Group Inc. is prohibited.

pervasive computing environments (PCE). That poses a very important interoperability issue, as it cannot be assumed that the various hardware and software components share common communication and data schemes. We argue that the use of Semantic Web technologies, namely the ontologies, present a intriguing way of resolving such issues and, therefore, their application in the deployment of PCE is a highly important research issue.

Introduction

The vision of pervasive computing presents many technical issues, such as scaling-up of connectivity requirements, heterogeneity of processors and access networks and poor application portability over embedded processors. These issues are currently being addressed by the research community; however the most serious challenges are not technological but structural, as embedded processors and sensors in everyday products imply an explosion in the number and type of organisations that need to be involved in achieving seamless interoperability (O'Sullivan, 2003). In a typical pervasive computing environment (PCE) there will be numerous devices with computing capabilities that need to interoperate (Nakajima, 2003). These devices might be of different vendors and may operate based on different protocols. Therefore, the key issue in deploying a PCE is achieving application level interoperability. The complexity of such a venture is considerable. It is extremely difficult to reach agreements when the players involved expand from all the hardware and software providers (e.g., IBM, HP, Microsoft) to all the organisations that will equip their products with computing and communication capabilities (e.g., coffee machines, refrigerators). Therefore, we cannot rely on shared a priori knowledge based on commonly accepted standards to resolve the issue. Instead, software components must adapt to their environment at runtime to integrate their functionality with other software components seamlessly. An intriguing way of resolving this issue is the use of semantics, namely the use of Semantic Web technologies such as ontologies. In this manner, software entities provide semantically enriched specifications of the services that they provide and the way they should be invoked. Moreover, the data that are exchanged are also semantically enriched, enabling the entities to reason and make effective decisions. This is particularly important for the description of contextual information, which is of main interest in a PCE. As context we identify any information that is, directly or indirectly, associated with any entity in the environment.

The novelty of the Semantic Web is that the data are required to be not only machine readable but also machine understandable, as opposed to today's Web which was mainly designed for human interpretation and use. According to Tim

Copyright © 2006, Idea Group Inc. Copying or distributing in print or electronic forms without written permission of Idea Group Inc. is prohibited.

Berners-Lee, the Director of World Wide Web Consortium, "the Semantic Web's goal is to be a unifying system which will (like the Web for human communication) be as un-restraining as possible so that the complexity of reality can be described" (Berners-Lee, 2001). With the realisation of a Semantic Web it would be easy to deploy a wide range of services that would be almost impossible to manage in the current Web. Semantics enable developers to create powerful tools for complex service creation, description, discovery, and composition. The application areas of the Semantic Web extend from knowledge repositories to e-commerce and from user profiling to PCE.

New standards are being developed as a first step in realising the Semantic Web. The Resource Description Framework (RDF), which is a Web mark-up language that provides basic ontological primitives, has been developed by the W3C (Beckett, 2004). RDF is a language for representing meta-information about resources in the World Wide Web. However, by generalising the concept of a "Web resource", RDF can also be used to represent information about things that can be identified on the Web, by means of URIs. The DARPA Agent Markup Language + Ontology Inference Layer (DAML+OIL) extends RDF with a much richer set of modelling primitives (Rapoza, 2000). The DAML+OIL have been submitted to W3C as a starting point for the Web Ontology Working Group and led to the creation of the standard for Web ontologies, namely the Web Ontology Language (OWL) (McGuinness, 2004; Noy et al., 2001).

In this chapter, we survey the research that has been carried out regarding ontology and knowledge engineering and try to map the key findings to the requirements of a PCE. We present standards and tools that are being used for the development of ontologies. Furthermore, we discuss efforts regarding the management of multiple, overlapping, and evolving ontologies and the semantic mappings among them. We also depict how ontologies can help in searching mechanisms and in profiling users and data sources. Moreover, due to the diverse nature of PCE using interoperability standards is not a sufficiently scalable approach. Hence, we show the important role that ontologies can play in achieving service interoperability, a key requirement in PCE. Finally, we present the efforts that have been carried out linking the pervasive computing paradigm with ontologies and illustrate with an example our vision that brings these two worlds together in terms of semantics usage.

The rest of this chapter is structured as follows: In section 2, we outline our vision of a service oriented PCE and some arguments in the use of the ontology paradigm. Section 3 deals with ontological engineering aspects by providing definitions related to ontologies, presenting current Semantic Web standards for developing ontologies and describing the work that has been carried out regarding their management and manipulation. Section 4 introduces issues related to how ontologies might be used to provide semantic searching mecha-

Copyright © 2006, Idea Group Inc. Copying or distributing in print or electronic forms without written permission of Idea Group Inc. is prohibited.

nisms related to information retrieval. In this section also, we give an overview of ontology-based profiling. In section 5, we portray how ontologies assist in providing semantic based service interoperability and how ontologies may be useful in a PCE. Section 6 outlines some several examples of semantic-based service interoperability through ontologies. Finally in section 7, we provide our conclusions and outline future research directions.

A Service-Oriented Pervasive Computing Environment

Service-oriented architectures focus on application-level interoperability, in terms of well-established service ontologies (Martin, 2003; Fensel, 2002). This is achieved by means of well-defined interfaces to various software components. In this fashion, rapid integration of existing functionality is achieved in the design and implementation of new applications. Object Management Group's (OMG) Common Object Request Broker Architecture (CORBA) (Orfali, 1998) is an example of such widespread service-oriented architecture. One of the most important issues in a PCE is the establishment of ad-hoc relationships between applications or between applications and devices. As a result, the use of a common mechanism, which will be able to provide interoperability in a dynamic fashion, is essential. Furthermore, the separation of the application's functionality from the adopted communication schemes is of considerable importance, as it enables system developers to distinguish between service functionality (e.g., functional and no-functional context-aware service features, as referred in McIlraith et al., 2001) and system functionality. Hence, the deployment of a PCE reduces to the problem of ad-hoc service discovery, composition (e.g., WSFL, service composition context-aware paradigm) and execution with minimum a-priori knowledge of the service's functionality. While OWL-S is based on the combination of OWL, WSDL and SOAP, WSMO uses F-Logic and XML-based features of Web Services (WS). Based on the previous languages, significant research has been devoted to semantic WS (e.g., ODE SWS [Gómez et al., 2004], METEOR [Aggarwa et al., 2004]).

However, interoperability is a difficult task to accomplish in a highly heterogeneous and volatile environment. Therefore, for service-oriented architectures to be applicable a well-defined mechanism is required to describe the communication schemes which are employed, provide data with semantics and describe the profiles and the policies of the entities that constitute the PCE. Furthermore, ontologies may be employed for defining entity-specific policies that govern the context access, usage, and manipulation. They may also provide abstract

Copyright © 2006, Idea Group Inc. Copying or distributing in print or electronic forms without written permission of Idea Group Inc. is prohibited.

descriptions of the physical and software components (i.e., conceptual modelling via ontology paradigm) that comprise the PCE. Hence, considerable flexibility in deploying such environments is provided.

Accurate and expressive context models are required to represent context in an efficient and effective way, as the manipulation of the context will be based on its model. The model will determine the variety of the actions that may be applied on it, as well as the accuracy of the results. Moreover, context knowledge and manipulation enable the system to be context aware and proactive and employ more efficient decision making mechanisms. Therefore, we consider the context representation to be of great importance, since formal context modelling (in terms of developing an object-centred or frame-based knowledge manipulation system) achieves multirepresentation knowledge, by rendering heterogeneous nature of PCEs. Moreover, model-mapping and context multirepresentation techniques should be employed to transform the whole context or part of it from one form (i.e., representation formulae or dissimilar semantic conceptual model) to another (e.g., [Ding et al., 2002]). As long as the appropriate mappings exist, the entities may exchange contextual information while preserving their own context model (i.e., localised ontologies [Bouquet, 2002]). Additionally, inference engines (e.g., rule-based systems [Friedman, 2005]) may be used to deduct previously unknown derivative context from primitive contextual information. In this fashion, a pervasive computing system is able to produce knowledge and provide proactivity, assisting the user in an unobstructive way. Last but not least, the profile and policy management for the descriptions of the physical and software entities and the definition of the rules that govern the context access, usage, and manipulation may be based on the ontological engineering paradigm. Therefore, we envisage an environment where all entities interact with services through abstract interfaces and argue that the use of Semantic Web technologies is useful in achieving semantic based service interoperability in PCEs.

Ontological Engineering

The word ontology comes from philosophy, where it means a systematic explanation of being. In the last decade, this word has become relevant to the knowledge engineering community. The authors in Guarino (1995) propose the words "Ontology" (with capital "o") and "ontology" to refer to the philosophical and knowledge engineering concepts, respectively. There are a lot of relevant definitions, but we keep only two terms that are relevant, to some extent, to the pervasive computing paradigm. In Neches et al. (1991) an "ontology" is defined as follows:

Copyright © 2006, Idea Group Inc. Copying or distributing in print or electronic forms without written permission of Idea Group Inc. is prohibited.

An ontology defines the basic terms and relations comprising the vocabulary of a topic area as well as the rules for combining terms and relations to define extensions to the vocabulary.

Such descriptive definition informs us what to do to build an ontology by giving some vague guidelines: This ontology description denotes how to identify basic terms and relations between terms and how to identify rules combining terms and their relationships. Through such a definition one can deduce that ontology does not include only terms that are explicitly defined in, but also the knowledge that can be inferred from it. In this point, the authors in Guarino (1995) consider an 'ontology' as:

A logical theory, which gives an explicit, partial account of a conceptualisation, where conceptualisation is basically the idea of the world that a person or a group of people can have. Consecutively, one may consider that the ontology distinguishes terms that are mainly taxonomies from dissimilar human senses and models a specific domain in a 'deeper' way by providing well-documented restrictions on such domain semantics.

In the pervasive computing area it can be claimed that ontology is used to provide a vocabulary with explicitly defined and machine comprehensible meaning. In multi-agent systems, ontologies are essential for providing a common vocabulary and a shared perception of domain knowledge to allow communication in an open environment. An ontology provides an explicit specification of the structure of a certain domain and includes a vocabulary for referring to the subject area, and a set of logical statements expressing the constraints existing in the domain and restricting the interpretation of the vocabulary. Thus, the developing of ontologies, especially for pervasive environments, provides the means by which distributed software components share common semantics of the terms being used for communication and knowledge representation.

Ontology Languages for the Semantic Web

As already mentioned, several standards for Web ontologies have emerged. We will provide a brief description of the most widely used standards, namely RDF, DAML+OIL and OWL.

The RDF is a language for representing information about resources in the WWW. It is particularly intended for representing metadata about Web resources, such as the title, author, and modification date of a Web page. However,

Copyright © 2006, Idea Group Inc. Copying or distributing in print or electronic forms without written permission of Idea Group Inc. is prohibited.

by generalising the concept of a "Web resource", RDF can also be used to represent information about things that can be identified on the Web, even when they cannot be directly retrieved on the Web. Examples include information about items available from online shopping facilities (e.g., information about specifications, prices, and availability). RDF is intended for situations in which this information needs to be processed by applications, rather than only being displayed to people. RDF provides a common framework for expressing this contextual information that it can be exchanged between applications without loss of meaning. It represents a simple graph model and uses a well-documented XML schema for datatypes by identifying all elements by means of URIs. The structure of any expression in RDF is a triplet (Figure 1), consisting of a subject, an object, and a predicate (also called property).

RDF provides a way to express simple statements about resources, using named properties and values. However, RDF user communities also need the ability to define the vocabularies (i.e., terms or taxonomies in ontology community) they intend to use in those statements, specifically, to indicate that they are describing specific kinds or classes of resources, and will use specific properties in describing those resources. RDF itself provides no means for defining such application-specific classes and properties. Instead, such classes and properties are described as an RDF vocabulary, using extensions to RDF provided by the RDF Schema (RDFS) (Brickley, 2004). The RDFS provides the facilities needed to describe such classes and properties, and to indicate which classes and properties are to be used together. In other words, RDFS provides a type system for RDF. The RDFS type system is somehow similar to the type systems of object-oriented programming languages, such as Java.

DAML+OIL is also a semantic markup language for Web resources. It builds upon earlier W3C standards such as RDF and RDFS, and extends these languages with richer modelling primitives (e.g., in terms of class subsupmtion

Figure 1. RDF triplet

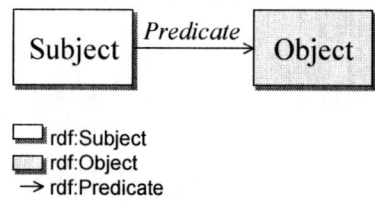

rdf:Subject
rdf:Object
→ rdf:Predicate

The SPO conceptual model of RDF

Copyright © 2006, Idea Group Inc. Copying or distributing in print or electronic forms without written permission of Idea Group Inc. is prohibited.

or object properties). It adds the familiar ontological primitives of object-oriented and frame-based systems, and the formality of a very expressive description logic (i.e., an ALC Description Logic language).

OWL is a semantic markup language for publishing and sharing ontologies on the World Wide Web. OWL is intended to be used when the information contained in documents needs to be processed by applications, as opposed to situations where the content only needs to be presented to humans. OWL can not only be used to explicitly represent the meaning of terms in vocabularies and the relationships between those terms, but also may serve implicitly representing inferred terms or whole taxonomies by means of logical reasoners (i.e., applications, such as RACER, FaCT, DLP (Haarslev, 1999), based on Description Logics axioms that provide inferred knowledge with respect to certain First Order Logic algorithms, such as Structural Subsumption and Tableau algorithms). OWL has more facilities for expressing meaning and semantics than XML, RDF, and RDFS, and thus OWL goes beyond these languages in its ability to represent machine interpretable content on the Web. OWL is an enhanced revision of DAML+OIL Web ontology language. OWL provides three increasingly expressive sublanguages designed for use by specific communities of implementers and ontology developers. The OWL class (i.e., Concepts and conceptual formulas in Description Logics) expressions are built with Knowledge Representation primitives that are properties. These will be organised in two groups. The first group includes primitives defined for OWL Lite while the second group includes primitives defined for OWL DL and OWL Full.

In the previous paragraph, we discussed a set of concept models (i.e., interpretations in the ABox that satisfy the TBox formulas) that are decidable and satisfiable, but with restricted expressiveness due to representation of an sublanguage of the ALC DL. In this context, there is no disjunction and negation concept formulas and the property restrictions (owl:Restriction) of objects (i.e., the instances of concepts in the ABox) are set to cardinality (owl:cardinality).

The latter group stands for fully expressive OWL languages, but the OWL DL is the only sublanguage that can be considered as a machine comprehensible ALCN Description Logic language for the Semantic Web. The OWL Full is not a decidable language in terms of Description Logics, since the ABox and the TBox may overlap (e.g., an instance of the ABox can be interpreted as a concrete concept of the TBox) according to their concept definitions and there is no such a reasoner (i.e., DL algorithm) that can infer whether an interpretation expressed in OWL Full is satisfiable or not. The OWL DL is a DL language enabling the definitions of conjunction, disjunction, and negation for any interpretation of the concept models and deals with certain value (i.e., owl:allValuesFrom, $\forall R.C$ in DL semantics), existential (owl:someValuesFrom, $\exists R.C$ in DL semantics) and number (e.g., owl:maxCardinality, $\geq_n$ in DL semantics) restrictions with

Copyright © 2006, Idea Group Inc. Copying or distributing in print or electronic forms without written permission of Idea Group Inc. is prohibited.

arbitrarily cardinalities of their fillers (i.e., instances that satisfy Roles in a Knowledge Base ABox).

Due to the dynamic nature of the contextual information in PCEs and the vague, temporal or probabilistic knowledge representation, there have been many extensions in OWL DL language in order to cope with such models (e.g., imprecise information representation, probabilistic knowledge concept definitions, temporal information representation). Such OWL DL languages deal with fuzzy context modelling (e.g., the proposed [Stracia, 1998]), temporal contextual concepts (e.g., the proposed [Artale, 1998]) and probabilistic knowledge representation in terms of Bayesian Networks (e.g., the proposed P-OWL language). Such OWL extensions are based on novelty extensions of the classical DLs (e.g., the OWL is a kind of a SHIN language) with the use of Modal Logics (Elgesem, 1997) that provide different world assumptions in well-defined DL Tboxes.

Moreover, in order to deduct new knowledge by reasoning over contextual facts (i.e., set of instances, such as the Abox interpretations), certain extensions in the ontological community by representing rules have been proposed and standardised. The SWRL (Horrocks et al., 2003) is a language for a Semantic Web Rule Language based on a combination of the OWL DL and OWL Lite sublanguages with the Unary/Binary Datalog RuleML sublanguages of the Rule Markup Language (Boley, 2001). Such language extends the set of OWL axioms to include Horn-like clauses. Hence, Horn-like clauses can be combined with an OWL knowledge base by defining rules for the ABox instances reasoning. A high-level abstract syntax is provided that extends the OWL abstract. An extension of the OWL model-theoretic semantics is also given to provide a formal meaning for OWL ontologies including rules written in this abstract syntax.

Managing the Ontology

Taking into account the distributed nature of a PCE, several key challenges emerge concerning the use of ontologies. One has to be able to deal with multiple and distributed ontologies (i.e., distributed knowledge representation in terms of ontological modularity of taxonomies) to enable reuse, maintainability, evolution, and interoperability. Furthermore, support is needed in managing the evolution of multiple and distributed ontologies, in order to ensure consistency. In fact, the ability of logical reasoners to link independent ontology modules, to allow them to be separately maintained, extended, and re-used is one of their most powerful features. In PCE we encounter large and distributed ontologies (i.e., in terms of number of concepts), representing semantically similar contexts (e.g., equivalent concepts or roles).

Copyright © 2006, Idea Group Inc. Copying or distributing in print or electronic forms without written permission of Idea Group Inc. is prohibited.

In highly distributed systems such as the Semantic Web, modularity naturally exists in terms of conceptual modelling. Large ontologies (e.g., medical or biological ontologies such as GALEN (Rector, 2003; Rector et al., 1999) may be split in smaller and reusable modules in order to meet the needs of ontology maintenance and distributed reasoning. Requirements for managing ontologies in such a distributed environment include the following:

- The semantic loose coupling, which refers to the conceptualisation sets of ontologies that are irrelevant (in terms of concept perception).

- The self-standing or self-containment of concepts is the idea of defining concepts that enable the automatic inference of other, classified concepts without having to access contextual information in other ontologies.

- The ontology consistency, which performs integrity checks to assess whether relevant knowledge in other systems has changed, and, updates the self-standing ontological module, if needed.

In Maedche (2003), the authors present a framework for managing multiple and distributed ontologies in a Semantic Web environment. They describe a conceptual modelling framework for single, multiple, and distributed ontologies, which renders feasible the realisation of ontology-based systems using well-established technologies such as relational databases. The authors make use of the object-oriented paradigm and extend it with simple deductive features. In their approach, information is organised in Ontology-Instance models, which contain both ontology entities and instances. They introduce the "single ontology consistency" and the "dependent ontology consistency" definitions and propose an evolution strategy that ensures the consistency of the evolving ontologies by unambiguously defining the way in which elementary changes will be resolved.

In another approach, discussed in Rector (2003), the key is modularity. Modularity is acknowledged as a key requirement for large ontologies in order to achieve reuse, maintainability, and evolution. Rector, motivated by the fact that mechanisms for normalisation are standard for databases, proposes the concept of normalisation for ontologies. He proposes a two-step normalisation implemented using OWL or related description logic-based formalisms. For the first step, namely the "ontological normalisation", he makes use of the analysis in Guarino (2000). For the second step (implementation normalisation) he proposes an approach based on decomposing the ontology into independent disjoint skeleton taxonomies restricted to simple trees, which can, then, be recombined using definitions and axioms to explicitly represent the relationships between them. The main requirement, though, for implementation normalisation is that the modules to be reused can be identified and separated from the whole, so as to evolve independently.

Copyright © 2006, Idea Group Inc. Copying or distributing in print or electronic forms without written permission of Idea Group Inc. is prohibited.

In Guarino (1999), the authors introduce the OntoSeek, a system that employs an ontology server for ontology management. The server provides an interface for applications willing to access or manipulate an ontology data model (i.e., a generic graph data structure), and facilities maintaining a persistent Lexical Conceptual Graph (LCG) database. End users and resource encoders can access the server to update the LCG database encoded in a markup language, (e.g., XML). The main function of the ontology server is to create, edit, evaluate, publish, maintain, and reuse ontologies. Particularly important is the ability to support the collaborative works through the Web.

Semantic Mappings Among Ontologies

Clearly, it would be desirable in a PCE that the various services are capable of interoperating with existing data sources and consumers. It is very difficult to build a consensus about the terminologies and structures that should be used. In order to introduce the significance of the ontology matching issue in the computing paradigm, we support the Semantic Web to make information on the World Wide Web more accessible using machine-readable metadata. Since the ontological development deals with conceptualising specific domains, the notion of context modelling could be imperative by using specific domain ontologies. Thus, an ontology is contextualised when its concepts are kept local (i.e., self-standing concepts that are not shared with other ontologies) and are associated only with the concepts of other ontologies, through well-defined explicit mappings (or implicit mappings, through distributed reasoning tasks, such as Compilation of Implied Knowledge with Queries).

An architecture is needed for building a Web of data by allowing incremental addition of sources, where each new source maps to the sources considered most convenient. Therefore, information processing across ontologies is only possible when semantic mappings between the ontologies exist. Technological infrastructure for semantic interoperability between semantically autonomous communities (i.e., ontological development that is compliant with the requirements in subsection 3.2) must be based on the capability of representing local ontologies and contextual mappings between them, rather than on the attempt to create a global and possibly shared conceptualisation. Manually finding such contextual mappings is tedious, error prone, and clearly not possible on a large scale. This is what researchers call semantic heterogeneity in the Semantic Web paradigm, namely, a situation in which different meanings use the same concept to mean the different things and use different granularity to describe the same domain forming explicit different perspectives of the world. In this context, the lack of uniformity prompts for the meaning negotiation among different agents, since they do not understand each others as they use languages with heterogeneous semantics.

Copyright © 2006, Idea Group Inc. Copying or distributing in print or electronic forms without written permission of Idea Group Inc. is prohibited.

A lot of work has been performed during the last years in this direction, which resulted in the development of systems that provide mappings semi-automatically (Doan, 2003; Halevy, 2003; Fowler et al., 1999; Chalupsky, 2000). In this way, given the semantic mappings between different ontologies, one may match them in order to be able to reason about a specific domain.

Given two ontologies, the ontology-matching problem is to find semantic mappings between them. The simplest type of mapping is the one-to-one mapping between the elements of the different ontologies, namely each element of one ontology is associated with at most one element of the second ontology and vice versa. Mappings between different types of elements are possible, meaning that a property may be mapped to a class. More complex mappings are possible, such that a class or an instance may map to a union of classes or instances. In general, a mapping could be specified as a query that transforms instances, classes and properties from one ontology into instances, classes, and properties of another.

To illustrate the above, suppose that we want to provide mappings between two WWW sites that sell books. In Figure 2, the ontology-matching problem between an author-centric ontology and a book-centric ontology is depicted. A matchmaking process must provide semantic mappings between the nodes in the two ontologies in an automatic manner.

In Doan (2003), the authors describe GLUE, a system that employs machine-learning techniques to determine such mappings. Given two ontologies, for each concept in one ontology the system finds the most similar concept in the other ontology. A key feature of the GLUE is that it uses multiple learning strategies, each of which exploits well a different type of information either in the data instances or in the taxonomic structure of the ontologies.

The first issue to address such conceptual matching is the notion of similarity between two concepts. The approach in Doan (2003) is based on the observation that many practical measures of similarity can be defined based solely on the joint probability distribution of the concepts involved. The second challenge is that of computing the joint distribution of any two given concepts A and B. Specifically, for any two concepts A and B, the joint distribution consists of $p(a \in A \cap B)$, $p(\alpha \in A \cap \neg B)$, $p(\alpha \in \neg A \cap B)$, $p(\alpha \in \neg A \cap \neg B)$, where a term such as $p(a:A \neg B)$, is the probability that an instance in the domain belongs to concept A but not to concept B. GLUE, also, uses a multi-strategy learning approach employing a set of learners and then combining their predictions using a metalearner. Furthermore, it exploits domain specific constraints using general heuristic algorithms to improve the matching accuracy.

In another approach, described in Halevy (2003), the authors provide a mediating schema between RDF and XML data sources. Most of these sources provide XML representations of the data and not RDF; XML and the applications that depend on it rely not only on the domain structure of the data, but also on its

Copyright © 2006, Idea Group Inc. Copying or distributing in print or electronic forms without written permission of Idea Group Inc. is prohibited.

Figure 2. Ontology matching problem

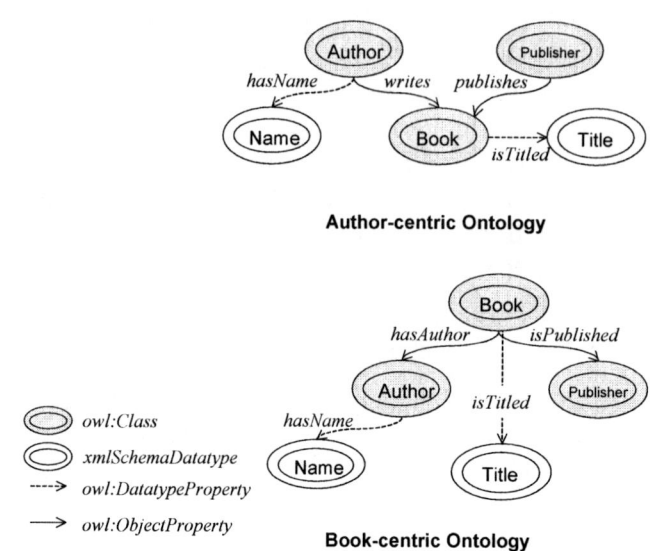

document structure. In order to provide interoperability between such sources, one must map between both their domain structures and their document structures. Second, data management practitioners often prefer to exchange data through local point-to-point data translations, rather than mapping to common mediated schemas or ontologies. Therefore, the authors built the Piazza system, which addresses these challenges. Piazza offers a language for mediating between data sources on the Semantic Web, which maps both the domain structure and document structure. Piazza also enables interoperation of XML data with RDF data that is accompanied by rich OWL ontologies. They proposed a language for mediating between nodes that allows mapping simple forms of domain structure and rich document structure. The language is based on Xquery (Fernández et al. 2005), the emerging standard for querying XML. They also showed that this language could map between nodes containing RDF data and nodes containing XML data.

In Fowler et al. (1999), the authors propose InfoSleuth, a system that can support construction of complex ontologies from smaller component ontologies so that tools tailored for one component ontology can be used in many application domains. Examples of reused ontologies include units of measure, chemistry knowledge, and geographic metadata. The mapping is explicitly specified among

Copyright © 2006, Idea Group Inc. Copying or distributing in print or electronic forms without written permission of Idea Group Inc. is prohibited.

these ontologies as relationships between terms in one ontology and related terms in other ontologies.

In another approach, the OntoMorph system of the Information Sciences Institute (ISI) of the University of Southern California aims to facilitate ontology merging and the rapid generation of knowledge base translators (Chalupsky, 2000). It combines two powerful mechanisms to describe Knowledge Base transformations. The first of these mechanisms is syntactic rewriting via pattern-directed rewrite rules that allow the concise specification of sentence-level transformations based on pattern matching, and the second mechanism involves semantic rewriting which modulates syntactic rewriting via semantic models and logical inference. The integration of ontologies can be based on any mixture of syntactic and semantic criteria.

Ontology-Based Profiling and Information Retrieval

In a typical PCE there exists a vast amount of contextual information available for retrieval. The traditional solution to the problem of information retrieval employs keyword-based searching techniques. This solution is inadequate and, therefore, new searching mechanisms must be developed. The context should be modelled in a semantically enriched manner, namely by means of ontologies, and searching techniques that make use of this semantic information must be applied in order to provide superior results. Furthermore, PCEs usually integrate sensor networks, which differ considerably from current networked and embedded systems. They combine the large scale and distributed nature of networked systems with the extreme energy constraints and physically coupled nature of embedded control systems. Sensors, however, produce raw data. Ontologies could be used to transform the raw data to a semantically enriched representation that could be queried by an advanced searching mechanism.

The proliferation of Semantic Web technologies, such as ontologies, has led to the development of more sophisticated semantic query mechanisms. Semantic querying techniques will exploit the semantics of content to provide results that are superior to those delivered by contemporary techniques, which rely mostly on the lexical and structural properties of a document. There are a number of proposals for querying data (e.g., Anyanwu et al., 2003) expressed by means of ontologies, but most of them are constrained in such a way that one is able to form queries only of the type "Get all elements that are related to element A through a relationship R" where R is, typically, specified as a path expression. Recently,

Copyright © 2006, Idea Group Inc. Copying or distributing in print or electronic forms without written permission of Idea Group Inc. is prohibited.

Figure 3. Domain specific concept hierarchy

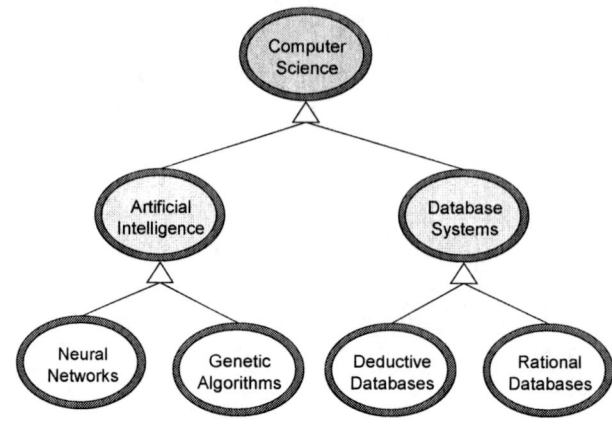

research is being conducted in order to be able to express queries of the type "How is element A related to element B?"

In Khan (2004), the authors propose a framework for querying databases of a specific context using an ontology-based index structure. The traditional solution employs keyword-based search. The only documents retrieved are those containing user specified keywords. But many documents convey desired semantic information without containing these keywords. One can overcome this problem by indexing documents according to context and meaning rather than keywords, although this requires a method of converting words to meanings and the creation of a meaning-based index structure. In the discussed work, the authors have solved the problem of an index structure through the design and implementation of a concept-based model using domain dependent ontologies. In other words, the ontology provides index terms/concepts that can be used to match with user requests. Furthermore, the generation of a database query takes place after the keywords in the user request are matched to concepts in the ontology. To achieve this, they employ a proprietary method for query expansion and SQL query generation.

In Wang et al. (2003), a system for classifying text documents based on ontologies for representing the domain specific concept hierarchy (an example is depicted in Figure 3) is presented. Many researchers have shown that similarity-based classification algorithms, such as k-nearest neighbour and centroid-based classification, are very effective for large document collections. However, these effective classification algorithms still suffer disadvantages

Copyright © 2006, Idea Group Inc. Copying or distributing in print or electronic forms without written permission of Idea Group Inc. is prohibited.

from high dimensionality that greatly limit their practical performance. Finding the nearest neighbours in high dimensional space is very difficult because the majority of points in high dimensional space are almost equi-distant from all other points. The k-nearest neighbour classifier is an instance-based classifier. This means that a training dataset of high quality is particularly important. An ideal training document set for each particular category will cover all the important terms, and their possible distribution in this category. With such a training set, a classifier can find the true distribution model of the target domain. Otherwise, a text that uses only some keywords out of a training set may be assigned to the wrong category. In practice, however, establishing such a training set is usually infeasible. Actually, a perfect training set can never be expected. By searching the concept hierarchy defined by a domain specific ontology, a more precise distribution model for a predefined classification task can be determined. The experiments that were carried out indicated that, by using this approach, the size of the feature sets can be effectively reduced and the accuracy of the classifiers can be increased.

An extension of the ontological reasoning, in terms of information retrieval, is discussed in Maedche et al. (2001). In this work, the authors, in order to define and exploit the semantics of domain ontologies, perform the context retrieval process by clustering the pre-classified concepts with several similarity measures. The metrics of the semantic distances among concepts, which are referred to as similarity measures, are the Taxonomy similarity, the Relation similarity, and the Attribute similarity. Such measures form the input to the hierarchical clustering algorithm. In this context, one may deduce about new relations among concepts, something that enriches the notion of the distributed domain ontologies in PCE. Furthermore, the classification process, with respect to the context retrieval, is not feasible with undefined relations among conceptual formulas (i.e., primitive concept, defined concepts, and complex concept definitions).

Moreover, such new relations may denote new activities (i.e., constraints), behaviours (i.e., axioms), and roles on entities or concepts in ontologies. In the pervasive computing paradigm we encounter a large number of interacting entities. The rules that govern that interaction are complex and should be based, among others, on the profiles of entities. According to the Cambridge Dictionary (1980), a profile is "a portrait, drawing, diagram, etc., of a side view of the subject". A model of an entity's characteristics, activity, behaviour and its role in the PCE is the entity's profile. Based on this knowledge, complex mechanisms manage interaction of entities in the most efficient and effective way.

Several methods have been proposed for using ontologies for user and data sources profiling, as well as for describing communication mechanisms. Hence,

Copyright © 2006, Idea Group Inc. Copying or distributing in print or electronic forms without written permission of Idea Group Inc. is prohibited.

complex queries over data sources may be exploited and auto-learning systems may propose content sources to users based on their preferences.

In Middleton et al. (2004) the authors explore an ontological approach to user profiling in the context of a recommender system. Their system, called Foxtrot, addresses the problem of recommending online research papers to academic researchers. Their ontological approach to user profiling is a novel idea to recommender systems. Researchers need to be able to search the system for specific research papers and have interesting papers recommended. Foxtrot uses a research-paper-topic ontology to represent the research interests of its users. A class is defined for each research topic and is-a relationships are defined where appropriate. Their ontology is based on the CORA (McCallum et al., 2000) digital library, as it classifies computer science topics and has example papers for each class. Research papers are represented as term vectors, with term frequency/total number of terms used for a terms weight; terms represent single words in a paper's text. Research papers in the central database are classified by an IBk (Kibler, 1991) classifier, a k-nearest neighbour type classifier, which is boosted by the AdaBoostM1 algorithm (Freund, 1996).

In Parkhomenko et al. (2003), the authors present an approach that uses ontologies to set up a peer profile containing all the data, necessary for peer-to-peer (P2P) interoperability (Milojicic et al., 2002). The use of this profile can help address important issues present in current P2P networks, such as security, resource aggregation, group management, etc. They also consider applications of peer profiling for Semantic Web built on P2P networks, such as an improved semantic search for resources not explicitly published on the Web but available in a P2P system. They developed the ontology-based peer profile in RDF format and demonstrated its manifold benefits for peer communication and knowledge discovery in both P2P networks and Semantic Web.

In another approach regarding P2P networks, the authors in (Nejdl et al., 2003), focus on peer profiling so as to enable complex user queries and routing algorithms. RDF-based P2P networks allow complex and extendable descriptions of resources, and they provide complex query facilities against these metadata instead of simple keyword-based searches. Furthermore, they are capable of supporting sophisticated routing and clustering strategies based on the metadata schemas, attributes and ontologies. Especially helpful in this context is the RDF functionality to uniquely identify schemas, attributes and ontologies. The resulting routing indices can be built using dynamic frequency counting algorithms and support local mediation and transformation rules.

Copyright © 2006, Idea Group Inc. Copying or distributing in print or electronic forms without written permission of Idea Group Inc. is prohibited.

Semantic-Based Service Interoperability Through Ontologies in Pervasive Environments

Although, the Web was once just a content repository, it is, now, evolving into a provider of services. Web-accessible programs, databases, sensors, and a variety of other physical devices realise such services. In the next decades, computers will most likely be ubiquitous and most devices will have some sort of computing functionality. Furthermore, the proliferation of intranets, ad-hoc, and mobile networks sharpen the need for service interoperation. However, the problem of service interoperability arises because today's Web is designed primarily for human use. Nevertheless, an increased automation of services interoperation, primarily in business-to-business and e-commerce applications, is being noted. Generally, such interoperation is realised through proprietary APIs that incorporate hard-coded functionality in order to retrieve information from Web data sources. Ontologies can prove very helpful in the direction of service description for automatic service discovery, composition, interoperation, and execution.

In McIlraith et al. (2001), the authors present an agent technology based on reusable generic procedures and customising user constraints that exploits and showcases WS markup. To realise their vision of Semantic WS, they created semantic markup of WS that makes them machine understandable and easy to use. They also developed an agent technology that exploits this semantic markup to support automated WS composition and interoperability. Driving the development of their markup and agent technology are the automation tasks that semantic markup of WS will enable, in particular, service discovery, execution, and composition and interoperation. Automatic WS discovery involves automatically locating WS that provide a particular service and adhere to requested properties. Automatic WS execution involves a computer program or agent automatically executing an identified WS. Automatic WS composition and interoperation involves the automatic selection, composition, and interoperation of appropriate WS to perform some task, given a high-level description of the task's objective. They use ontologies to encode the classes and subclasses of concepts and relations pertaining to services and user constraints. The use of ontologies enables sharing common concepts, specialisation of these concepts, and vocabulary for reuse across multiple applications, mapping of concepts between different ontologies, and the composition of new concepts from multiple ontologies.

In Gibbins et al. (2003), the authors describe the design and implementation of an ontologically enriched WS system for situational awareness. Specifically,

Copyright © 2006, Idea Group Inc. Copying or distributing in print or electronic forms without written permission of Idea Group Inc. is prohibited.

they discuss the merits of using techniques from the multi-agent systems community for separating the intentional force of messages from their content and the implementation of these techniques within the DAML Services model. They identify that the world of WS may be characterised as a world of heterogeneous and loosely coupled distributed systems where adaptability to ad-hoc changes in the services offered by system components is considered advantageous. The term loosely coupled, means that the interactions between system components are not rigidly specified at design time. Instead, system components may opportunistically make use of new services that become available during their lifetime without having been explicitly told of their existence from the outset. The task of searching for a system component which can perform some given service (i.e., service discovery) is the enabling technique that makes loosely-coupled systems possible, and provides a process by which system components may find out about new services being offered. Service descriptions are more complex expressions, which are based on terms from agreed vocabularies and attempt to describe the meaning of the service, rather than simply ascribing a name to it. An essential requirement to service discovery is service description. A key component in the semantics-rich ap-proach is the ontology. In the conventional WS approach exemplified by Web Services Definition Language (WSDL) or even by DAML Services, the communicative intent of a message (e.g., whether it is a request or an assertion) is not separated from the application domain. This is quite different from the convention from the Multi-Agent Systems world, where there is a clear separation between the intent of a message, which is expressed using an Agent Communication Language (ACL), and the application domain of the message, which is expressed in the content of the message by means of domain-specific ontologies. In this approach, there is a definite separation of the intent of the messages and the application domain.

In Medjahed et al. (2003), the authors propose an ontology-based framework for the automatic composition of WS. An important issue in the automatic compo-sition of WS is whether those services are composable (Berners-Lee, 2001). Composability refers to the process of checking if WS to be composed can actually interact with each other. A composability model is proposed for comparing syntactic and semantic features of WS (Figure 4). The second issue is the automatic generation of composite services. A technique is proposed to generate composite service descriptions while obeying the aforementioned composability rules. Such technique uses as input a high-level specification of the desired composition. This specification contains the list of operations to be performed through composition without referring to any composite service. Based on their composability model, they propose an approach for the automatic composition of WS. This approach consists of four conceptually separate phases: specification, matchmaking, selection, and generation. The authors

Copyright © 2006, Idea Group Inc. Copying or distributing in print or electronic forms without written permission of Idea Group Inc. is prohibited.

Figure 4. Composability model for Web services

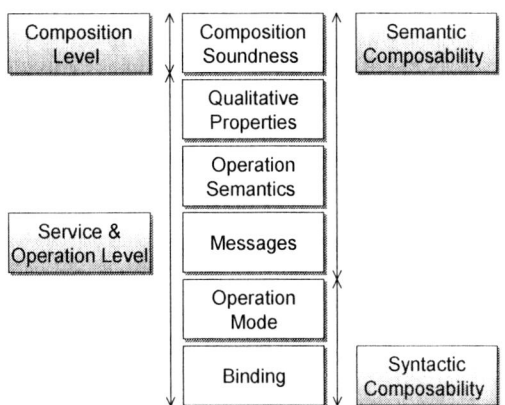

define an XML language, which adopts an ontology-based model to cater for Semantic Web-enabled WS, for the specification of composite services. Once specifications are provided, the next step is to generate appropriate composition plans using a matchmaking algorithm. The general principle of the matchmaking algorithm is to map each operation of the composite service to one or more operations of existing services. At the end of the matchmaking phase, several composition plans may have been generated. To facilitate the selection of relevant plans, they define three parameters for the quality of composition: ranking, relevance, and completeness. The ranking of a composition gives an approximation of its "importance", the relevance gives an approximation of the composition soundness and the completeness gives the proportion of composite service operations that are composable with other services. The last phase in their approach aims at generating a detailed description of a composite service. This description includes the list of outsourced services, mappings between composite services and component service operations, mappings between messages and parameters and flow of control and data between component services.

Let us imagine a PCE where all devices interoperate through services. In this way, a service-oriented architecture for functionality integration is being employed. Therefore, the problem of deploying a PCE reduces to the problem of service description and ad-hoc service discovery, composition, and execution.

We propose that the services/agents that represent the various devices share a common communication scheme, possibly by means of an ACL or a common ontology that describes the communication details. However, they do not share any a priori knowledge about the semantics of the functionality that these devices

Copyright © 2006, Idea Group Inc. Copying or distributing in print or electronic forms without written permission of Idea Group Inc. is prohibited.

provide. As described above, through the use of the device-specific ontologies one can achieve runtime service discovery and execution based on an abstract description of his objectives. In other words, a device will query its environment, giving the description of its objectives and employing complex semantic querying mechanisms. After obtaining a service reference, it will proceed with its execution. As new devices join the PCE and while others abandon it, the cycle of service description, discovery, and execution will continue forever with minimum external interference.

Furthermore, the work that has been carried out in the field of ontology management and ontology mapping ensures that the consistency of the ontologies is maintained as they evolve throughout time. Furthermore, webs of complex services are being created as devices join in or withdraw from the PCE and as new composite services are being created based on noncomposite (i.e., primitive) or even other composite services. This is only possible when one can provide semantic context mappings between the different services and automatically produce semantic descriptions of the composite services. Hence, a complicated hierarchy of services may be created. Such a hierarchy should be both manageable and highly flexible.

Examples of Semantic-Based Service Interoperability Through Ontologies in PCE

To illustrate the key issues, consider the following simple scenario (Figure 5). Bob has an important professional meeting (i.e., human society context) in his office (OfficeA) with Alice. For this meeting it is important that Bob has a certain postscript file (FileA) printed (i.e., application context). Unfortunately, the printer in his office (PrinterA) is not capable of printing postscript files. However, there is one such printer (PrinterB) in another office (OfficeB), but Bob is unaware of its existence (i.e., service registry context). In addition, the two offices are adjacent (i.e., spatial context). Bob is engaged in the activity of editing (i.e., human activity-oriented context) the file and finishes just 15 minutes before his appointment (e.g., temporal context). The PCE, by recording the time at which Bob finishes editing the file and knowing that the appointment is scheduled in 15 minutes, (i.e., contextual history) infers that Bob wants this file for the appointment (e.g., applying deduction rules over certain contexts). Furthermore, it judges that there is sufficient time for Bob to go to the next office and print the file and, consequently, issues the suggestion (i.e., ambient notification to human).

Copyright © 2006, Idea Group Inc. Copying or distributing in print or electronic forms without written permission of Idea Group Inc. is prohibited.

Figure 5. An example ontology

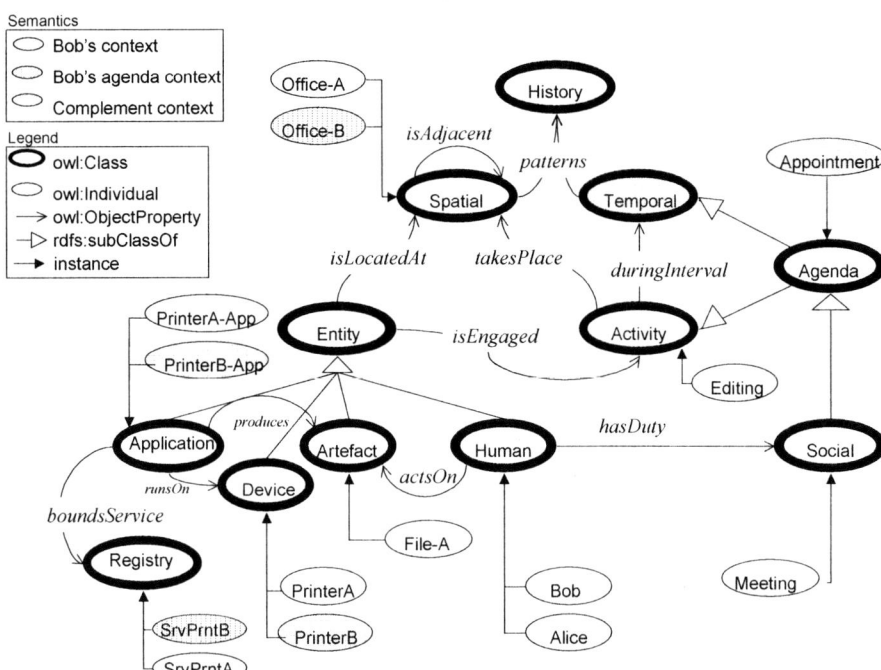

Contextualised Ontology

There are several types of semantically enriched contexts in this scenario. First of all the system knows the location of Bob, as well as the locations of the offices within the building. It also has knowledge about the equipment that is present at the offices and its capabilities (i.e., terminal profile). The system has access to his personal agenda and monitors the activities that he is engaged in (i.e., user profile). It is, also, capable of associating any activity with specific entities, which, in turn have specific attributes (i.e., inference over context facts). Furthermore, it employs an inference engine to deduce new contextual information (Bob wants the files for his appointment) and uses this information to issue suggestions.

In another scenario, a system for observing and assisting elderly people, who are alone at home, may be considered. Ontologies may be used to define simple cases (e.g., the person is sleeping for the past fifteen hours or has been in the bathroom for too long) in which the system must take specific actions, wake the patient by ringing a bell or informing the attending doctor. Furthermore, with the addition of a rule engine, complex conditions may be defined that were,

Copyright © 2006, Idea Group Inc. Copying or distributing in print or electronic forms without written permission of Idea Group Inc. is prohibited.

previously, required to be diagnosed by a trained doctor. For instance, the system may constantly monitor the patient's Electro Cardiogram (ECG) and, based on time depended composite rules, may decide that the patient requires immediate medical attention.

Research has been reported in the association of Semantic Web technologies and pervasive computing. In Chen et al. (2004), the authors describe a framework for an agent-based PCE. Central to their framework is the presence of an intelligent context broker that accepts context related information from devices and agents in the environment as well as from other sources, including information available on the Web describing the space and events and activities scheduled for it. The context broker integrates and reasons over this information to maintain a coherent model of the space, the devices, agents, and people in it, and their associated services and activities. A key to the realisation of this architecture is the use of a set of common ontologies (e.g., specialising an Upper Level Ontology or semantics similarity measures). They identify ontologies as a key component for building open and dynamically distributed PCEs in which agents and devices share contextual information. They describe the use of OWL and other tools for building the foundation ontology for the Context Broker Architecture (CoBrA). The CoBrA ontology models the basic concepts of people, agents, place, and presentation events in an intelligent meeting room environment. It provides a vocabulary of terms for classes and properties suitable for building practical systems that model context in PCEs.

In Wang (2003), the author describes "The context broker", a broker-centric agent architecture, which aims to support context-aware computing in intelligent spaces. In this architecture, a broker maintains and manages a shared context model on the behalf of a community of devices and agents. It provides necessary common services including ontology mediation and teamwork support, and enforces policies for security, trust, and privacy. A reasoning engine is responsible for reasoning with ontological and contextual knowledge. As ontological knowledge, the author identifies the static knowledge derived from the underlying ontology model (expressed in OWL), and as contextual knowledge, the dynamic knowledge that is inferred from acquired situational information.

In Ranganathan et al. (2003) and Ranganathan-GAIA (2003), the authors describe GAIA, an infrastructure for Smart Spaces, which are ubiquitous computing environments that encompass physical spaces. GAIA converts physical spaces and the ubiquitous computing devices they contain into a programmable computing system. It offers services to manage and program a space and its associated state. GAIA is similar to traditional operating systems in that it manages the tasks common to all applications built for physical spaces. Each space is self-contained, but may interact with other spaces or modular ontologies (e.g., via context mappings). GAIA provides core services, including

Copyright © 2006, Idea Group Inc. Copying or distributing in print or electronic forms without written permission of Idea Group Inc. is prohibited.

events, entity presence (e.g., devices, users and services), discovery, and naming. By specifying well-defined interfaces to services, applications may be built in a generic way so that they are able to run in arbitrary active spaces. The core services are started through a bootstrap protocol that starts the GAIA infrastructure. Finally, GAIA allows application developers to specify different behaviours of their applications for different contexts. The use of contextualised ontologies makes easier for developers to specify context-sensitive behaviour.

One of the main uses of ontologies is that it allows developers to define all the terms that can be used in the environment. Ontological engineering allows the attachment of precise semantics to various terms and the clear definition of the relationships between different terms. Hence, it prevents semantic ambiguities where different entities in the environment have different ideas from what a particular term means. Different entities in the environment can refer to the ontology to get a definition of a term, in case they are not sure (e.g., due to the imprecise conceptual modelling or the nature of the contextual information). Furthermore, developers employ ontologies for describing both entities (e.g., concepts) and contextual information (domain specific conceptual modelling). In this manner, they can determine whether these descriptions are valid with respect to the schema defined by the ontology. When a new entity is introduced into the system, its description can be checked against the existing ontology to see whether it is satisfiable (Roman et al., 2002). If the description is not consistent with the concepts described in the ontology, then the description is faulty. Moreover, ontologies are used to provide semantic discovery of objects and services. Conventional object registries provide a limited capability for object discovery and the existing discovery protocols support limited ability to spontaneously discover entities on a network. However, in PCEs it must be possible to discover only the "relevant" entities, without knowing, in advance, what will be relevant (Amann et al., 2003). This process has also been termed "matchmaking". Finally, a key feature of applications in PCEs is that they are context-aware, i.e., they are able to obtain the current context and adapt their behaviour to different situations.

Conclusion and Future Work

The vision of pervasive computing introduces structural issues, as in a typical ubiquitous environment there will be numerous devices with computing capabilities that need to interoperate. Therefore, we cannot rely on shared a priori knowledge based on commonly accepted standards to resolve such issues. Software components must adapt to their environment at runtime to integrate

Copyright © 2006, Idea Group Inc. Copying or distributing in print or electronic forms without written permission of Idea Group Inc. is prohibited.

their functionality with other software components seamlessly. We argued that the concept of Semantic Web, by the means of ontologies, could prove to be a satisfactory solution to the aforementioned problem, as it enables the system and the software providers to deal with semantically enriched specifications of services and semantically enriched data. "Models" based on ontologies may be used to provide objects, and mainly contextual information, with semantics. The context representation is thought to be of the utmost importance, as context knowledge and manipulation enables a system to be context aware and proactive and to employ more effective and efficient decision-making mechanisms. In addition, profile and policy management could be based on ontology models, in order to provide maximum flexibility for describing entities and rules regarding resource usage and manipulation.

Moreover, we surveyed the research that has been carried out regarding ontologies and tried to map the key findings to the requirements of a PCE. We have discussed efforts dealing with the management of multiple, overlapping, and evolving ontologies and the semantic mappings among them. We, also, depicted how ontologies can help in searching mechanisms and in profiling of users and data sources. Furthermore, due to the diverse nature of PCEs, the use of interoperability standards is not a sufficiently scalable approach. Hence, we showed the important role that ontologies could play in achieving service interoperability, a key requirement in PCEs, and depicted how one may use ontologies in the deployment of PCE.

Research on the association of ontologies and PCEs is rather restricted. Mainly the artificial intelligence community is carrying out the research regarding ontologies, whilst the research regarding pervasive computing is mostly conducted by the network community. Thus, the need of collaboration between the two communities is clearly visible. Nevertheless, a significant trend is identified. More and more researchers acknowledge the necessity of incorporating ontologies into the development of systems for supporting PCE. In this way, the emerging standard for Web ontologies, such as the (OWL), may prove to be a significant factor in the road ahead.

However, there are several future research directions that should be considered. Description logics may perform well in defining domain vocabularies and concepts but lacks in describing modal entities, such as temporal entities, vague knowledge, and probabilistic knowledge. Several extensions have been proposed (Baader et al., 2003) in that direction. These extensions include the following: the ability to predefine specific predicates over concrete domains (e.g., the greater-than or less-than operators over the domain of integers), the use of modal extensions in order to define concepts like belief and obligation, temporal extensions for describing time dependence, representation of uncertain and vague knowledge (probabilistic and fuzzy logic extensions) and nonmonotonic

Copyright © 2006, Idea Group Inc. Copying or distributing in print or electronic forms without written permission of Idea Group Inc. is prohibited.

reasoning. These extensions should be incorporated into the Semantic Web standards in order to achieve maximum expressiveness. Last but not least, collaboration between various scientific communities should be employed, so as to see the Semantic Web reach its full potential, pervading in everyday life.

References

Aggarwal, R., Verma, K., Miller, J., & Milnor, W. (2004). *Constraint driven Web service composition in METEOR-S.* University of Georgia, LSDIS Lab.

Amann, P., & Quirchmayr, G. (2003). Foundation of a framework to support knowledge management in the field of context-aware and pervasive computing. *Proceedings of the Australasian Information Security Workshop Conference on ACSW03* (Vol. 21), Adelaide, Australia.

Anyanwu, K., & Sheth, A. (2003). ρ-Queries: Enabling querying for semantic associations on the Semantic Web. *Proceedings of the 12th International Conference on World Wide Web,* Budapest, Hungary.

Artale, A., & Fraconi, E. (1998). A temporal description logic for reasoning about actions and plans. *Journal of Artificial Intelligence Research, 9,* 463-506.

Baader, F., Kusters, R., & Wolter, F. (2003). Extensions to description logics. In *The description logics handbook* (pp. 226-268). Cambridge University Press.

Beckett, D. (2004). *RDF/XML syntax specification. Resource Description Framework.* W3C Recommendation.

Berners-Lee, T. (2001). Services and semantics: Web architecture. *The 10th International World Wide Web Conference Hong Kong.* W3C Document. Retrieved from http://www.w3.org/2001/04/30-tbl

Boley, H. (2001). *RuleML in RDF. Version 0.2.*

Bouquet, P., Dona, A., Serafini, L., & Zanobini, S. (2002). ConTeXtualized local ontology specification via CTXML. Supported by the project EDAMOK Enabling Distributed and Autonomous Management of Knowledge, American Association for Artificial Intelligence.

Brickley, D., & Guha, R. (2004). *RDF vocabulary description language 1.0: RDF Schema.* W3C Recommendation.

Chalupsky, H. (2000). OntoMorph: A translation system for symbolic knowledge. *Proceedings of 7th International Conference on Principles of*

Copyright © 2006, Idea Group Inc. Copying or distributing in print or electronic forms without written permission of Idea Group Inc. is prohibited.

Knowledge Representation and Reasoning (KR2000), Breckenridge, CO.

Chen, H., Finin, T., & Anupam, J. (2004). An ontology for context-aware pervasive computing environments. *Special Issue on Ontologies for Distributed Systems, Knowledge Engineering Review.*

Ding, Y., & Foo, S. (2002). Ontology research and development. Part 2 — A review of ontology mapping and evolving. *Journal of Information Science, 28*(5), 375-388.

Doan, A., Madhavan, J., Dhamankar, R., Domingos, P., & Halevy, A. (2003). Learning to match ontologies on the Semantic Web. *The VLDB Journal — The International Journal on Very Large Data Bases, 12*(4), 303-319.

Elgesem, D. (1997). The modal logic of agency. *Nordic Journal of Philosophical Logic, 2*(2), 1-46.

Fensel, D., & Bussler, C. (2002). The Web service modeling framework WSMF. *Electronic Commerce: Research and Applications, 1,* 113-137.

Fernández, M., Malhotra, A., Marsh, J., Nagy, M., & Walsh, N. (2005). *Xquery 1.0 and XPath 2.0 Data Model.* W3C Working Draft 4.

Fowler, J., Nodine, M., Perry, B., & Bargmeyer, B. (1999). Agent based semantic interoperability in Infosleuth. *Sigmod Record, 28.*

Freund, Y., & Schapire (1996). Experiments with a new boosting algorithm. *Proceedings of the 13th International Conference on Machine Learning.*

Friedman, E. J. (2005). *Jess™, The Rule Engine for the Java™ Platform.* Version 7.0a6. Sandia National Laboratories.

Gibbins, N., Harris, S., & Shadbolt, N. (2003). Agent-based Semantic Web Services. *Proceedings of the 12th International Conference on World Wide Web,* Budapest, Hungary.

Gómez, P., Cabero, G., & Penín, L. (2004). *ODE SWS: A Semantic Web service development tool. Version 1.0.*

Guarino, N. (1995). Formal ontology, conceptual analysis and knowledge representation. *International Journal Human-Computer Studies, 43*(5), 625-640.

Guarino, N., Masolo, C., & Vetere, G. (1999, May/June). OntoSeek: Content based access to the Web. *IEEE Intelligent Systems,* 70-80.

Guarino, N., & Welty, C. (2000). *Towards a methodology for ontology-based model engineering.* ECOOP-2000 Workshop on Model Engineering, Cannes, France.

Copyright © 2006, Idea Group Inc. Copying or distributing in print or electronic forms without written permission of Idea Group Inc. is prohibited.

Haarslev, V., & Möller, R. (1999). RACER, A Semantic Web inference engine.

Halevy, A. Y., Halevy, Z. G., Ives, P. M., Tatarinov, I. (2003). Piazza: Data management infrastructure for Semantic Web Applications. *Proceedings of the 12th International Conference on World Wide Web*, Budapest, Hungary.

Horrocks, I., Patel-Schneider, P., Boley, H., Tabet, S., Grosof, B., & Dean, M. (2003). *SWRL: A Semantic Web Rule Language Combining OWL and RuleML. Version 0.5.*

Khan, L., McLeod, D., & Hovy, E. (2004). Retrieval effectiveness of an ontology-based model for information selection. *The VLDB Journal — The International Journal on Very Large Data Bases, 13*(1), 71-85.

Kibler, D., & Albert, M. (1991). Instance-based learning algorithms. *Machine Learning, 6*, 37-66.

Lei, L., & Horrocks, I. (2003). A software framework for matchmaking based on Semantic Web technology. *Proceedings of the 12th International Conference on World Wide Web*, Budapest, Hungary.

Maedche, A., & Zacharias, V. (2001). *Clustering ontology-based metadata in the Semantic Web.* Supported by the project HRMore project, FZI Research Center for Information Technologies, University of Karlsruhe, Germany.

Maedche, A., Motik, B., & Stojanovic, L. (2003). Managing multiple and distributed ontologies on the Semantic Web. *The VLDB Journal — The International Journal on Very Large Data Bases, 12*(4), 286-302.

Martin, D. (2003). OWL-S 1.0 Release. OWL-based Web Service Ontology.

McCallum, A., Nigam, N., Rennie, K., & Seymore, J. (2000). Automating the construction of Internet portals with machine learning. *Information Retrieval, 3*(2), 127-163.

McGuinness, D., & van Harmelen, F. (2004). *OWL Web Ontology Language overview.* W3C Recommendation.

McIlraith S., Son, T., & Zeng, H. (2001). *Semantic Web services.* 1094-7167/ 01/2001. IEEE Intelligent Systems.

Medjahed B., Bouguettaya, A., & Elmagarnid, A. (2003). Composing Web services on the Semantic Web. *The VLDB Journal — The International Journal on Very Large Data Bases, 12*(4), 333-351.

Middleton, S., Shadbolt, N., & De Roure, D. (2004). Capturing interest through inference and visualization: Ontological user profiling in recommender systems. *ACM Transactions on Information Systems (TOIS), 22*(1).

Copyright © 2006, Idea Group Inc. Copying or distributing in print or electronic forms without written permission of Idea Group Inc. is prohibited.

Milojicic, D., Kalogeraki, S., Lukose, V., Nagaraja, Pruyne, K, Richard, J., et al. (2002). *Peer-to-peer computing.* HP Laboratories Technical Report.

Nakajima, T. (2003). Pervasive servers: A framework for creating a society of appliances. *Personal and Ubiquitous Computing, 7*(3), 182-188.

Neches, R., Fikes, R., Finin, R., Gruber, T., Senator, T., & Swartout, W. (1991). Enabling technology for knowledge sharing. *AI Magazine, 12,* 36-56.

Nejdl, W., Wolpers, M., Siberski, W., Schmitz, C., Schlosser, M., Brunkhorst, I., et al. (2003). Super-peer-based routing and clustering strategies for RDF-based peer-to-peer network. *Proceedings of the 12th International Conference on World Wide Web,* Budapest, Hungary.

Noy, N., & McGuinness, D. (2001). *Development 101: A guide to creating your first ontology.* Stanford Knowledge Systems Laboratory Technical Report KSL-01-05 and SMI-2001-0880.

Orfali, R., & Harkey, D. (1998). *Client/server programming with Java and CORBA.* Wiley.

O'Sullivan, D., & Lewis, D. (2003). Semantically driven service interoperability for pervasive computing. *Proceedings of the 3rd ACM International Workshop on Data Engineering for Wireless and Mobile Access,* San Diego, CA.

Parkhomenko, O., Lee, Y., & Park, E. (2003). Ontology-driven peer profiling in peer-to-peer enabled Semantic Web. *Proceedings of the 12th International Conference on Information and Knowledge Management,* New Orleans, LA.

Ranganathan, A., McGrath, R., Campbell, R., & Mickunas, M. (2003). Ontologies in a pervasive computing environment. *Proceedings of the IJCAI-03 Workshop on Ontologies and Distributed Systems (ODS 2003),* Acapulco, Mexico.

Ranganathan-GAIA. (2003). An infrastructure for context-awareness based on first order logic. *Personal and Ubiquitous Computing, 7*(6), 353-364.

Rapoza. (2000). *DAML could take search to a new level.* PC Week Labs.

Rector. (2003). Modularization of domain ontologies implemented in description logics and related formalisms including OWL. *Proceedings of the International Conference on Knowledge Capture,* Sanibel Island, FL.

Rector, A., & Rogers, J. (1999). *Ontological issues in using a description logic to represent medical concepts: Experience from GALEN.* Appeared at IMIA WG6 Workshop: Terminology and Natural Language in Medicine, Phoenix, AZ.

Copyright © 2006, Idea Group Inc. Copying or distributing in print or electronic forms without written permission of Idea Group Inc. is prohibited.

Roman, M., Hess, C., Cerqueira, R., Ranganathan, A., Cambell, R., & Nahrstedt, K. (2002). A middleware infrastructure for active spaces. *IEEE Personal Communications, 1*(4), 74-83

Stracia, U. (1998). A fuzzy description logic. *Proceedings of the 15th National/10th Conference on Artificial Intelligence/Innovative Applications of Artificial Intelligence.*

The Cambridge Dictionary. (1980). *The Cambridge dictionary of English.* Daniel Lawrence Corporation.

Wang. (2003). *The context gateway: A pervasive computing infrastructure for context aware services.* Report submitted to School of Computing, National University of Singapore & Context-Aware Dept, Institute for Infocomm Research.

Wang, B., Mckay, R., Abbass, H., & Barlow, M. (2003). A comparative study for domain ontology guided feature extraction. *Proceedings of the 26th Australasian Computer Science Conference on Conference in Research and Practice in Information Technology,* Adelaide, Austria.

Copyright © 2006, Idea Group Inc. Copying or distributing in print or electronic forms without written permission of Idea Group Inc. is prohibited.

Chapter XII

Description of Policies Enriched by Semantics for Security Management

Félix J. García Clemente, University of Murcia, Spain

Gregorio Martínez Pérez, University of Murcia, Spain

Juan A. Botía Blaya, University of Murcia, Spain

Antonio F. Gómez Skarmeta, University of Murcia, Spain

Abstract

Policies, which usually govern the behavior of networking services (e.g., security, QoS, mobility, etc.) are becoming an increasingly popular approach for the dynamic regulation of Web information systems. By appropriately managing policies, a system can be continuously adjusted to accommodate variations in externally imposed constraints and environmental conditions. The adoption of a policy-based approach for controlling a system requires an appropriate policy representation regarding both syntax and semantics, and the design and development of a policy management framework. In the context of the Web, the use of languages enriched with semantics has been limited primarily to represent Web content and services. However the

Copyright © 2006, Idea Group Inc. Copying or distributing in print or electronic forms without written permission of Idea Group Inc. is prohibited.

capabilities of these languages, coupled with the availability of tools to manipulate them, make them well suited for many other kinds of applications, as policy representation and management. In this chapter, we present an evaluation of the ongoing efforts to use ontological (Semantic Web) languages to represent policies for distributed systems.

Introduction

The heterogeneity and complexity of computer systems is increasing constantly, but their management techniques are not changing and so system configuration processes are getting more complicated and error-prone. Therefore, there is a clear need for standardized mechanisms to manage advanced services and applications. Policy-based management (PBM) frameworks (Verma, 2000; Kosiur, 2001; Strassner, 2003) define languages, mechanisms and tools through which computer systems can be managed dynamically and homogeneously.

One of the main goals of policy-based management is to enable service and application control and management on a high abstraction level. The administrator specifies rules that describe domain-wide policies independent of the implementation of the particular service and/or application. It is then the policy management architecture that provides support to transform and distribute the policies to each node and thus to enforce a consistent configuration in all involved elements, which is a prerequisite for achieving end-to-end security services, or consistent access control configuration in different Web servers, for example.

The scope of policy-based management is increasingly going beyond its traditional applications in significant ways. The main functions of policy management architectures are:

- **Enforcement:** to implement a desired policy state through a set of management commands.
- **Monitoring:** ongoing active or passive examination of the information system, its services and applications for checking its status and whether policies are being satisfied or not.
- **Decision-taking:** to compare the current state of the communication system to a desired state described by a policy (or a set of them) and to decide how the desired state can be achieved or maintained.

In the information systems security field, a policy (i.e., security policy) can be defined as a set of rules and practices describing how an organization manages,

Copyright © 2006, Idea Group Inc. Copying or distributing in print or electronic forms without written permission of Idea Group Inc. is prohibited.

protects, and distributes sensitive information. The research community, the industry, and standardization bodies have proposed different secure policy specification languages (Martínez Pérez, 2005) that range from formal policy languages that can be processed and interpreted easily and directly by a computer, to rule-based policy notation using *if-then* rules to express the mandatory behavior of the target system, and to the representation of policies as entries in a table consisting of multiple attributes. There are also ongoing standardization efforts toward common policy information models and frameworks such as CIM (Common Information Model) from the DMTF (Distributed Management Task Force, 2005).

In the Web services world, standards for SOAP-based message security and XML-based languages for access control are now appearing, as it is the case of XACML (OASIS, 2004). However the immaturity of the current tools along with the limited scope and total absence of explicit semantics of new languages make them less than-ideal candidates for sophisticated Web-based services or applications.

Ontological languages like OWL (Connolly et al., 2003) and others can be used to incorporate semantic expressiveness into the management information specifications and some reasoning capabilities which definitely would help in handling the management tasks aforementioned (i.e., enforcement, monitoring, and decision-taking). For example, in this kind of systems it is an important improvement to allow simple operations on different policies like testing equality, inclusion, equivalency and so. Ontological languages allow this kind of operations for the entities expressed with them.

This chapter provides some initial concepts related with policy management and their formal representation complemented with a case study and specific examples in the field of Web systems security where the reader can see the requirements regarding syntax and semantics that any policy representation should satisfy.

This analysis also provides a comparative study of main semantic security policy specification languages against a set of "traditional" non-semantic policy languages, such as Ponder (Damianou et al., 2001), XACML and XML/CIM. Some of the criteria used for this analysis include their ability to express semantic information, the representation technique they use, and the concepts they are able to express.

Then, this chapter examines the main existing policy-related technologies based on specifications with semantics. In particular, it presents the results of our in-depth analysis of four candidates: KaoS (Uszok et al., 2003), RuleML (The Rule Markup Initiative, August 2004), Rei (Kagal et al., 2003) and SWRL (The Rule Markup Initiative, May 2004) along with their associated schemas and ontology specifications. The analysis focuses on the expressiveness of each language, and

Copyright © 2006, Idea Group Inc. Copying or distributing in print or electronic forms without written permission of Idea Group Inc. is prohibited.

is presented along several dimensions and summarized in a comparison table. Moreover, the chapter shows the implementation of our case study by the four languages analyzed and, to some extent, a criticism of each of the resulting specifications.

The state-of-the-art vision is also complemented with some future trends in semantic-based policy languages, mainly the open issues that should be resolved as a prerequisite to widespread adoption. Some new designs and developments from the research community considered of relevance for the reader are also being presented.

Background and Related Work

This section is intended to provide the reader with an overview of the main elements and technologies existing in semantic security policy specifications and frameworks from both a theoretical and a practical perspective. It starts by describing some useful concepts around policy management paradigm and formal representation techniques. Then, it presents a case study related with Business-to-Consumer (B2C) Web-based e-commerce systems, which will be used throughout this chapter to illustrate different concepts like, for example, how different security policy languages enriched by semantics express the same high-level policy. Next, it describes two of the main policy languages in use today. Finally, this section provides a comparative study between these two "traditional" non-semantic policy languages and other semantic security policy languages; it is based on our own research work, but also incorporating the view of others. This study is intended to help the reader to understand the need of semantic policy languages.

Policy-Based Management: A Brief Overview

Introduction

Policy rules define in abstract terms a desired behavior. They are stored and interpreted by the policy framework, which provides a heterogeneous set of components and mechanisms that are able to represent, distribute, and manage policies in an unambiguous, interoperable manner, thus providing a consistent behavior in all affected policy enforcement points (i.e., entities where the policy decisions are actually enforced when the policy rule conditions evaluate to "true").

Copyright © 2006, Idea Group Inc. Copying or distributing in print or electronic forms without written permission of Idea Group Inc. is prohibited.

Security policies can be defined to perform a wide variety of actions, from IPsec/IKE management (example of network security policy) to access control over a Web service (example of application-level policy). To cover this wide range of security policies, this chapter aims to examine the current state of policy engines and policy languages, and how they can be applied to both Web security and information systems security to protect information at several levels. Some future directions of interest for the reader are provided as well.

Requirements of a Policy Language and Policy Framework

Administrators use policy languages assuring that the representation of policies will be unambiguous and verifiable. Other important requirements of any policy language are:

- **Clear and well-defined semantics.** The semantics of a policy language can be considered as well-defined if the meaning of a policy written in this language is independent of its particular implementation.

- **Flexibility and extensibility.** A policy language has to be flexible enough to allow new policy information to be expressed, and extensible enough to allow new semantics to be added in future versions of this language.

- **Interoperability with other languages.** There are usually several languages that can be used in different domains to express similar policies, and interoperability is a must to allow different services or applications from these different domains to communicate with each other according to the behavior stated in these policies.

Once the policy has been defined for a given administrative domain, a policy management architecture is needed to transfer, store, and enforce this policy in that domain. The main requirements for such policy management architecture are:

- **Well-defined model.** Policy architectures need to have a well-defined model independent of the particular implementation in use. In it, the interfaces between the components need to be clear and well-defined.

- **Flexibility and definition of abstractions to manage a wide variety of device types.** The system architecture should be flexible enough to allow addition of new types of devices with minimal updates and recoding of existing management components.

Copyright © 2006, Idea Group Inc. Copying or distributing in print or electronic forms without written permission of Idea Group Inc. is prohibited.

- **Interoperability with other architectures (inter-domain).** The system should be able to interoperate with other architectures that may exist in other administrative domains.

- **Conflict detection.** It has to be able to check that a given policy does not conflict with any other existing policy.

- **Scalability.** It should maintain quality performance under an increased system load.

Introduction to Formal Representation Techniques

The Concept of Ontology

An ontology consists on a number of terms precisely defined. It is actually a vocabulary which can be shared between humans and/or applications. The concepts in the vocabulary are usually arranged as a hierarchy of elements. Moreover, each element, called a concept, comes with a set of properties and another set of instances pertaining to the concept. A simple definition of what is really an ontology can be found in Gruber et al. (1993) and says "an ontology is an explicit specification of a conceptualization". From the point of view of knowledge representation in computers, what is represented is what exists, in a declarative way. And what is represented is done by using an abstract model of a concrete phenomenon, in where the model represents the most relevant concepts of the phenomenon.

Traditionally, ontologies have been used for knowledge sharing and reuse in the context of knowledge-based systems like distributed problem solvers (Avouris et al., 1992) and more recently multi-agent systems (Wooldridge, 2001). In turn, this kind of systems has been used for intelligent information integration, cooperative information systems, information gathering, electronic commerce, and knowledge management.

In the classification of the different ontologies introduced by Fensel et al. (2004) and using as a dimension for classification the kind of phenomenon they try to represent, we are interested, in the context of this work, in the *domain ontology*. A domain ontology represents the knowledge extracted from a particular application domain like the one we are working on this chapter: policy management for securing Web services and environments.

Copyright © 2006, Idea Group Inc. Copying or distributing in print or electronic forms without written permission of Idea Group Inc. is prohibited.

The Role of Formal Logics in Semantic Web Technologies

Ontology technology finds its roots in traditional logic. Formal logic provides us with two important things used in ontological research:

- Languages for knowledge representation, and
- Functional models for the implementation of reasoning processes.

Knowledge representation is half of the story with respect to ontology technology. The other part of the story gets to do with reasoning. With a proper reasoning mechanism, we can use an ontological specification to infer interesting details about our model of the world, in other words, deduce or discover properties or relations between concepts that we did not explicitly define.

There are three different and important logics for representing knowledge. Two basic logics: propositional and first order logic. The third one, description logic, is considered as an extension of first order logic, and nowadays is seen as the most suitable to manage ontologies in the context of the Semantic Web.

They all are distinct kind of logics for representing distinct kind of conceptualizations. If we talk about semantics in propositional logic, it has to be understood in terms of truth values. The semantic of each well formed formula depends of how the truth values are evolved with using the above mentioned connectives. With respect to reasoning, in the context of propositional logic it is of deductive nature. In this kind of process, there are some premises which in turn are well-formed formulae, and a conclusion. In order to deduce the conclusion from the premises, these must imply the conclusion.

Propositional logic compound a very poor expressiveness language as it does not allow for properties-based or general-relations-based reasoning. As general knowledge can not be represented, no general reasoning can be performed. As the reader may have supposed, this kind of logic is not suitable for representing policy specifications. For a more detailed explanation, please see Russell et al. (1995) and Genesereth et al. (1986).

First order predicate logic is a more advanced kind of logic, in the sense that the basic representational element is a predicate. With a predicate, we can represent properties of an entity or relations between different entities. This is accomplished by using variables inside predicates for representing objects of individuals. Moreover, we can specify the scope of variables by using quantifiers like the universal quantifier (i.e., for all) and the existential one (i.e., it exists at least one).

If semantics in propositional logic were based on a truth value for the propositions, in first order predicate logic we have the interpretation for giving meaning

Copyright © 2006, Idea Group Inc. Copying or distributing in print or electronic forms without written permission of Idea Group Inc. is prohibited.

to expressions. Talking about reasoning, predicate logic is consistent with the appropriate inference rules and it is semi-decidable, in other words, if a logic expression can be deduced from a set of other logic expressions, then a demonstration of the first one exist but you can not always find the demonstration that a logic expression can not be deduced from a set of logic expressions.

A more advanced logic is description logic (DL). This kind of logic uses concepts (or classes) instead of predicates. If predicate logic is oriented to truth values of true and false, descriptive logic is oriented to concepts and concept belonging relations. For example, we can represent the fact that someone has a child with the binary predicate:

$$\{x|(\exists y)(hasChild(x,y) \wedge Person(y))\}$$

but in descriptive logic the approach is slightly different as we use properties for that and represent it with:

$$\exists hasChild.Person.$$

Other of the interesting features of DL is the availability of a number of constructors to build more complex classes from basic and other complex classes.

Languages for Ontology Representation

Ontology can be, in principle, considered as independent of the language used to serialize it. As argued in Gruber et al. (1993), in systems where knowledge must be shared between different and independent entities, it is needed a common agreement at three different levels: format of languages for knowledge representation, knowledge sharing communication protocol and specification of the knowledge to be shared. In this way, the task of defining an ontology is totally decoupled of the task of using a concrete language for representing knowledge outside of the entity using it internally. That is why we can use languages like XML or RDF, which can be seen as not related to ontologies at all, for knowledge exchanging.

RDF results in an interesting approach because it offers a direct and easy way to state facts (W3C, 1999). On the other hand, RDFS (Brickley et al., 2004), the schema of RDF, organizes the modeling elements in a very convenient way through classes and properties. For example, you can define a hierarchy of

Copyright © 2006, Idea Group Inc. Copying or distributing in print or electronic forms without written permission of Idea Group Inc. is prohibited.

classes in terms of subsumption of classes inside other. You can also add semantics by means of adding ranges, domains, and cardinality to properties. OWL (Connolly et al., 2003) is the cutting edge language nowadays for dealing with semantics, more specifically in the context of the Web. With OWL we will be able of doing things like:

- State that two different kinds of policies are disjoint classes.
- Declare a new policy as the inverse for other policy.
- State new policies by adding restrictions to the properties of other policies.

A Case Study

Once we have described the basic concepts of policy management and formal representation techniques and the requirements of policy languages and frameworks, this background analysis will be complemented with one case study, which will be used throughout the chapter to show the advantages, and also some of the current open issues of using policy specification languages with semantics.

Business-to-consumer (B2C) e-commerce systems are an example of Web Information System (WIS) where security is a fundamental aspect to be considered. A secure e-commerce scenario requires transmission and reception of information over the Internet such that:

- It is accessible to the sender and the receiver only (privacy/confidentiality).
- It cannot be changed during transmission (integrity).
- The receiver is sure it came from the right sender (source authenticity).
- The sender can be sure the receiver is genuine (authenticity of the destination).
- The sender cannot deny he/she sent it (non-repudiation).

As stated before, security policies can be specified at different levels of abstraction. The process starts with the definition of a business security policy. This can be the case of the next authorization security policy, which is defined in natural language: *"Permit the access to the e-payment service, if the user is in the group of customers registered for this service."*

Next, the security policy is usually expressed by a policy administrator as a set of IF-THEN policy rules as, for example:

Copyright © 2006, Idea Group Inc. Copying or distributing in print or electronic forms without written permission of Idea Group Inc. is prohibited.

IF ((<Requester> is member of Payment Customers) AND (<Server> is member of Payment Servers)) THEN (<Requester> granted access to <Server>)

Policy languages to be analyzed in this chapter (both semantic and non=semantic) will be used to represent this specific policy, so the reader will be able to understand their descriptions and compare them.

Non-Semantic Security Policy Languages

This section describes two of the most relevant "traditional" non=semantic policy languages in use nowadays: Ponder and XACML (eXtensible Access Control Markup Language), together with their main advantages and disadvantages.

Ponder

Ponder (Damianou et al., 2001) is a declarative, object-oriented language developed for specifying management and security policies. Ponder permits to express authorizations, obligations (stating that a user must or must not execute or be allowed to execute an action on a target element), information filtering (used to transform input or output parameters in an interaction), refrain policies (which define actions that subjects must not perform on target objects), and delegation policies (which define what authorizations can be delegated to whom).

Figure 1 presents an example of policy expressed in Ponder using our case study.

It shows how to express with an authorization policy the business level security policy and IF-THEN rules defined in our case study. The positive authorization policy defines that the subject PayCustomer is permitted to access to the target PayServer.

Ponder can describe any rule to constrain the behavior of components in a simple and declarative way. However, it does not take care of the description of the content of the policy (e.g., description of the specified components, the system, etc.). The adoption of a Semantic Web language can clearly overcome this limitation.

Copyright © 2006, Idea Group Inc. Copying or distributing in print or electronic forms without written permission of Idea Group Inc. is prohibited.

Figure 1. Example of policy representation in PONDER

```
inst auth+ PaymentAuthPolicy1 {
            subject s = people/PayCustomer ;
            target t = servers/PayServer ;
            action t.access (s);
        }
```

XACML

The eXtensible Access Control Markup Language (XACML) (OASIS, 2004) describes both an access control policy language and a request/response language. The policy language provides a common means to express subject-target-action-condition access control policies and the request/response language expresses queries about whether a particular access should be allowed (requests) and describes answers to those queries (responses).

Figure 2 presents an example of policy expressed in XACML using our case study.

From our point of view, the main failing of XACML is that the policy is rather verbose and not really aimed at human interpretation. In addition, the language model does not include any way of grouping policies.

Discussion

Now the basic concepts of policy-based management and formal representation techniques have been presented, a case study for WIS security has been introduced, and two of the main "traditional" non-semantic languages have been introduced and generally compared with semantic languages, this section shows how the administrator models business security policies using a policy representation based on ontology to create a formal representation of the business policy, and which are the frameworks allowing the management of such policies.

Although many semantic security policy specifications exist, we have selected four of them as they are considered nowadays as the most promising options: KAoS, RuleML, SWRL, and Rei. They are now described in detail, while a comparison of their main pros and cons is presented at the end of this section.

Copyright © 2006, Idea Group Inc. Copying or distributing in print or electronic forms without written permission of Idea Group Inc. is prohibited.

Figure 2. Example of policy representation in XACML

```
<Policy PolicyId="PaymentAuthPolicy1">
  <Target>
     <Subjects><AnySubject/></Subjects>
     <Resources> <Resource>
         <ResourceMatch MatchId="function:anyURI-equal">
          <AttributeValue>
                    http://payserver.ourcompany.com
          </AttributeValue>
          <ResourceAttributeDesignator/>
          </ResourceMatch>
     </Resource> </Resources>
     <Actions><AnyAction/></Actions>
  </Target>
  <Rule RuleId="ReadRule" Effect="Permit">
     <Target>
          <Subjects><AnySubject/></Subjects>
          <Resources><AnyResource/></Resources>
          <Actions> <Action>
           <ActionMatch MatchId="function:string-equal">
             <AttributeValue>access</AttributeValue>
             <ActionAttributeDesignator/>
           </ActionMatch>
          </Action></Actions>
     </Target>
     <Condition FunctionId="function:string-equal">
          <Apply FunctionId="function:string-one-and-only">
          <SubjectAttributeDesignator AttributeId="group"/>
          </Apply>
          <AttributeValue>PayCustomer</AttributeValue>
     </Condition>
  </Rule>
</Policy>
```

Semantic Policy Languages and Frameworks for Managing System Security

KAoS

KAoS (Uszok et al., 2003) is a collection of services and tools that allow for the specification, management, conflict resolution, and enforcement of policies within domains describing organizations of human, agent, and other computational actors. While initially oriented to the dynamic and complex requirements of software agent applications, KAoS services are also being extended to work equally well with both general-purpose grid computing and Web service environments.

Copyright © 2006, Idea Group Inc. Copying or distributing in print or electronic forms without written permission of Idea Group Inc. is prohibited.

KAoS uses ontology concepts encoded in OWL to build policies. The KAoS Policy Service distinguishes between authorization policies (i.e., constraints that permit or forbid some action) and obligation policies (i.e., constraints that require some action to be performed when a state- or event-based trigger occurs, or else serve to waive such a requirement). The applicability of the policy is defined by a set of conditions or situations whose definition can contain components specifying required history, state and currently undertaken actions. In the case of the obligation policy the obligated action can be annotated with different constraints restricting possibilities of its fulfillment.

The current version of the KAoS Policy Ontologies (KPO) defines basic ontologies for actions, conditions, actors, various entities related to actions, and policies, as depicted in Figure 3. It is expected that for a given application, the ontologies will be further extended with additional classes, individuals, and rules.

Figure 4 shows an example of the type of policy that administrators can specify using KAoS. It is related with the case study described earlier.

KAoS defines a Policy Framework (see Figure 5) that includes the following functionality:

- Creating/editing of policies using KAoS Policy Administration Tool (KPAT). KPAT implements a graphical user interface to policy and domain management functionality.

- Storing, deconflicting and querying policies using KaoS Directory Service.

- Distribution of policies to Guard, which acts as a policy decision point.

- Policy enforcement/disclosure mechanism, i.e., finding out which policies apply to a given situation.

Figure 3. KAoS policy ontology

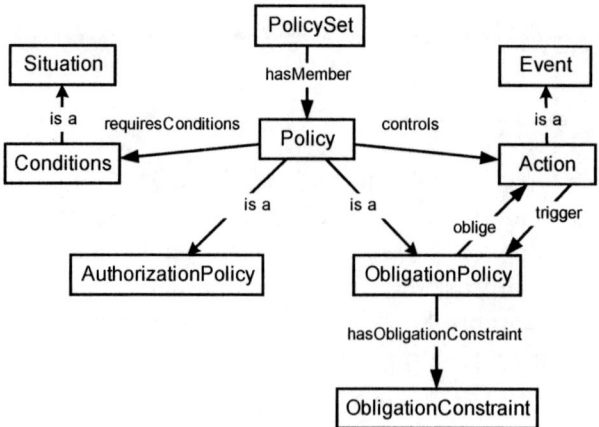

Copyright © 2006, Idea Group Inc. Copying or distributing in print or electronic forms without written permission of Idea Group Inc. is prohibited.

Figure 4. Example of policy representation in KAoS

```
<owl:Class rdf:ID="PaymentAuthAction">
<owl:intersectionOf rdf:parseType="owl:collection">
 <owl:Class rdf:about="&action;AccessAction"/>
 <owl:Restriction>
 <owl:onProperty rdf:resource="&action;#performedBy"/>
 <owl:toClass rdf:resource="&domains;MembersOfPayCustomer"/>
 </owl:Restriction>
 <owl:Restriction>
 <owl:onProperty rdf:resource="&action;#performedOn"/>
 <owl:toClass rdf:resource="&domains;MembersOfPayServer"/>
 </owl:Restriction>
</owl:intersectionOf>
</owl:Class>
<policy:PosAuthorizationPolicy rdf:ID="PaymentAuthPolicy1">
 <policy:controls rdf:ID="PaymentAuthAction"/>
 <policy:hasSiteOfEnforcement rdf:resource="#TargetSite"/>
 <policy:hasPriority>1</policy:hasPriority>
</policy:PosAuthorizationPolicy>
```

Figure 5. KAoS framework

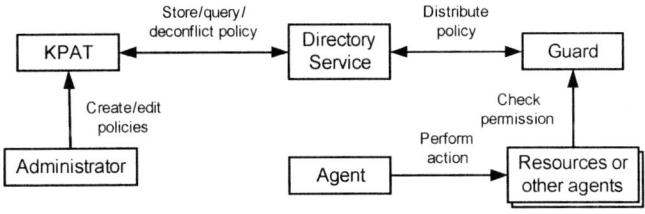

Every agent in the system is associated with a Guard. When an action is requested, the Guard is automatically queried to check whether the action is authorized based on the current policies and, if not, the action is prevented by various enforcement mechanisms. Policy enforcement requires the ability to monitor and intercept actions, and allow or disallow them based on a given set of policies. While the rest of the KAoS architecture is generic across different platforms, enforcement mechanisms are necessarily specific to the way the platform works.

The use of OWL as a policy representation enables runtime extensibility and adaptability of the system, as well as the ability to analyze policies relating to entities described at different levels of abstraction. The representation facilitates

Copyright © 2006, Idea Group Inc. Copying or distributing in print or electronic forms without written permission of Idea Group Inc. is prohibited.

Figure 6. RuleML family of sublanguages

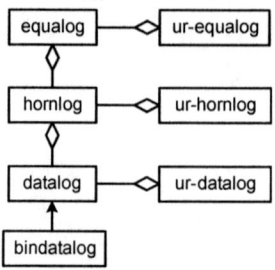

careful reasoning about policy disclosure, conflict detection, and harmonization about domain structure and concepts.

RuleML

The Rule Markup Language (RuleML) initiative defines a family of XML-based rule languages (The Rule Markup Initiative, 2004 August). The model of the RuleML family of sublanguages is shown in Figure 6.

For these families, XML Schemes are provided. The main RuleML sublanguages are described as follows:

- Datalog RuleML sublanguage is entirely composed of element modules: core, description, clause, boolean, atom, role, and data. Each module declares a set of XSD elements.

- Binary Datalog RuleML sublanguage redefines Datalog so that atoms are binary.

- Horn-Logic RuleML sublanguage adds the Horn elements module.

- Equational-Logic RuleML sublanguage adds the equation elements module to Hornlog.

- 'UR' Datalog RuleML sublanguage redefines datalog to permit href attributes. The same way is used to define the 'UR' Horn-Logic RuleML sublanguage and the 'UR' Equational-Logic RuleML sublanguage.

RuleML encompasses a hierarchy of rules, including reaction rules (event-condition-action rules), transformation rules (functional-equation rules), derivation rules (implicational-inference rules), also specialized to facts ("premise-less"

Copyright © 2006, Idea Group Inc. Copying or distributing in print or electronic forms without written permission of Idea Group Inc. is prohibited.

Figure 7. Graphical view of RuleML rules

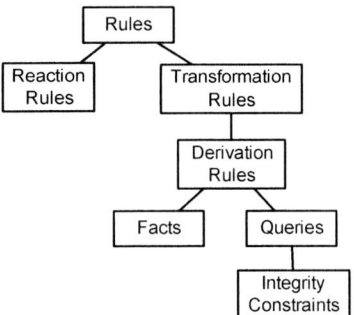

derivation rules) and queries ("conclusion-less" derivation rules), as well as integrity-constraints (consistency-maintenance rules). A graphical view of RuleML rules is a reduction tree rooted in general rules is showed in the Figure 7.

RuleML permits both forward (bottom-up) and backward (top-down) rules and facts in XML to describe constrains. We distinguish between the following facts/rules for policy representation:

- **Structural/organizational** facts and rules. These rules are used to encode domain specific ontologies.

- **Service definition** facts and rules, provided with links to the structural rules and facts.

- **Task-specific** rules and facts, provided by the service clients.

The benefit of categorizing the policy rules in this way is that an organization can utilize a common ontology that can be shared amongst services and service clients. The rules/facts (i.e., policies) inform to clients about the usage of services in achieving business objectives.

Let consider an example to illustrate the policy representation in RuleML with the ur-datalog sublanguage. It is related with the case study described earlier (see Figure 8).

One or more rule engines will be needed for executing RuleML rulebases (Friedman-Hill, 2004; Spencer, 2004). RuleML provides tools for deduction, rewriting, and further inferential-transformational tasks. However, the main motivation is to employ RuleML as a semantic interoperable vehicle for heterogeneous policy languages, standards, protocols, and mechanisms. RuleML

Copyright © 2006, Idea Group Inc. Copying or distributing in print or electronic forms without written permission of Idea Group Inc. is prohibited.

Figure 8. Example of policy representation in RuleML

```
<rulebase direction="backward">
<imp>
 <_head>
  <atom>
   <_opr href="#GrantedAccess"/>
   <var>requester</var>
   <var>server</var>
  </atom>
 </_head>
 <_body>
  <and>
  <atom>
   <_opr href="#isMember"/>
   <var>requester</var>
   <ind>PayCustomer</ind>
  </atom>
  <atom>
   <_opr href="#isMember"/>
   <var>server</var>
   <ind>PayServer</ind>
  </atom>
  </and>
 </_body>
</imp>
</rulebase>
```

provides intermediate markup syntax, with associated deep knowledge representation semantics for interchange between those languages, standards, and mechanisms. For interchange between policy languages/standards/protocols/mechanisms that are already XML-based, this can, for instance, be achieved using XSL transformations (XSLT), e.g., to translate into and then out of RuleML.

SWRL

SWRL, acronym of Semantic Web Rule Language (The Rule Markup Initiative, May 2004) is based on a combination of the OWL DL and OWL Lite sublanguages of the OWL with the Unary/Binary Datalog RuleML sublanguages.

Copyright © 2006, Idea Group Inc. Copying or distributing in print or electronic forms without written permission of Idea Group Inc. is prohibited.

SWRL extends the OWL abstract syntax to include a high-level abstract syntax for Horn-like rules. A model-theoretic semantics is given to provide the formal meaning for OWL ontologies including rules written in this abstract syntax.

In the same way that in RuleML, SWRL rules are used for policy representation. Atoms in SWRL rules can be of the form C(x), P(x,y), sameAs(x,y) or differentFrom(x,y), where C is an OWL description, P is an OWL property, and x,y are either variables, OWL individuals or OWL data values.

SWRL is defined by an XML syntax based on RuleML and the OWL XML Presentation Syntax. The rule syntax is illustrated with the following example (see Figure 9) related with the case study described earlier.

A useful restriction in the form of the rules is to limit antecedent and consequent classAtoms to be named classes, where the classes are defined purely in OWL.

Figure 9. Example of policy representation in SWRL

```
<ruleml:imp>
 <ruleml:_head>
        <swrlx:individualPropertyAtom
        swrlx:property="GrantedAccess">
                <ruleml:var>requester</ruleml:var>
                <ruleml:var>server</ruleml:var>
        </swrlx:individualPropertyAtom>
 </ruleml:_head>
 <ruleml:_body>
        <swrlx:classAtom>
                <owlx:Class owlx:name="User" />
                <ruleml:var>requester</ruleml:var>
        </swrlx:classAtom>
        <swrlx:classAtom>
                <owlx:Class owlx:name="Server" />
                <ruleml:var>server</ruleml:var>
        </swrlx:classAtom>
        <swrlx:individualPropertyAtom swrlx:property="Member">
                <ruleml:var>requester</ruleml:var>
                <owlx:Individual owlx:name="#PayCustomer" />
        </swrlx:individualPropertyAtom>
        <swrlx:individualPropertyAtom swrlx:property="Member">
                <ruleml:var>server</ruleml:var>
                <owlx:Individual owlx:name="#PayServer" />
        </swrlx:individualPropertyAtom>
 </ruleml:_body>
</ruleml:imp>
```

Copyright © 2006, Idea Group Inc. Copying or distributing in print or electronic forms without written permission of Idea Group Inc. is prohibited.

Adhering to this format makes it easier to translate rules to or from existing or future rule systems, including Prolog.

Rei

Rei (Kagal et al., 2003) is a policy framework that integrates support for policy specification, analysis, and reasoning. Its deontic-logic-based policy language allows users to express and represent the concepts of rights, prohibitions, obligations, and dispensations. In addition, Rei permits users to specify policies that are defined as rules associating an entity of a managed domain with its set of rights, prohibitions, obligations, and dispensations.

Rei provides a policy specification language in OWL-Lite that allows users to develop declarative policies over domain specific ontologies in RDF, DAML+OIL, and OWL.

A policy (see Figure 10 for the Rei Ontology) primarily includes a list of grantings and a context used to define the policy domain. A granting associates a set of constraints with a deontic object to form a policy rule. This allows reuse of deontic objects in different policies with different constraints and actors. A deontic object represents permissions, prohibitions, obligations, and dispensations over entities in the policy domain. It includes constructs for describing what action (or set of actions) the deontic is described over, who the potential actor (or set of actors) of the action is and under what conditions is the deontic object applicable.

Figure 10. Rei ontology

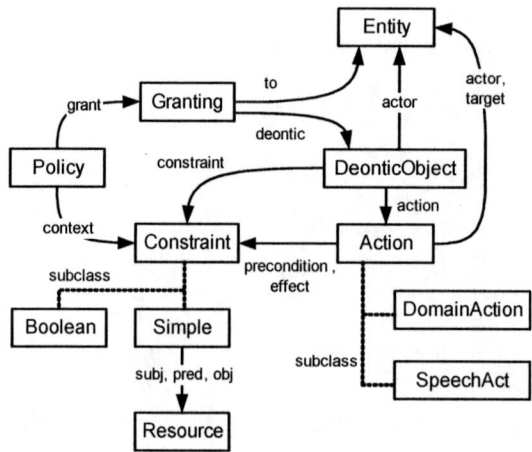

Copyright © 2006, Idea Group Inc. Copying or distributing in print or electronic forms without written permission of Idea Group Inc. is prohibited.

Actions are one of the most important components of Rei specifications as policies are described over possible actions in the domain. The domain actions describe application or domain specific actions, whereas the speech acts are primarily used for dynamic and remote policy management.

There are six subclasses of SpeechAct: Delegate, Revoke, Request, Cancel, Command, and Promise. A valid delegation leads to a new permission. Similarly, a revocation speech act nullifies an existing permission (whether policy based or delegation based) by causing a prohibition. An entity can request another entity for a permission, which if accepted causes a delegation, or to perform an action on its behalf, which if accepted causes an obligation. An entity can also cancel any previously made request, which leads to a revocation and/or a dispensation. A command causes an obligation on the recipient and the promise causes an obligation on the sender.

To enable dynamic conflict resolution, Rei also includes meta-policy specifications, namely setting the modality preference (negative over positive or vice versa) or stating the priority between rules within a policy or between policies themselves.

Figure 11 shows an example illustrating the policy representation in Rei. It is related with the case study described earlier.

Figure 11. Example of policy representation in Rei

```
<constraint:SimpleConstraint rdf:ID="IsPayCustomer"
  constraint:subject="#RequesterVar"
  constraint:predicate="&example;memberOf"
  constraint:object="&example;payCustomer"/>
<constraint:SimpleConstraint rdf:ID="IsPayServer"
  constraint:subject="#PayServerVar"
  constraint:predicate="&example;memberOf"
  constraint:object="&example;payServer"/>
<constraint:And rdf:ID="ArePayCustomerAndPayServer"
  constraint:first="#IsPayCustomer"
  constraint:second="#IsPayServer"/>
<deontic:Permission rdf:ID="PayServerPermission">
  <deontic:actor rdf:resource="#RequesterVar"/>
  <deontic:action rdf:resource="&example;access"/>
  <deontic:constraint
        rdf:resource="#ArePayCustomerAndPayServer"/>
</deontic:Permission>
<policy:Policy rdf:ID="PaymentAuthPolicy1">
  <policy:grants rdf:resource="#PayServerPermission"/>
</policy:Policy>
```

Copyright © 2006, Idea Group Inc. Copying or distributing in print or electronic forms without written permission of Idea Group Inc. is prohibited.

The Rei framework provides a policy engine that reasons about the policy specifications. The engine accepts policy specifications in both the Rei language and in RDFS, consistent with the Rei ontology. Specifically, the engine automatically translates the RDF specification into triplets of the form (subject, predicate, object). The engine also accepts additional domain-dependent information in any semantic language that can then be converted into this recognizable form of triplet. The engine allows queries according to the Prolog language about any policies, meta-policies, and domain dependent knowledge that have been loaded in its knowledge base.

The Rei framework does not provide an enforcement model. In fact, the policy engine has not been designed to enforce the policies but only to reason about them and answer to queries.

Comparative Analysis Between Semantic and Non-Semantic Security Policy Languages

A comparative analysis between PONDER, Rei, and KAoS is presented in Tonti et al. (2003). In this analysis, it is discussed the differences between PONDER, a non-semantic policy language, and Rei, KAoS that are based on OWL. Based on this work and our research, which includes an extensive analysis of XACML, RuleML, and SWRL, we can identify the next topics as the most relevant ones for the comparison:

- **Abstraction.** Levels of abstraction (high, medium, and low) that a policy specification can represent.

- **Extensibility.** The capability to include new concepts and elements of the system in the representation.

- **Representability.** Types of environments (complex or specific) that the specification can represent.

- **Readability.** The quality of specification that makes it easy to read and to understand.

- **Conflict Detection.** A policy framework should be able to check that a given policy does not conflict with any other existing policy.

- **Policy Access.** A policy framework has to be able to access to policy repositories.

- **Interoperation.** A policy framework has to be able to interoperate with other policy frameworks.

- **Enforcement.** A policy framework has to be able to enforce a given policy.

Copyright © 2006, Idea Group Inc. Copying or distributing in print or electronic forms without written permission of Idea Group Inc. is prohibited.

Table 1. Comparative analysis between semantic and non-semantic policy languages

	Semantic Languages	Non-Semantic Languages
Abstraction	Multiple levels	Medium and low level
Extensibility	Easy and at runtime	Complex and at compile-time
Representability	Complex environments	Specific environments
Readability	Specialized tools	Direct
Conflict Detection	Easy, directly supported	Complex
Policy Access	Querying the ontology	By API
Interoperation	By common Ontology	By Interfaces
Enforcement	Complex	Easy

Table 1 shows a summary of the comparison between semantic and non-semantic policy languages using these criteria, and how the first ones provide some really interesting features to be used in policy-based management frameworks.

Semantic approaches have the ability to analyze policies related with entities described at a high level of abstraction, while in non-semantic ones, such as Ponder and XACML, it is necessary the identification of resources and therefore, in some cases, they have to pay attention to some implementation details. For example, in the policy expressed in XACML using our case study (see Figure 2) the tag <Resources> is used to identify the resources, whereas in SWRL (see Figure 9) the entities are described at a high level using an ontology. On the contrary, this capability to express policy at multiples levels of abstraction makes automation mechanisms for policy enforcement more complex in the case of semantic languages.

The use of ontologies permits the extensibility at runtime and the representability of complex environments in the semantic languages, although it is necessary the use of specialized tools for the policy specification. For example, the reader can compare the readability of the Ponder policy in Figure 1 with the Rei policy in Figure 11. Moreover, the use of ontologies facilitates the reasoning over the policies; therefore it makes easy the detection of conflicts.

In addition, the access to policy information and its sharing between different entities is easier when using ontologies than when using APIs and exchange interfaces in traditional non-semantic languages. For example, the interoperability between two heterogeneous applications, one using Ponder and other one using XACML should be based on the creation of complex interfaces to permit the exchange of policies between them, whereas one application using KAoS and another one using SWRL just need a common ontology for being interoperable between them.

Copyright © 2006, Idea Group Inc. Copying or distributing in print or electronic forms without written permission of Idea Group Inc. is prohibited.

Table 2. Comparative analysis between KAoS, RuleML, SWRL, and Rei

	KAoS	RuleML	SWRL	Rei
Approach	Deontic Logic	Rules	Rules	Deontic Logic + Rules
Specification language	DAML/OWL	Prolog-like syntax + XML	RuleML + OWL	Prolog-like syntax + RDFS
Tools for specification	KPAT	No	No	No
Reasoning	KAoS engine	External Engine	External Engine	Prolog engine
Enforcement	Supported	External Engine	External Engine	External Functionality

Comparative Analysis of the Semantic Policy Languages Described

Table 2 shows a comparison of the aforementioned policy languages. Many aspects can be identified as part of this comparison, although the most relevant are:

- Approach. Two types of approaches have been identified: Rule-based and deontic logic-based.
- Specification language. It can be XML, RDFS, or OWL.
- Tools for policy specification.
- Reasoning engine for policy analysis and verification.
- Enforcement support to the policy deployment.

KAoS presents a full solution that includes from the policy language to the policy enforcement, while the rest of approaches lack some components (e.g., tools for policy specification).

On the other hand, OWL has a limited way of defining restrictions using the tag *owl:Restriction*. This limitation also appears in KAoS, but SWRL overcomes it by the extending the set of OWL axioms including horn-like rules.

Moreover, SWRL is not limited to deontic policies as it happens in Rei and KAoS. For example, in the case of the next operational security policy, which is defined in natural language as *"If an intrusion is detected in the e-payment service, stop this service"*, it cannot be described in KAoS or Rei, but it can in SWRL as it is illustrated in Figure 12.

Copyright © 2006, Idea Group Inc. Copying or distributing in print or electronic forms without written permission of Idea Group Inc. is prohibited.

Figure 12. Operational security policy in SWRL

```
<ruleml:imp>
  <ruleml:_head>
        <swrlx:individualPropertyAtom
        swrlx:property="Status">
                <ruleml:var>server</ruleml:var>
                <owlx:Individual owlx:name="#Stopped" />
        </swrlx:individualPropertyAtom>
  </ruleml:_head>
  <ruleml:_body>
        <swrlx:classAtom>
                <owlx:Class owlx:name="Server" />
                <ruleml:var>server</ruleml:var>
        </swrlx:classAtom>
        <swrlx:individualPropertyAtom swrlx:property="Intrusion">
                <ruleml:var>server</ruleml:var>
                <owlx:Individual owlx:name="#Detected" />
        </swrlx:individualPropertyAtom>
  </ruleml:_body>
</ruleml:imp>
```

Conclusion and Future Work

This chapter has provided the current trends of policy-based management enriched by semantics applied to the protection of information systems. It has described some initial concepts related with policy management and formal representation techniques complemented by a case study where the reader can see the advantages and also some of the current open issues of using this approach to management versus traditional "non-semantic" languages and frameworks.

This chapter has also provided some discussions of the most relevant security aware semantic specification languages and information models. Our perspective on the main issues and problems of each of them has also been presented, based on different criteria such as their approach or the specification technique they use.

According to our analysis KAoS presents a full collection of services and tools allowing the specification, reasoning, and enforcement of policies, whereas Rei lacks in the definition of a full framework with tools and enforcement mechanisms. RuleML and SWRL (which is based on RuleML) are ideal as interoperable

Copyright © 2006, Idea Group Inc. Copying or distributing in print or electronic forms without written permission of Idea Group Inc. is prohibited.

semantic language formats; in fact, they can be integrated in other external frameworks with a compatible engine.

In the field of security policy languages enriched by semantics the future work is focused on its application in different contexts such as grid-computing environments, coalition search and rescue, and semantic firewalls. More specifically, and regarding the policy languages analyzed in this chapter, currently a graphical user interface is being developed to specify policies for Rei (UMBC eBiquity, 2005), and KAoS services are being adapted to general-purpose grid computing and Web services environments (Institute of Human and Machine Cognition, 2005).

Regarding our future work, it is being devoted to investigate how the Common Information Model (CIM) defined by DMTF can be expressed in OWL and used as common ontology for SWRL or other specific-purpose semantic security policy languages.

Acknowledgments

This work has been partially funded by the EU POSITIF (Policy-based Security Tools and Framework) IST project (IST-2002-002314) and by the ENCUENTRO (00511/PI/04) Spanish Seneca project.

References

Avouris, N. M., & Gasser, L. (1992). *Distributed artificial intelligence: Theory and praxis*. Kluwer Academic Publishers.

Brickley, D., & Guha, R. V. (2004, January). *Rdf vocabulary description language 1.0: Rdf schema*. Technical Report, W3C Working Draft.

Connolly, D., Dean, M., van Harmelen, F., Hendler, J., Horrocks, I., McGuinness, D. L., et al. (2003, February). *Web ontology language (owl) reference version 1.0*. Technical Report, W3C Working Draft.

Damianou, N., Dulay, N., et al. (2001). *The ponder policy specification language*. Policy 2001: Workshop on Policies for Distributed Systems and Networks. Springer-Verlag.

Distributed Management Task Force, inc. (2005). *Common Information Model (CIM), Version 2.9.0*.

Copyright © 2006, Idea Group Inc. Copying or distributing in print or electronic forms without written permission of Idea Group Inc. is prohibited.

Fensel, D. (2004) *Ontologies: Silver bullet for knowledge management and electronic commerce.* Springer-Verlag.

Friedman-Hill, E. (2004, December). *Jess Manual, Version 7.0.* Distributed Systems Research, Sandia National Labs.

Genesereth, M. R., & Nilsson, N. (1986) *Logical foundation of artificial intelligence.* Morgan Kaufmann.

Gruber, T. R. (1993). *A translation approach to portable ontology specifications.* Knowledge Acquisition.

Institute of Human and Machine Cognition. (2005). *IHMC ontology and policy management.* University of West Florida.

Kagal, L., Finin, T., & Johshi, A. (2003). *A policy language for pervasive computing environment.* Policy 2003: Workshop on Policies for Distributed Systems and Networks. Springer-Verlag.

Kosiur, D. (2001). *Understanding policy-based networking.* John Wiley & Sons.

Martínez Pérez, G., García Clemente, F. J., & Gómez Skarmeta, A. F. (2005) *Policy-based management of Web and information systems security: An emerging technology.* Hershey, PA: Idea Group.

OASIS. (2004, November). *Extensible Access Control Markup Language (XACML), Version 2.0.*

Russell, S., & Norvig, P. (1995). *Artificial intelligence: A modern approach.* Prentice Hall.

Spencer, B. (2004). *A Java deductive reasoning engine for the Web (jDREW), Version 1.1.*

Strassner, J. (2003). *Policy-based network management: Solutions for the next generation.* Morgan Kaufmann.

The Rule Markup Initiative. (2004, May). *SWRL: A Semantic Web rule Language combining OWL and RuleML, Version 0.6.*

The Rule Markup Initiative. (2004, August). *Schema specification of RuleML 0.87, Version 0.87.*

Tonti, G., Bradshaw, J. M., Jeffers, R., Montanari, R., Suri, N., & Uszok, A. (2003). Semantic Web languages for policy representation and reasoning: A comparison of KAoS, Rei, and Ponder. *The Semantic Web — ISWC 2003. Proceedings of the Second International Semantic Web Conference.* Springer-Verlag.

UMBC eBiquity. (2005). *Project Rei: A policy specification language.* University of Maryland, Baltimore County (UMBC).

Copyright © 2006, Idea Group Inc. Copying or distributing in print or electronic forms without written permission of Idea Group Inc. is prohibited.

Uszok, A., Bradshaw, J., Jeffers, R., Suri, N., et al. (2003). *KAoS Policy and domain services: Toward a description-logic approach to policy representation, deconfliction, and enforcement*. Policy 2003: Workshop on Policies for Distributed Systems and Networks. Springer-Verlag.

Verma, D. C. (2000). *Policy-based networking: Architecture and algorithms*. Pearson Education.

W3C. (1999, February). *Resource Description Framework (RDF), data model and syntax*. W3C Recommendation.

Wooldridge, M. (2001). *An introduction to multiagent systems*. John Wiley & Sons.

Copyright © 2006, Idea Group Inc. Copying or distributing in print or electronic forms without written permission of Idea Group Inc. is prohibited.

About the Authors

David Taniar received a PhD in databases from Victoria University, Australia, in 1997. He is now a senior lecturer at Monash University, Australia. He has published more than 100 research articles and edited a number of books in the Web technology series. He is in the editorial board of a number of international journals, including *Data Warehousing and Mining, Business Intelligence and Data Mining, Mobile Information Systems, Mobile Multimedia, Web Information Systems,* and *Web and Grid Services*. He has been elected as a fellow of the Institute for Management of Information Systems (UK). Dr. Taniar is editor-in-chief of the *International Journal of Data Warehousing and Mining* (www.idea-group.com/ijdwm).

Wenny Rahayu is an associate professor at the department of computer science and computer engineering La Trobe University, Australia. She has been actively working in the areas of database design and implementation covering object-relational databases, Web and e-commerce databases, semistructured databases and XML, Semantic Web and ontology, data warehousing, and parallel databases. She has worked with industry and expertise from other disciplines in a number of projects including bioinformatics databases, parallel data mining, and e-commerce catalogues. She has been invited to give seminars in the area of e-commerce databases in a number of international institutions. She has edited three books which forms a series in Web applications, including Web databases, Web information systems and Web Semantics. She has also published over 70 papers in international journals and conferences.

*   *   *

Copyright © 2006, Idea Group Inc. Copying or distributing in print or electronic forms without written permission of Idea Group Inc. is prohibited.

Muhammad Abulaish is a faculty in the Department of Mathematics, Jamia Millia Islamia, India (A Central University) (since January 2004). Prior to this he has served Hamdard University and UPTEC Computer Consultancy Ltd. (3.5 and 2.5 years, respectively). He obtained his postgraduate degree, MCA, from Motilal Nehru National Institute of Technology, Allahabad and is presently pursuing his PhD from IIT, Delhi. He has qualified the UGC-NET examination in Computer Applications. He is member of professional societies like CSI, ISTE, IETE and ISCA. His research interests include intelligent information retrieval, Semantic Web and data mining.

Rama Akkiraju is a senior software engineer at IBM T.J. Watson Research Center in Hawthorne, NY, USA. She holds a master's degree in computer science and an MBA from New York University. Since joining IBM Research in 1995, she has worked on agent-based decision support systems, electronic marketplaces, and business process integration technologies. Ms. Akkiraju is interested in applying artificial intelligence techniques to solving business problems. Her current focus is on Semantic Web services.

Christos Anagnostopoulos received a BSc from the Department of Informatics and Telecommunications, University of Athens (2002) and an MSc from the Department of Informatics and Telecommunications at the University of Athens (2003). He is currently a PhD candidate in the Department of Informatics at the University of Athens, Greece. He is currently participating in the Pervasive Computing Research Unit of CNL. His research interests include pervasive and mobile computing, context awareness, and ontological engineering in mobile environments.

Marie Aude Aufaure obtained her PhD in computer science from the University of Paris 6 (1992). From 1993 to 2001, she was an associate professor at the University of Lyon. Now, she is a professor at Supélec (France). Her research interests deal with the combination of data mining techniques and ontology to improve the retrieval process of complex data. She applied her research work to complex data such as spatial and multimedia data and to the Semantic Web in order to automate ontology construction. Her work was published in international journals and conferences.

Aïcha-Nabila Benharkat is a graduated engineer from the Computer Science Institute of ANNABA University (Algeria, 1984). She received an MS and PhD from the National Institute of Applied Sciences of Lyon (1985 and 1990, respectively). Since 1992, she has been an associate professor at the Computer

Copyright © 2006, Idea Group Inc. Copying or distributing in print or electronic forms without written permission of Idea Group Inc. is prohibited.

Science Department (INSA de Lyon, France). From 1984 to 1991, her research interests included compilation/meta compilation, software engineering environment. From 1991 to 1993, she worked on the design of information systems, object oriented modeling, and computer aided software engineering. Since 1993, she works on the federation of heterogeneous databases and on the use of reasoning mechanisms in order to define a methodology of integration of databases schemes. Currently, her research interest includes in the domain of systems and applications interoperability, matching techniques, and ontologies.

Nacera Bennacer is an engineer in industrial engineering and doctor in computer science at the Conservatoire National des Arts et Métiers of Paris (CNAM). She was a lecturer in CNAM for two years and an assistant professor in Ecole Internationale des Sciences du Traitement de l'Information (EISTI). Currently, she is an assistant professor in the Computer Department of Ecole Supérieure d'Electricité (Supélec, France). Her research domain concerns Semantic Web, knowledge extraction using data-mining methods and knowledge representation using Web languages and formal languages like description logics.

Juan A. Botía Blaya (juanbot@um.es) received MSc and PhD degrees in computer science from the University of Murcia (1996 and 2002, respectively). From 1996 to 1997, he was a senior engineer with Tecnatom, S.A. From 1997 to 1999, he worked as an associate professor in the University of Alcala. Since 2003, he is a full professor in the Department of Information and Communications Engineering at the University of Murcia, Spain. His main research interests include machine learning and distributed artificial intelligence.

María Laura Caliusco was born in Argentina in 1974. She received her degree in information system engineering in 1999 and her PhD in Information Systems from Universidad Tecnológica Nacional - Facultad Regional Santa Fe (FRSF), Argentina, in 2005. She is now a Professor Assistant at FRSF, Argentina. She has published a number of papers on Knowledge and Software Engineering. Furthermore, she has a written three-chapter book. She has been elected as a fellow of the CONICET (Argentina). She is working as research scientist in CIDISI Research Center at FRSF. She is interested in software agents, information retrieval, knowledge discovery, e-collaboration and contextual ontology.

Silvana Castano is full professor of computer science at the University of Milan, Italy, where she currently chairs the Information Systems & Knowledge

Copyright © 2006, Idea Group Inc. Copying or distributing in print or electronic forms without written permission of Idea Group Inc. is prohibited.

Management (ISLab) Group. She received a PhD in computer and automation engineering from Politecnico di Milano (1993). Her main research interests are in the area of databases and information systems and ontologies and Semantic Web, with current focus on knowledge sharing in peer-based systems, information integration, semantic interoperability, database, and XML security. She has published her research results in the major journals and in the refereed proceedings of the major conferences of the field. On these topics, she has been working in several national and international research projects, including the recent EU FP6 NoE INTEROP (Interoperability Research for Networked Enterprises Applications and Software) project, where she is responsible of the University of Milano Group, and the FIRB WEB MINDS project, where she is responsible of the Working Group on Metadata and Ontologies. She is an author of the book *Database Security* (Addison Wesley-ACM Press, 1995).

Omar Chiotti was born in Argentina in 1959. He received his degree in chemical engineering in 1984 from Universidad Tecnológica Nacional (UTN) - Facultad Regional Villa María (Córdoba) and his PhD in chemical engineering from Universidad Nacional del Litoral (Santa Fe) in 1989. Since 1984, he has been working for the Argentina's National Council of Scientific Research (CONICET) as a researcher. He is a professor of information system engineering at UTN - Facultad Regional Santa Fe (Argentina) since 1986. Currently, he is the director of CIDISI Research Center in Information System Engineering. His current research interests include decision support systems, data warehouse, e-collaboration, and multi-agent systems.

Lipika Dey is a faculty member in the Department of Mathematics at Indian Institute of Technology, Delhi, India (since 1995). She has done integrated MSc in mathematics, MTech in computer science and data processing and a PhD in computer science and engineering, all from IIT, Kharagpur, India. Her research interest spans across various areas in artificial intelligence and cognitive science, especially in the areas of pattern recognition, text information retrieval and mining interesting exceptions from data bases. She is working on text-classification models based on soft computing techniques like rough sets, fuzzy sets and evolutionary algorithms.

Tharam S. Dillon is the dean of the Faculty of Information Technology at the University of Technology, Sydney (UTS) (Australia). His research interests include data mining, Internet computing, e-commerce, hybrid neuro-symbolic systems, neural nets, software engineering, database systems and computer networks. He has also worked with industry and commerce in developing systems in telecommunications, healthcare systems, e-commerce, logistics,

Copyright © 2006, Idea Group Inc. Copying or distributing in print or electronic forms without written permission of Idea Group Inc. is prohibited.

power systems, and banking and finance. He is editor-in-chief of the *International Journal of Computer Systems Science and Engineering* and the *International Journal of Engineering Intelligent Systems*, as well as co-editor of the *Journal of Electric Power and Energy Systems*. He is on the advisory editorial board of *Applied Intelligence* (Kluwer) and *Computer Communications* (Elsevier). He has published more than 400 papers in international and national journals and conferences and has written four books and edited five other books. He is a fellow of the IEEE, fellow of the Institution of Engineers (Australia), and fellow of the Australian Computer Society.

Alfio Ferrara is assistant professor in computer science at the University of Milan, Italy. His research interest include database and semistructured data integration, Web-based information systems, ontology engineering, and knowledge sharing and evolution in strong distributed systems, such as P2P and Super-peer networks. On these topics, he participates to the INTEROP Network of Excellence IST Project n. 508011 - 6th EU Framework Programme and to the "Wide-scalE, Broadband, MIddleware for Network Distributed Services (WEB-MINDS)" FIRB Project funded by the Italian Ministry of Education, University, and Research. Previously, he has participated with the group of the University of Milan to the D2I (Integration, Warehousing and Mining of Heterogeneous Sources) project, founded by the Italian Ministry of Education, University, and Research. He received a master's degree in methodologies of computer science and communication for humanities (2001), and the Laurea degree in philosophy (2000).

María Rosa Galli is a professor of operation research of Universidad Tecnológica Nacional - Facultad Regional Santa Fe (Argentina). She received her degree in chemical engineering in 1983 from Universidad Nacional de Mar del Plata, Argentina, and a PhD in chemical engineering from Universidad Nacional del Litoral (Santa Fe), in1989. Since 1984, she has been working for the Argentina's National Council of Scientific Research (CONICET) as a researcher. Currently, she is the directory member of CIDISI Research Center in Information System Engineering. Her current research interests include decision support systems, e-collaboration, knowledge management systems, ontologies, and multi-agents systems.

Félix J. García Clemente (fgarcia@dif.um.es) is a lecturer at the University of Murcia, Spain. He received the MSc in computer engineering from the same university. His research interests include ontology, Web service security and policy-based management systems.

Copyright © 2006, Idea Group Inc. Copying or distributing in print or electronic forms without written permission of Idea Group Inc. is prohibited.

Richard Goodwin leads the Semantic e-Business Middleware Research Group at the IBM T.J. Watson Research Center in Hawthorne, NY, USA. He received his PhD and MS degrees in computer science from Carnegie Mellon University. His graduate work focused on artificial intelligence, planning, and machine-learning. Since joining IBM, Dr. Goodwin has worked on agent-based optimization (asynchronous teams of agents), electronic marketplaces, and decision support systems. His current focus is on Semantic Web representations and Semantic Web services.

Antonio F. Gómez Skarmeta (skarmeta@dif.um.es) is an assistant professor at the University of Murcia, Spain. He received an MSc in computer science from the University of Granada, and a BSc (Honors) and PhD in computer science from the University of Murcia. His research interests include distributed artificial intelligence, computer support for collaborative learning, and advanced networking services and applications over IP networks.

Stathes Hadjiefthymiades received his BSc, MSc and PhD degrees in informatics from the Department of Informatics and Telecommunications, University of Athens (UoA). He also received a joint engineering-economics MSc from the National Technical University of Athens. In 1992, he joined the Greek consulting firm Advanced Services Group, Ltd. In 1995, he joined the Communication Networks Laboratory (CNL) of UoA. From 2001-2002, he served as a visiting assistant professor at the University of Aegean, Department of Information and Communication Systems Engineering. In the summer of 2002 he joined the faculty of the Hellenic Open University, Patras, Greece, as an assistant professor. Since December 2003, he is in the faculty of the Department of Informatics and Telecommunications, University of Athens, Greece, where he is presently an assistant professor. He has participated in numerous EU projects and national initiatives. His research interests are in the areas of Web engineering, mobile/pervasive computing, and networked multimedia. He has contributed to more than 90 publications in these areas. Since 2004, he has coordinated the Pervasive Computing Research Unit of CNL.

Hyoil Han is an assistant professor in the College of Information Science and Technology at Drexel University, USA. She graduated in 2002 from the University of Texas at Arlington with a PhD in computer science and engineering. Her research lies in the areas of merging techniques from databases (DB) and artificial intelligence (AI) with emphasis on data/text mining, ontology learning, and Semantic Web for biomedicine and Web. She has 19 referred research papers. She has also industry research experience in the Telecommu-

Copyright © 2006, Idea Group Inc. Copying or distributing in print or electronic forms without written permission of Idea Group Inc. is prohibited.

nication Network Laboratory at Korea Telecom and Samsung Electronics. She is a member of ACM and IEEE.

Pankaj Kamthan (Concordia University, Canada) has been teaching in academia and industry for several years. He has also been a technical editor and participated in standards development. His professional interests and experience are in software quality and markup languages.

Izambo Karali is a lecturer at the Department of Informatics and Elecommunications, University of Athens, Greece. From 1989 to 1999, she was a research associate in the department and was employed in European Union research projects. She obtained her first degree in mathematics in 1986 (Department of Mathematics, University of Athens), her MSc in computer science in 1988 (University College, University of London) and her PhD in Informatics in 1995 (Department of Informatics, University of Athens). Her research interests include knowledge representation, logic programming, as well as Web information extraction and representation. She has published a number of papers and has been reviewer in various journals and conferences related to the above areas.

Bénédicte Le Grand is an associate professor at University Pierre and Marie Curie, in the Computer Science Laboratory of Paris 6 (LIP6) (France). After receiving her engineer diploma in telecommunications from the National Institute of Telecommunications, she earned her PhD in computer science (2001). Her research interests deal with information retrieval in complex systems, the Web in particular. She works on Semantic Web standards, especially topic maps, in order to propose Semantic Web visualization solutions. She also works on conceptual analysis techniques to extract knowledge from data and automate ontology construction. She published her work in international conferences and contributed to several books.

Juhnyoung Lee is a research staff member at the IBM T.J. Watson Research Center in Hawthorne, NY, USA. He received his PhD from the University of Virginia, Charlottesville, and MS and BS degrees from Seoul National University, Korea (all in computer science). His graduate work focused on database systems, transaction processing and real-time systems. Dr. Lee joined IBM Research in 1997 and has worked on e-commerce intelligence, electronic marketplaces, and decision support systems. His current research focus is on ontology management systems and model-driven business transformation.

Copyright © 2006, Idea Group Inc. Copying or distributing in print or electronic forms without written permission of Idea Group Inc. is prohibited.

César Maidana was born in Argentina in 1978. He is pursuing his information system engineering degree at Universidad Tecnológica Nacional - Facultad Regional Santa Fe, Argentina. He has been working for CIDISI Research Center since April 2003. His research interest is related to XML Technologies, ontologies and Semantic Web.

Gregorio Martínez Pérez (gregorio@dif.um.es) is a lecturer at the University of Murcia, Spain. He received MSc and PhD degrees in computer engineering from the same university. His research interests include security and management of IPv4/IPv6 communication networks.

Stefano Montanelli is a PhD student in computer science at the University of Milan, Italy. He received the Laurea degree in computer science from the University of Milan (2003). He participates with the Group of the University of Milan to the INTEROP Network of Excellence IST Project n. 508011 - 6th EU Framework Programme and to the "Wide-scalE, Broadband, MIddleware for Network Distributed Services (WEB-MINDS)" FIRB Project funded by the Italian Ministry of Education, University, and Research. His main research interests include Semantic Web, ontology engineering, and semantic interoperability in open distributed systems.

Hsueh-leng Pai received her master's degree from McGill University, Canada, and is currently pursuing a doctoral degree at Concordia University, Canada. Her research interests include intelligent Web searching and knowledge management on the Semantic Web.

Rajugan Rajagopalapillai holds a bachelor's degree in information systems from La Trobe University (Australia). He worked in industry as a chief application/database programmer in developing sports planing and sports fitness & injury management software and as database administrator. He was also involved in developing an e-commerce solution for a global logistics (logistics, cold-storage and warehousing) company as a software engineer/architect. He is currently a PhD student at the University of Technology, Sydney (UTS) Australia. His published research articles have appeared in international refereed conference and journal proceedings. His research interests include object-oriented conceptual models, XML, data warehousing, software engineering, database and e-commerce systems. He is a member of IEEE and ACM. He is also an associate member of the Australian Computer Society (AACS).

Copyright © 2006, Idea Group Inc. Copying or distributing in print or electronic forms without written permission of Idea Group Inc. is prohibited.

Lawrence Reeve is a doctoral student in the College of Information Science and Technology at Drexel University, USA. He has a master's degree in computer and software engineering from the School of Engineering at Widener University, and a bachelor's degree in computer science from Rutgers University. His research interests are in applying semantic annotation, semantic indexing and retrieval, and information visualization, as well as the supporting fields of information extraction, natural language processing, text mining, and machine learning methods, to the biomedical domain. He has four referred research papers.

Rami Rifaieh received his PhD in computer science from the National Institute of Applied Sciences of Lyon, his MS degree from Claude Bernard University, and his graduate degree from Lebanese University Lebanon (2004, 2000, and 1999, respectively). His PhD was sponsored by the French National Association of Technical Research (ANRT) and TESSI Group. Meanwhile, he was working as a research engineer in the R&D division of TESSI Computing & Consulting St-Etienne, France (2001-2004). He taught computer science in INSA of Lyon, Claude Bernard University, Jean-Moulin University, and Lumière University (2000-2005). Currently, he is pursuing a post doctoral research position with the San Diego Supercomputer Center, University of California San Diego, USA. His interest focus on managing and integrating biological data.

Michel Soto received his PhD in computer networks from University Pierre and Marie Curie, Paris (1990). He was a research associate in the Multimedia Communication Research Laboratory at Ottawa University, Ontario, Canada (October 1991 to September 1992). He is currently an associate professor at René Descartes University, Paris and a researcher in the Computer Science Laboratory of Paris 6 (LIP6) (France). His current research interests are knowledge management and virtual reality with special emphasis on Semantic Web visualization, ontology engineering, semantic-oriented standards, and networked virtual worlds interoperability.

Martin Tomko is a postgraduate student at The University of Melbourne, Australia, working on a thesis on granular route directions. He has a MSc in Geodesy from the University of Technology Bratislava, Slovakia (2003). His master thesis was done in cooperation with TU Delft, The Netherlands, on spatial databases for mobile GIS. Martin Tomko was also representative for Slovakia in the CEN TC 287, Geographic Information.

Copyright © 2006, Idea Group Inc. Copying or distributing in print or electronic forms without written permission of Idea Group Inc. is prohibited.

Athanasios Tsounis has received a BSc from the Department of Informatics and Telecommunications, University of Athens (2002) and an MSc from the Department of Informatics and Telecommunications, University of Athens (2003). He is currently a PhD candidate in the Department of Informatics at the University of Athens, Greece. His research interests include pervasive computing, smart agents, expert systems, and ontological engineering in mobile environments.

Stephan Winter is a senior lecturer at The University of Melbourne, Australia. He previously held positions at the University of Bonn, Germany (PhD, 1997), and the Technical University Vienna, Austria (Habilitation, 2001). His research focuses on spatial information theory, spatial cognition and spatial communication, and spatial interoperability, all with a specific interest in wayfinding. Parts of his findings were included in the OpenGIS specifications. Dr. Winter is co-chair of the ISPRS WG II/6: Geo-spatiotemporal semantics and interoperability, and chaired of the AGILE WG in Interoperability until 2003. He chaired also the EuroConference on Ontology and Epistemology for Spatial Data Standards in France, 2000. For a full CV and publication list, visit http://www. geom.unimelb.edu.au/winter.

Carlo Wouters is a PhD candidate in the Department of Computer Science and Computer Engineering at La Trobe University, Melbourne (Australia). His research interests include ontologies and extraction of materialized ontology views. He obtained a BSc in audio-visual sciences (RITS, Belgium), a GradDipSc in computer science (La Trobe, Australia) and an MSc in information technology (La Trobe, Australia).

Copyright © 2006, Idea Group Inc. Copying or distributing in print or electronic forms without written permission of Idea Group Inc. is prohibited.

Index

Copyright © 2006, Idea Group Inc. Copying or distributing in print or electronic forms without written permission of Idea Group Inc. is prohibited.

Copyright © 2006, Idea Group Inc. Copying or distributing in print or electronic forms without written permission of Idea Group Inc. is prohibited.

Copyright © 2006, Idea Group Inc. Copying or distributing in print or electronic forms without written permission of Idea Group Inc. is prohibited.

Copyright © 2006, Idea Group Inc. Copying or distributing in print or electronic forms without written
permission of Idea Group Inc. is prohibited.

Single Journal Articles and Case Studies Are Now Right at Your Fingertips!

Purchase any single journal article or teaching case for only $18.00!

Idea Group Publishing offers an extensive collection of research articles and teaching cases that are available for electronic purchase by visiting www.idea-group.com/articles. You will find over 980 journal articles and over 275 case studies from over 20 journals available for only $18.00. The website also offers a new capability of searching journal articles and case studies by category. To take advantage of this new feature, please use the link above to search within these available categories:

- ◆ Business Process Reengineering
- ◆ Distance Learning
- ◆ Emerging and Innovative Technologies
- ◆ Healthcare
- ◆ Information Resource Management
- ◆ IS/IT Planning
- ◆ IT Management
- ◆ Organization Politics and Culture
- ◆ Systems Planning
- ◆ Telecommunication and Networking
- ◆ Client Server Technology

- ◆ Data and Database Management
- ◆ E-commerce
- ◆ End User Computing
- ◆ Human Side of IT
- ◆ Internet-Based Technologies
- ◆ IT Education
- ◆ Knowledge Management
- ◆ Software Engineering Tools
- ◆ Decision Support Systems
- ◆ Virtual Offices
- ◆ Strategic Information Systems Design, Implementation

You can now view the table of contents for each journal so it is easier to locate and purchase one specific article from the journal of your choice.

Case studies are also available through XanEdu, to start building your perfect coursepack, please visit www.xanedu.com.

For more information, contact cust@idea-group.com or 717-533-8845 ext. 10.

www.idea-group.com

IDEA GROUP INC.

International Journal of
Information Technology and Web Engineering

ISSN 1554-1045
eISSN 1554-1053
Published quarterly
www.idea-group.com/ijitwe

ABOUT THIS JOURNAL

Organizations are continuously overwhelmed by a variety of new information technologies, many are Web based. These new technologies are capitalizing on the widespread use of network and communication technologies for seamless integration of various issues in information and knowledge sharing within and among organizations. This emphasis on integrated approaches is unique to this journal and dictates cross platform and multidisciplinary strategy to research and practice.

MISSION

The main objective is to publish refereed papers in the area covering IT concepts, tools, methodologies, and ethnography, in the contexts of global communication systems and Web engineered applications. In accordance with this emphasis on the Web and communication systems, the journal publishes papers on IT research and practice that support seamless end-to-end information and knowledge flow among individuals, teams, and organizations.

TOPICS OF INTEREST:

• Semantic Web Studies
• Data and Knowledge Capture
• IT Education and Training
• RFID Research
• and many more! Visit www.idea-group.com/ijitwe

> **Full submission guidelines available at:**
> **http://www.idea-group.com**

Now when your institution's library subscribes to any IGP journal, it receives the print version as well as the electronic version for one inclusive price. For more information call 717/533-8845, ext. 14

For subscription information, contact:

Idea Group Publishing
701 E Chocolate Ave., Ste 200
Hershey PA 17033-1240, USA
cust@idea-group.com
www.idea-group.com

For paper submission information:
Ghazi Alkhatib
<alkhatib@asu.edu.jo> &
<ghazi_al@wanadoo.jo>
David C. Rine, drine@gmu.edu;
davidcrine@aol.com